Surveys in Differential Geometry

Vol. 1: Lectures given in 1990
edited by S.-T. Yau and H. Blaine Lawson

Vol. 2: Lectures given in 1993
edited by C.C. Hsiung and S.-T. Yau

Vol. 3: Lectures given in 1996
edited by C.C. Hsiung and S.-T. Yau

Vol. 4: Integrable systems
edited by Chuu Lian Terng and Karen Uhlenbeck

Vol. 5: Differential geometry inspired by string theory
edited by S.-T. Yau

Vol. 6: Essays on Einstein manifolds
edited by Claude LeBrun and McKenzie Wang

Vol. 7: Papers dedicated to Atiyah, Bott, Hirzebruch, and Singer
edited by S.-T. Yau

Vol. 8: Papers in honor of Calabi, Lawson, Siu, and Uhlenbeck
edited by S.-T. Yau

Vol. 9: Eigenvalues of Laplacians and other geometric operators
edited by A. Grigor'yan and S-T. Yau

Vol. 10: Essays in geometry in memory of S.-S. Chern
edited by S.-T. Yau

Vol. 11: Metric and comparison geometry
edited by Jeffrey Cheeger and Karsten Grove

Vol. 12: Geometric flows
edited by Huai-Dong Cao and S.-T. Yau

Vol. 13: Geometry, analysis, and algebraic geometry
edited by Huai-Dong Cao and S.-T.Yau

Vol. 14: Geometry of Riemann surfaces and their moduli spaces
edited by Lizhen Ji, Scott A. Wolpert, and S.-T. Yau

Volume III

Surveys in Differential Geometry

Proceedings of the conference
on geometry and topology
held at Harvard University
May 3-5, 1996

edited by
C.C. Hsiung and S.-T. Yau

International Press
www.intlpress.com

Surveys in Differential Geometry, Vol. 3
Proceedings of the conference on geometry and topology held at Harvard University, May 3-5, 1996

Editors:
C.C. Hsiung, Lehigh University
S.-T. Yau, Harvard University

ISBN 978-1-57146-214-5

Paperback reissue 2010. Previously published in 1998 under ISBN 1-57146-067-5 (clothbound).

Printed in the United States of America.

Preface

This year is the 30th anniversary of the Journal of Differential Geometry and the 80th birthday of the founder of the journal , Chuan-Chih Hsiung of Lehigh University, who also initiated the idea of expanding the classical (or traditional) coverage of the subject of "differential geometry" to be as broad as today's. So this conference was held together with a joint celebration of these two big events, and these proceedings are published in honor of the celebration.

C. C. Hsiung
Lehigh University

S.-T. Yau
Harvard University

CHUAN-CHIH HSIUNG

Table of Contents

Local index theory, eta invariants and holomorphic torsion : a survey

by

JEAN-MICHEL BISMUT

Département de Mathématique
Université Paris-Sud
URA D1169 du CNRS
Bâtiment 425
91405 ORSAY Cedex
FRANCE
bismut@topo.math.u-psud.fr

Abstract. The purpose of this paper is to review various results related to the local families index theorem, eta invariants, and the Ray-Singer holomorphic torsion.

Acknowledgments. The support of Institut Universitaire de France (IUF) is gratefully acknowledged.

Also the author is indebted to Jeff Cheeger and Paolo Piazza for very helpful comments.

The purpose of this paper is to survey various recent developments in local index theory, including applications to eta invariants, Quillen metrics on determinant bundles, and analytic torsion forms. In this refined index theory, the construction of secondary objects plays an important role. On the analytic side, these will be the $\widetilde{\eta}$ forms (which are extensions of eta invariants), analytic torsion forms (which extend the Ray-Singer torsion). These objects refine on the construction of the index bundle [**A**], [**AS2**]. On the geometric side, they include Chern-Simons forms [**ChSi**], Bott-Chern forms [**BoCh**] and Bott-Chern currents [**BGS4,5**], which refine on the classical cohomological objects of index theory like $\widehat{A}$, Td, ch. The ultimate purpose of the new theory is to relate the analytic secondary invariants to corresponding geometric secondary invariants.

This line of thought is most clearly illustrated by the Riemann-Roch formula in Arakelov geometry of Gillet-Soulé [**GS3,4**], which applies to arithmetic varieties. In its simplest form, it evaluates the artihmetic degree of a determinant bundle (whose evaluation involves Quillen metrics at places at infinity) in terms of integrals of arithmetic characteristic classes [**GS1,2**] (whose construction involves Bott-Chern classes and Bott-Chern currents at places at infinity).

Needless to say, we will always work here in the context of real or complex geometry. Still the idea that the above objects fit "naturally" in an algebraic context has been a powerful motivation for their development.

Let us now briefly discuss in more detail the content of this survey. As we said before, this paper is organized around the local index theorem [**P1,2**], [**Gi1,2**], [**ABoP**]. Let Z be a compact even dimensional Riemannian oriented spin manifold, let D^Z be a Dirac operator acting on smooth sections of the twisted spinors $S^{TZ} \otimes \xi$. The Atiyah-Singer index theorem [**AS1**] asserts that the index $\operatorname{Ind}(D^Z_+) \in \mathbf{Z}$ of D^Z_+ (which is D^Z restricted to twisted positive spinors) is given by

$$\operatorname{Ind}(D^Z_+) = \int_Z \widehat{A}(TZ)\operatorname{ch}(\xi) \tag{0.1}$$

where the right-hand side is an integral of a characteristic class, which is a cohomology class.

Let $P_t(x,y)$ be the heat kernel for $\exp(-tD^{Z,2})$. The Mc-Kean Singer formula [**McKS**] says that for $T > 0$,

$$\operatorname{Ind}(D^Z_+) = \operatorname{Tr_s}\left[\exp(-tD^{Z,2})\right] = \int_Z \operatorname{Tr_s}\left[P_t(x,x)\right] dv_Z(x) \tag{0.2}$$

(in (0.2), $\operatorname{Tr_s}$ is our notation for a supertrace, which is a graded trace).

The local index theorem, conjectured in [**McKS**] and proved in [**P**], [**Gi1,2**], [**ABoP**], asserts that as $t \to 0$, we have "fantastic" cancellations in $\operatorname{Tr_s}[P_t(x,x)]$ (which means that as $t \to 0$, $\operatorname{Tr_s}[P_t(x,x]$ is non singular), and that

$$\operatorname{Tr_s}\left[P_t(x,x)\right] dv_X(x) \to \left\{\widehat{A}(TZ,\nabla^{TZ})\operatorname{ch}(\xi,\nabla^{\xi})\right\}^{\max}, \tag{0.3}$$

where in (0.3), the corresponding characteristic forms in Chern-Weil theory are calculated using the Levi-Civita connection ∇^{TZ} on TZ, and the given connection ∇^{ξ} on ξ. Of course, from (0.2), (0.3), we recover (0.1).

In the above form, the local index theorem was used in **[ABoP]** to give a new proof of the Atiyah-Singer index theorem.

In **[APS1]**, Atiyah-Patodi-Singer developed an index theory for manifolds with boundary. If Z is a compact even dimensional oriented spin manifold with boundary, the index problem on the Dirac operator D^Z on Z imposes global boundary conditions on ∂Z. The index formula of **[APS1]** takes the form

$$\operatorname{Ind}(D^Z_+) = \int_Z \widehat{A}(TZ, \nabla^{TZ}) \operatorname{ch}(\xi, \nabla^{\xi}) - \overline{\eta}^{D^{\partial Z}}(0)\,. \tag{0.4}$$

In (0.4), $\overline{\eta}^{D^{\partial Z}}(s)$ is a meromorphic function of S, which is calculated in terms of the spectrum of a Dirac operator $D^{\partial Z}$ on the boundary ∂Z . The quantity $\overline{\eta}^{D^{\partial Z}}(0)$ is called a reduced eta invariant. To establish (0.4), the local index theorem plays an essential role. In fact, the first term in the right-hand side of (0.4) is the integral of a closed differential form.

Formula (0.4) is quite important. In effect, $\int_Z \widehat{A}(TZ, \nabla^{TZ}) \operatorname{ch}(\xi, \nabla^{\xi})$ is a Chern-Simons invariant. On the other hand, $\overline{\eta}^{D^{\partial Z}}(0)$ is a global spectral invariant of ∂Z, which is a prototype of the analytic secondary invariants which will be considered later. Then (0.4) implies that mod($\mathbf{Z}$), $\overline{\eta}^{D^{\partial Z}}(0)$ is equal to a Chern-Simons invariant. This is a simple prototype of a refined index theorem. Such a theorem was formulated first in the context of differential characters by Cheeger and Simons **[CSi]**.

Section 1 is devoted to a short exposition of local index theory and eta invariants.

Let now $\pi : X \to B$ be a fibration with compact fibres as before. Let $(D^Z_b)_{b\in B}$ be the corresponding family of Dirac operators. In **[AS2]**, Atiyah and Singer have shown how to associate to this family an (analytic) index bundle $\operatorname{Ind}(D^Z_+) \in K^0(B)$. They also defined a topological index, and they proved a corresponding families index formula. When mapping $K^0(B)$ in $H(B, \mathbf{Q})$ by the Chern character ch, the formula of **[AS2]** takes the form

$$\operatorname{ch}(\operatorname{Ind}(D^Z_+)) = \pi_* \left[\widehat{A}(TZ) \operatorname{ch}(\xi)\right] \text{ in } H(B, \mathbf{Q})\,. \tag{0.5}$$

The local index theorem of **[B2]** refines on the right hand-side of (0.5), by replacing it by an explicit geometrically constructed differential form $\pi_*[\widehat{A}(TZ, \nabla^{TZ}) \operatorname{ch}(\xi, \nabla^{\xi})]$. When $\operatorname{Ind}(D^Z_+)$ is a honest vector bundle, the theory of **[B2]** replaces the analytic index by an analytically constructed differential form $\operatorname{ch}(\ker D^Z, \nabla^{\ker D^Z, u})$. Then an essential by-product of **[B2]**, **[BeV]**, **[BeGeV]**, **[BC1]** is the construction of an explicit differential form $\widetilde{\eta}$ on B such that

$$d\widetilde{\eta} = \pi_* \left[\widehat{A}(TZ, \nabla^{TZ}) \operatorname{ch}(\xi, \nabla^{\xi})\right] - \operatorname{ch}(\ker D^Z, \nabla^{\ker D^Z, u})\,. \tag{0.6}$$

Quillen's superconnections **[Q1]** are an important tool to obtain the above results. Superconnections provide a useful extension of Chern-Weil theory to $\mathbf{Z}_2$-graded (and possibly infinite dimensional) vector bundles. In **[B2]**, the form $\pi_*[\widehat{A}(TZ,\nabla^{TZ})\operatorname{ch}(\xi,\nabla^{\xi})]$ is produced by refining the local index theorem (0.3) in a relative context, hence its name of a local families index theorem.

Equation (0.5) has been extended in **[BC2,3]**, **[MeP1,2]** to a families index theorem for Dirac operators on manifolds with boundary, by using either the techniques of Cheeger **[C1,2,3]** on manifolds with conical singularities, or the *b*-calculus of Melrose **[Me]**. The important concept of a spectral section **[MeP1,2]** has emerged in this context.

If (E,g^E) is a holomorphic Hermitian vector bundle on a complex manifold, E is naturally equipped with the holomorphic Hermitian connection ∇^E. We will denote by $\operatorname{Td}(E,g^E)$ the form $\operatorname{Td}(E,\nabla^E)$. It is a sum of forms of type (p,p).

If $\pi: X\to S$ is a holomorphic submersion with compact fibre Z, if TZ is equipped with a Hermitian metric g^{TZ}, and, if (ξ,g^{ξ}) is a holomorphic Hermitian vector bundle on X such that $R\pi_*\xi$ is locally free, a natural equation related to (0.6) is

$$\frac{\bar\partial\partial}{2i\pi}T=\operatorname{ch}(R\pi_*\xi,g^{R\pi_*\xi})-\pi_*\left[\operatorname{Td}(TZ,g^{TZ})\operatorname{ch}(\xi,g^{\xi})\right], \tag{0.7}$$

where $g^{R\pi_*\xi}$ is the metric on $R\pi_*\xi$ obtained from g^{TZ},g^{ξ} by Hodge theory along the fibres.

Assume that X is Kähler and let ω^X be the corresponding Kähler from. When g^{TZ} is obtained from ω^X by restriction to TZ, the forms $T(\omega^X,g^{\xi})$ were constructed in **[BGS3]**, **[BK]** and were called analytic torsion forms, because in degree 0, $T^{(0)}(\omega^X,g^{\xi})$ coincides with the Ray-Singer analytic torsion **[RS]** of the corresponding Dolbeault complex.

In fact of special interest are $\widetilde{\eta}^{(1)}$ and $T^{(0)}$. In **[BF1,2]**, **[BGS3]**, $\widetilde{\eta}^{(1)}$ appears as a connection form on the determinant line bundle $(\det\ker D^Z)^{-1}$, and $T^{(0)}(\omega^X,g^{\xi})$ is the natural correction to the obvious Hodge metric on the line bundle $(\det R\pi_*\xi)^{-1}$ introduced by Quillen **[Q2]** to construct the Quillen metric on $(\det R\pi_*\xi)^{-1}$. When suitably interpreted, in degree 2, equations (0.6) and (0.7) appear as curvature theorems for natural connections on the line bundles $(\det\ker D^Z)^{-1}$ and $(\det R\pi_*\xi)^{-1}$. Of course, one of the points of **[BF1,2]**, **[BGS3]** is that such curvature theorems still hold without any assumption on $\ker D^Z$ or $R\pi_*\xi$.

Objects like $\widetilde{\eta}$ and $T(\omega^X,g^{\xi})$ are secondary invariants which refine on the family index theorem of Atiyah-Singer **[AS2]** or on Riemann-Roch-Grothendieck. These last theorems are naturally functorial, in the sense they are compatible to the composition of maps. It is natural to ask whether $\widetilde{\eta}$ or $T(\omega^X,g^{\xi})$ have related functorial properties.

A first simple question is to ask how $T(\omega^X,g^{\xi})$ depends on ω^X,g^{ξ}. It was shown in **[BK]** that $T(\omega^X,g^{\xi})$ depend on ω^X,g^{ξ} via obvious Bott-Chern classes **[BoC]**, **[BGS1]**.

Another question is related to the behaviour on $\widetilde{\eta}$ or $T(\omega^X,g^{\xi})$ by the composition of two submersions. This question was solved in **[BC1]**, **[BeB]**,

[D], **[Ma]**, using the idea of adiabatic limits. Let us give an elementary application of this idea. In fact, if

$$0 \to L \to M \underset{j}{\to} N \to 0 \tag{0.8}$$

is an exact sequence of holomorphic vector bundles, and if g^M, g^N are Hermitian metrics on M, N for $\varepsilon > 0$, set

$$g_\varepsilon^M = g^M + \frac{1}{\varepsilon} j^* g^N . \tag{0.9}$$

Then if Q is any characteristic polynomial, one can easily show that if g^L is the metric on L induced by g^M,

$$Q(M, g_\varepsilon^M) \to Q(L, g^L) Q(N, g^N) . \tag{0.10}$$

Now (0.10) can be applied to the exact sequence

$$0 \to TZ \to TX \to \pi^* TS \to 0 .$$

Following a terminology introduced by physicists **[W]**, studying geometric or spectral objects depending on a metric $g_\varepsilon^{TX} = g^{TX} + \frac{1}{\varepsilon}\pi^* g^{TS}$ as $\varepsilon \to 0$ is called passing to the adiabatic limits.

As was observed in **[BF2]**, there is an analogue of (0.10) for the Levi-Civita connection of a fibered manifold.

On the other hand, the Leray spectral sequence for the de Rham or Dolbeault complexes of a fibered manifold makes the left-hand side of Riemann-Roch-Grothendieck compatible with the composition of submersions.

The behaviour of $\widetilde{\eta}$ and $T(\omega^X, g^\xi)$ under composition of submersions was obtained by adiabatic limit techniques. In **[BC1]**, **[D]**, **[BerB]**, the case where the last submersion maps to a point was considered, and the results were expressed as results on the adiabatic limit of eta invariants, or on Quillen metrics. In **[Ma]**, corresponding results were obtained for the composition of arbitrary holomorphic submersions. The above results also rely on an observation of Mazzeo-Melrose **[MazMe]** relating the adiabatic limit of the spectrum of the Dirac operator to the Leray spectral sequence.

Similar questions can be asked in the case of embeddings. We will explain the problem in the context of complex geometry. Let $i : Y \to X$ be an embedding of complex manifolds. Let η be a vector bundle on Y. If X is projective, there is a resolution of $i_* \eta$ by a complex (ξ, v) of holomorphic vector bundles on X. Then by definition, the direct image $i_! \eta \in K(X)$ is given by

$$i_! \eta = [\xi] \text{ in } K(X) , \tag{0.11}$$

the point being to show that $[\xi]$ does not depend on ξ. Now by Riemann-Roch-Grothendieck,

$$\mathrm{ch}(i_! \eta) = i_* \left(\mathrm{Td}^{-1}(N_{Y/X}) \, \mathrm{ch}(\eta) \right) \text{ in } H^{\mathrm{even}}(X, \mathbf{Q}) , \tag{0.12}$$

so that

$$\mathrm{ch}(\xi) = i_* \left(\mathrm{Td}^{-1}(N_{Y/X})\,\mathrm{ch}(\eta)\right) \text{ in } H^{\mathrm{even}}(X, \mathbf{Q}). \tag{0.13}$$

Let $g^\xi, g^\eta, g^{N_{Y/X}}$ be Hermitian metrics on $\xi, \eta, N_{Y/X}$. By analogy with (0.7) it is natural to ask whether one can refine (0.13). Namely one can ask for the existence of a current $T(\xi, g^\xi)$ on X such that

$$\frac{\overline{\partial}\partial}{2i\pi} T(\xi, g^\xi) = \mathrm{Td}^{-1}(N_{Y/X}, g^{N_{Y/X}})\,\mathrm{ch}(\eta, g^\eta)\delta_Y - \mathrm{ch}(\xi, g^\xi). \tag{0.14}$$

The current $T(\xi, g^\xi)$ has been constructed in **[B3]**, **[BGS4]**. Again it is natural to study the compatibility of the currents $T(\xi, g^\xi)$ to the composition of embeddings. This has been done in detail in **[BGS5]**.

Having now constructed objects $T(\omega^X, g^\xi)$ and $T(\xi, g^\xi)$ associated to a submersion or an embedding, which are compatible to the composition of submersions or of embeddings, the last obvious final step is to study the compatibility of these objects to the composition of an embedding and a submersion. This has been done in **[BL]** when the submersion maps to a point, and in **[B5,6]** in the general case. In **[BL]**, the main result is formulated naturally in terms of Quillen metrics. In the proof of **[BL]**, **[B5,6]**, a mysterious secondary invariant associated to a short exact sequence of holomorphic Hermitian vector bundles appeared, whose construction was somewhat puzzling. A preliminary step for the proof of **[BL]**, **[B5,6]** was the explicit evaluation of this class in **[B4]**.

The most elaborate formula in **[B5,6]** expresses a combination of analytic torsion forms as a sum of integrals along the fibre of analytic torsion currents and of Bott-Chern classes. This indicates that the refined objects introduced above fit in a refined Riemann-Roch algebra. As explained before, Gillet and Soulé **[GS3]** have explained how, in the case of arithmetic varieties, these results can be used to prove a Riemann-Roch-Grothendieck formula in Arakelov geometry. They have proved such a formula in **[GS4]** for the first Chern class, and their proof for higher Chern classes is pending.

This paper is organized as follows. Section 1 is devoted to the local index theorem and the eta invariant. In Section 2, we review various results on the local families index theorem and the $\tilde{\eta}$ forms. Finally, in Section 3, we consider analytic torsion forms and analytic torsion currents.

Part of the material contained in this survey already has been reviewed in **[B7]**.

I. The local index theorem and the eta invariant.

In this Section, we review a few well-known results on the local index theorem for Dirac operators closed manifolds, on the index theorem for manifolds with boundary, and we also give related results on eta invariant.

This Section is organized as follows. In a), we state the local Atiyah-Singer index theorem for closed manifolds of Patodi **[P1,2]**, Gilkey **[Gi1,2]**, Atiyah-Bott-Patodi **[ABoP]**. In b), we state the Atiyah-Patodi-Singer index theorem on manifolds with boundary **[APS1]** and we introduce the associated eta invariant. In c), we give the formula for the signature of a manifold with boundary obtained by Atiyah-Patodi-Singer **[APS1]**. In d), we describe result by Cheeger **[C1,2]** and Chou **[Ch]** on the index theorem on manifolds with conical singularities. In e), we recall the result by Cheeger **[C3]** on the L_2 signature of such manifolds. Finally in f), we review briefly the approach by Melrose **[Me]** to the Atiyah-Patodi-Singer index theorem, using the b-calculus.

For a detailed approach to the local index theorem, we refer to Berline-Getzler-Vergne **[BeGeV]**.

a) The local Atiyah-Singer index theorem for Dirac operators on closed manifolds.

Let Z be an even dimensional compact oriented spin manifold. Let g^{TZ} be a Riemannian metric on TZ. Let $S^{TZ} = S_+^{TZ} \oplus S_-^{TZ}$ be the Hermitian $\mathbf{Z}_2$-graded vector bundle of (TZ, g^{TZ}) spinors.

Let ∇^{TZ} be the Levi-Civita connection on (TZ, g^{TZ}). Let $\nabla^{S^{TZ}}$ be the connection induced by ∇^{TZ} on S^{TZ}. Let (ξ, g^ξ, ∇^ξ) be a Hermitian vector bundle equipped with a unitary connection. Let $\nabla^{S^{TZ}\otimes\xi}$ be the obvious connection on $S^{TZ} \otimes \xi$.

Let $c(TZ)$ be the bundle of Clifford algebras of (TZ, g^{TZ}). It is generated over $\mathbf{R}$ by 1, $X \in TZ$, and the commutation relations

$$XY + YX = -2\langle X, Y\rangle\,. \tag{1.1}$$

Then S^{TZ} is a $c(TZ)$-Clifford module. If $X \in TZ$, we denote by $c(X)$ the action of $X \in c(TZ)$ on S^{TZ}. Then $c(X)$ acts like $c(X) \otimes 1$ on $S^{TZ} \otimes \xi$.

Let D^Z be the Dirac operator associated to (g^{TZ}, ∇^ξ). If $e_1, \ldots, e_n$ is an orthonormal basis of TZ,

$$D^Z = \sum_1^n c(e_i)\nabla_{e_i}^{S^{TZ}\otimes\xi}\,. \tag{1.2}$$

Then D^Z is an odd operator, i.e. it exchanges $C^\infty(Z, S_+^{TZ}\otimes\xi)$ and $C^\infty(Z, S_-^{TZ}\otimes\xi)$. Also D^Z it is a first-order self-adjoint elliptic operator. Let $D_\pm^Z$ be the restriction of D^Z to $C^\infty(Z, S_\pm^{TZ} \otimes \xi)$, so that

$$D^Z = \begin{bmatrix} 0 & D_-^Z \\ D_+^Z & 0 \end{bmatrix}. \tag{1.3}$$

Since D_+^Z is elliptic, it is a Fredholm operator. By definition, the index $\operatorname{Ind}(D_+^Z) \in \mathbf{Z}$ is given by

$$\operatorname{Ind}(D_+^Z) = \dim\ker(D_+^Z) - \dim\ker(D_-^Z)\,. \tag{1.4}$$

Put

$$\widehat{A}(x) = \frac{x/2}{\sinh(x/2)} \ , \tag{1.5}$$
$$\mathrm{Td}(x) = \frac{x}{1-e^{-x}} \, .$$

We identify $\widehat{A}$ and Td with the corresponding multiplicative genera, the Hirzebruch genus and the Todd genus. Similarly the Chern character ch is the additive genus associated with the function $\exp(x)$.

If $\pi : F \to Z$ is a complex vector bundle, and if P is a real invariant polynomial, let $P(F) \in H^{\mathrm{even}}(Z, \mathbf{R})$ be the corresponding characteristic class. If ∇^F is a connection on F, and $R^F = \nabla^{F,2}$ is its curvature, we denote by $P(F, \nabla^F)$ the closed even form $P\left(\frac{-R^F}{2i\pi}\right)$ on Z, which represents $P(F)$ in cohomology.

Then, by the Atiyah-Singer index theorem **[AS1]**,

$$\mathrm{Ind}\,(D_+^Z) = \int_Z \widehat{A}(TZ)\,\mathrm{ch}(\xi)\, . \tag{1.6}$$

If $E = E_+ \oplus E_-$ is a $\mathbf{Z}_2$-graded vector space, let $\tau = \pm 1$ on $E_\pm$ define the $\mathbf{Z}_2$-grading. If $A \in \mathrm{End}\,(E)$, we define its supertrace $\mathrm{Tr_s}\,[A]$ by

$$\mathrm{Tr_s}\,[A] = \mathrm{Tr}[\tau A]\, . \tag{1.7}$$

The algebra $\mathrm{End}\,(E)$ is naturally $\mathbf{Z}_2$-graded, the even (resp. odd) elements commuting (resp. anticommuting) with τ. If $A, B \in \mathrm{End}\,(E)$, we define the supercommutator $[A, B]$ by the formula

$$[A, B] = AB - (-1)^{\deg A \deg B} BA. \tag{1.8}$$

By **[Q1]**, if $A, B \in \mathrm{End}\,(E)$,

$$\mathrm{Tr_s}\,[[A, B]] = 0\, . \tag{1.9}$$

Observe that since D^Z is elliptic, for $t > 0$, $\exp(-tD^{Z,2})$ is trace class. We then have the formula of Mc Kean-Singer **[McKS]**.

Proposition 1.1. *For $t > 0$,*

$$\mathrm{Ind}\,(D_+) = \mathrm{Tr_s}\left[\exp(-tD^{Z,2})\right]\, . \tag{1.10}$$

PROOF : By spectral theory,

$$\lim_{t\to+\infty} \mathrm{Tr_s}\left[\exp(-tD^{Z,2})\right] = \mathrm{Ind}\,(D_+^Z)\, . \tag{1.11}$$

Also we have the "Bianchi" identity

$$\left[D^Z, D^{Z,2}\right] = 0\, . \tag{1.12}$$

Using (1.9), (1.12), we get

$$\frac{\partial}{\partial t}\mathrm{Tr_s}\left[\exp(-tD^{Z,2})\right] = -\,\mathrm{Tr_s}\left[D^{Z,2}\exp(-tD^{Z,2})\right] = \tag{1.13}$$
$$-\tfrac{1}{2}\,\mathrm{Tr_s}\left[\left[D^Z, D^Z\exp(-tD^{Z,2})\right]\right] = 0\, .$$

Note that in the last steps of (1.13), one should express the various quantities in terms of smooth heat kernels to justify the use of (1.9). □

Let $P_t(x,y)$ be the smooth kernel of $\exp(-tD^{Z,2})$ with respect to the Riemannian volume dy. Then by (1.10),

$$\text{Ind}\,(D_+^Z) = \int_Z \text{Tr}_s\,[P_t(x,x)]\,dx\,. \tag{1.14}$$

Put $n = \dim Z$. By general results on elliptic differential operators, we know that for $x \in Z$, as $t \to 0$,

$$\text{Tr}_s\,[P_t(x,x)] = \sum_{k=0}^{p} a_{-n/2+k}(x)t^{-n/2+k} + \mathcal{O}_x(t^{-n/2+p+1})\,, \tag{1.15}$$

and $\mathcal{O}_x(t^{-n/2+p+1})$ is uniform in $x \in X$. Moreover the $a_j(x)$ only depend on the complete symbol of D^Z near x. By (1.14), (1.15),

$$\begin{aligned} &\int_Z a_j(x)dx = 0 \qquad \text{for } j \neq 0\,, \\ &\int_Z a_0(x)dx = \text{Ind}\,(D_+^Z)\,. \end{aligned} \tag{1.16}$$

Let ∇^{TZ} be the Levi-Civita connection on (TZ, g^{TZ}). In **[McKS]**, Mc Kean and Singer conjectured that "fantastic cancellations" should occur in (1.16) so that for the considered operator D^Z,

$$\begin{aligned} &a_j = 0\,,\, j < 0\,, \\ &a_0(x) = \left\{\widehat{A}(TZ, \nabla^{TZ})\,\text{ch}\,(\xi, \nabla^\xi)\right\}^{\max}\,. \end{aligned} \tag{1.17}$$

Needless to say, (1.16), (1.17) would imply the index formula (1.6).

Theorem 1.2. *Equation (1.17) holds.*

PROOF : There are two kinds of proofs of (1.17).

• The algebraic proofs of Patodi **[P1,2]**, Gilkey **[Gi1,2]**, Atiyah-Bott-Patodi **[ABoP]** describe explicitly the a_j $(j \leq 0)$ as polynomial functions of the metric g^{TZ}, the connection ∇^ξ and their derivatives. For $j < 0$, arguments of Gilkey show that there are no such polynomials other than 0. Also a_0 is shown to be a universal combination of certain Chern-Weil forms. One then only needs to verify the identity (1.17) for a_0 on sufficiently many examples, given by the $\mathbf{P}^n(\mathbf{C})$.

• The direct proofs of Getzler **[Ge1,2]**, Bismut **[B1]**, Berline-Vergne **[BeV]**. These proofs were stimulated by arguments by Alvarez-Gaumé **[Al]**, using functional integration, which suggested that there should be some explicit algebraic mechanism forcing the above local cancellations.

The proof of Getzler **[Ge2]** uses a powerful rescaling technique on the Clifford algebra. It is explained in detail in **[BeGeV**, Chapter 4]. Of particular importance is the Getzler operator **[Ge1,2]**, **[BeGeV**, Proposition 4.19], which appears when doing the Getzler rescaling on the Clifford algebra. This operator is an harmonic oscillator, and its heat kernel produces the genus $\widehat{A}$ in (1.17).

Theorem 1.2 is often called the local index theorem for Dirac operators. □

b) The Atiyah-Patodi-Singer index theorem.

Let Z be a compact manifold of dimension n with boundary $Y = \partial Z$. Let $\partial Z \times [0,1]$ be a tubular neighborhood of ∂Z in Z, u being the inward normal coordinate. Let g^{TZ} be a Riemannian metric on TZ, which is product near ∂Z, i.e.

$$g^{TZ} = g^{T\partial Z} + |du|^2 \,. \tag{1.18}$$

Let E, F be complex Hermitian vectors bundles on Z, which, near Y, are pull-backs of vector bundles on Y. Let $D : C^\infty(Z,E) \to C^\infty(Z,F)$ be an elliptic first order differential operator. Let $\sigma : E_{|Y} \to F_{|Y}$ be a bundle isometry. Let $A : C^\infty(Y, E_{|Y}) \to C^\infty(Y, F_{|Y})$ be an elliptic first order differential operator, which is self-adjoint with respect to the obvious L_2 Hermitian product. We assume that on $\partial Z \times [0,1]$,

$$D = \sigma\left(\frac{\partial}{\partial u} + A\right) . \tag{1.19}$$

Let $P_{\geq 0}$ (resp. $P_{<0}$) be the orthogonal projection operator on the direct sum of the eigenspaces of A associated to nonnegative (resp. negative) eigenvalues.

Let $C^\infty(Z,E,P)$ be the vector space of smooth sections of E over Z, such that

$$P_{\geq 0} f_{|Y} = 0 \,. \tag{1.20}$$

In **[APS1]**, Atiyah-Patodi-Singer showed that D defines a Fredholm operator. More precisely for $\ell \geq 0$, let $H^\ell(Z,E)$ be the ℓth Sobolev space of sections of E, and for $\ell \geq 1$, let $H^\ell(Z,E,P_{\geq 0})$ be the set of $s \in H^\ell(Z,E)$ such that $P_{\geq 0} s_{|Y} = 0$. Then in **[APS1]**, the authors proved that for any $\ell \geq 1$, $D : H^\ell(Z,E,P) \to H^{\ell-1}(Z,F)$ is Fredholm.

By definition, the index of D is given by

$$\operatorname{Ind}(D) = \dim\ker D - \dim\ \operatorname{coker} D \,. \tag{1.21}$$

Let D^* be the formal adjoint of D. We make D^* act on $C^\infty(Z,F,P_{<0})$ or on $H^\ell(Z,F,P_{<0})$. Then

$$\operatorname{Ind}(D) = \dim\ker D - \dim\ker D^* \,. \tag{1.22}$$

Let R_t, S_t be the heat kernels of $\exp(-tD^*D)$, $\exp(-tDD^*)$ acting on the smooth sections of E, F over the double of Z. If $x \in Z \setminus \partial Z$, as $t \to 0$, we have the asymptotic expansion

$$\operatorname{Tr}\left[(R_t - S_t)(x,x)\right] = \sum_{k=0}^{p} a_{-n/2+k}(x) t^{-n/2+k} + O_x(t^{-n/2+p+1}) \,. \tag{1.23}$$

Observe that the $a_{-n/2+k}(x)$ vanish identically near ∂Z. Also the constant coefficient $a_0(x)$ in the expansion (1.24) vanishes if n is odd.

Let $\mathrm{Sp}(A)$ be the spectrum of the operator A. For $s \in \mathbf{C}$, $\mathrm{Re}(s) > \dim Y$, put

$$\eta^A(s) = \sum_{\substack{\lambda \in \mathrm{Sp}(A) \\ \lambda \neq 0}} \frac{\mathrm{sign}\,(\lambda)}{|\lambda|^s}. \tag{1.24}$$

Then $\eta^A(s)$ is holomorphic in s on its domain of definition, and moreover

$$\eta^A(s) = \frac{1}{\Gamma(\frac{s+1}{2})} \int_0^{+\infty} t^{(s-1)/2}\, \mathrm{Tr}\left[Ae^{-tA^2}\right] dt. \tag{1.25}$$

As $t \to 0$,

$$\mathrm{Tr}\left[Ae^{-tA^2}\right] = \sum_{k=0}^{p} b_{-(\frac{\dim Y}{2}+1)+k}\, t^{-(\frac{\dim Y}{2}+1)+k}. \tag{1.26}$$

From (1.26), we deduce that $\eta^A(s)$ extends to a meromorphic function of $s \in \mathbf{C}$, with simple poles. In particular 0 appears to be a simple pole of $\eta^A(s)$.

Now we state the index formula of Atiyah-Patodi-Singer [**APS1,** Theorem 3.10].

Theorem 1.3. *The function $\eta^A(s)$ is holomorphic at 0. Moreover*

$$\mathrm{Ind}\,(D) = \int_Z a_0(x)dx - \tfrac{1}{2}\left(\eta^A(0) + \dim\ker(A)\right). \tag{1.27}$$

Proof : The proof of both statements in [**APS1**] is briefly sketched. The proof can be divided into three main steps :

- <u>The construction of a parametrix on $\partial Z \times [0, +\infty[$ for $\pm \frac{\partial}{\partial u} + A$</u>

By using the spectral decomposition for A, one constructs a parametrix for $\pm\frac{\partial}{\partial u} + A$ with the required boundary conditions. The operator A is then replaced by a scalar $\lambda \in \mathbf{R}$ in $\mathrm{Sp}(A)$, and the point is to show that these finite dimensional parametrices patch into a global parametrix Q.

By patching these parametrices with an inner parametrix for D and D^*, Atiyah-Patodi-Singer construct a parametrix for D, which demonstrates the Fredholm property of D.

- <u>The heat equation method on Z</u>

By patching the heat kernel for DD^* and D^*D on $\partial Z \times [0, +\infty[$ (which is explicitly obtained by the functional calculus) with the heat kernel on the double of Z, one obtains a global heat kernel on Z, from which one obtains a Mc Kean-Singer formula for $\mathrm{Ind}\,(D)$. Namely for $t > 0$,

$$\mathrm{Ind}\,(D) = \mathrm{Tr}\left[\exp(-tD^*D)\right] - \mathrm{Tr}\left[\exp(-tDD^*)\right]. \tag{1.28}$$

Using the corresponding heat kernels, one gets

$$\text{(1.29)} \qquad \operatorname{Ind}(D) = \int_Z \left(\operatorname{Tr}[P_t(x,x)] - \operatorname{Tr}[Q_t(x,x)] \right) dt\,.$$

One then makes $t \to 0$ in (1.29). By taking the constant term in $\int_{Z \setminus \partial Z \times [0,1]}$ as $t \to 0$, one obtains the easy term $\int_{Z \setminus \partial Z \times [0,1]} a_0(x) dx$. The contribution of $\int_{\partial Z \times [0,1]}$ as $t \to 0$ is obtained from the explicit form of the heat kernel on $\partial Z \times [0, +\infty[$. In particular, in **[APS1]**, the holomorphy at 0 of $\eta^A(0)$ is a consequence of the heat equation formula (1.29). The quantity $-\frac{1}{2}(\eta^A(0) + \dim \ker(A))$ is shown to be the contribution of $\int_{\partial Z \times [0,1]}$ to $\operatorname{Ind}(D)$. □

Let Z, g^{TZ} be as before. We assume that Z is even dimensional, oriented and spin. Let $S^{TZ} = S_+^{TZ} \oplus S_-^{TZ}$ be the Hermitian $\mathbf{Z}_2$-graded vector bundle of (TZ, g^{TZ}) spinors. Let (ξ, g^ξ, ∇^ξ) be a complex Hermitian vector bundle with unitary connection.

Let ∇^{TZ} be the Levi-Civita connection on (TZ, g^{TZ}). Then ∇^{TZ} lifts to a unitary connection on $S^{TZ} = S_+^{TZ} \oplus S_-^{TZ}$. Let $\nabla^{S^{TZ} \otimes \xi}$ be the obvious connection on $S^{TZ} \otimes \xi$.

We orient $T\partial Z$ so that if $e_1, \ldots, e_{n-1}$ is an oriented basis of $T\partial Z$, $(e_1, \ldots, e_{n-1},$
$\frac{\partial}{\partial u})$ is an oriented basis of $TZ_{|\partial Z}$.

Let $S^{T\partial Z}$ be the Hermitian vector bundle on ∂Z of the $(T\partial Z, g^{T\partial Z})$ spinors.Then $S^{T\partial Z} \simeq S^{TY}_{+|\partial Z} \simeq S^{TZ}_{-|\partial Z}$. The bundle $S^{T\partial Z} \otimes \xi_{|\partial Z}$ is naturally equipped with a connection $\nabla^{S^{T\partial Z} \otimes \xi_{|\partial Z}}$.

Let D^Z be the Dirac operator associated to the metric g^{TZ} and the connection ∇^ξ. We can write D^Z as a matrix operator,

$$\text{(1.30)} \qquad D^Z = \begin{bmatrix} 0 & D_-^Z \\ D_+^Z & 0 \end{bmatrix}.$$

Also D^Z is an elliptic first order differential operator.

Let $D^{\partial Z}$ be the Dirac operator associated to $g^{T\partial Z}$ and ∇^ξ. Then $D^{\partial Z}$ acts on $C^\infty(\partial Z, S^{T\partial Z} \otimes \xi)$ as a self-adjoint first order elliptic operator.

To the operator $D = D_+^Z$, we can apply the general approach of Section 1 b), with A replaced by $D^{\partial Z}$. So D_+^Z is restricted to act on $C^\infty(Z, S_+^{TZ} \otimes \xi, P_{\geq 0})$ and D_-^Z on $C^\infty(Z, S_-^{TZ} \otimes \xi, P_{<0})$.

Let $\eta^{D^{\partial Z}}(s)$ be the eta function of $D^{\partial Z}$.

As an application of Theorem 1.3, Atiyah-Patodi-Singer [**APS1**, Theorem 4.2] obtain the following result.

Theorem 1.4. *The fonction $\eta^{D^{\partial Z}}(s)$ is holomorphic at 0. Moreover*

$$\begin{aligned}\operatorname{Ind}(D^Z_+) = &\int_Z \widehat{A}(TZ,\nabla^{TY})\operatorname{ch}(\xi,\nabla^\xi)\\ &-\tfrac{1}{2}(\eta^{D^{\partial Z}}(0)+\dim\ker(D^{\partial Z}))\,.\end{aligned} \tag{1.31}$$

PROOF : By using the "fantastic cancellations" conjectured by Mc Kean-Singer in **[McKS]**, it is shown in **[APS1]** that in (1.23),

$$a_0(x) = \left\{\widehat{A}(TZ,\nabla^{TZ})\operatorname{ch}(\xi,\nabla^\xi)\right\}^{\max}. \tag{1.32}$$

Theorem 1.4 follows from Theorem 1.3. □

REMARK 1.5. The local index theorem, Theorem 1.2, has been used by **[ABoP]** as a tool to prove the Atiyah-Singer index theorem for Dirac operators, which by an argument of K-theory, is enough to prove the full Atiyah-Singer index theorem. However the "locality" of the index does not appear in the final answer. This is in dramatic contrast with Theorem 1.4, where $\widehat{A}(TZ,\nabla^{TZ})\operatorname{ch}(\xi,\nabla^\xi)$ is viewed as a form, and not as a representative of a cohomology class.

c) The signature of manifolds with boundary.

Let Z be an oriented manifold with boundary of dimension $4k$. Let $H^\cdot(Z)$, $H^\cdot(Z,\partial Z)$ be the absolute and relative cohomology groups of Z. Then by Poincaré duality, $H^{2k}(Z)$ and $H^{2k}(Z,\partial Z)$ are Poincaré dual.

Let $\widehat{H}(Z)$ be the image of $H(Z,\partial Z)$ in $H(Z)$. Then $\widehat{H}^{2k}(Z)$ is naturally equipped with the symmetric intersection form

$$(\alpha,\beta)\in\widehat{H}^{2k}(Z)\to\int_Z\alpha\wedge\beta\,. \tag{1.33}$$

Let $\operatorname{sign}(Z)$ be the signature of this form.

Let (V,g^V) be a real euclidean vector space of odd dimension $2\ell-1$. Let S^V be the vector space of (V,g^V) spinors.

The star operator $*^V$ induces an isomorphism $\Lambda^{\text{even}}(V^*)\simeq\Lambda^{\text{odd}}(V^*)$. Moreover

$$\Lambda^{\text{even}}(V^*)\simeq\Lambda^{\text{odd}}(V^*)\simeq S^V\otimes S^{V*}. \tag{1.34}$$

Since S^V and S^{V^*} are $c(V)$-Clifford modules, $\Lambda^{\text{even}}(V^*)\simeq\Lambda^{\text{odd}}(V^*)$ are left and right Clifford modules, and the corresponding Clifford actions commute. More precisely if $\alpha\in\Lambda^p(V^*)$, $e\in V$, if $e^*\in V^*$ corresponds to V by the metric, put

$$\begin{aligned}c(e)\alpha &= i^{\ell+p(p-1)}\left((-1)^V e^*\wedge *^V - *^V e^*\wedge\right)\alpha\,,\\ \widehat{c}(e)\alpha &= i^{\ell+p(p-1)}\left((-1)^V e^*\wedge *^V + *^V e^*\wedge\right)\alpha\,.\end{aligned} \tag{1.35}$$

Then $c(e), \widehat{c}(e)$ induce the left and right Clifford module structure on $\Lambda(V^*)$. They both preserve $\Lambda^{\text{even}}(V^*)$ and $\Lambda^{\text{odd}}(V^*)$. Let ρ be the one to one map from $\Lambda(V^*)$ into itself such that if $\alpha \in \Lambda^p(V^*)$,

$$\rho(\alpha) = i^{\ell+p(p-1)} *^V \alpha \in \Lambda^{2\ell-1-p}(V^*) . \tag{1.36}$$

Then ρ exchanges $\Lambda^{\text{even}}(V^*)$ and $\Lambda^{\text{odd}}(V^*)$ and moreover

$$\begin{aligned} c(e)\rho &= \rho c(e) \\ \widehat{c}(e)\rho &= \rho\widehat{c}(e) . \end{aligned} \tag{1.37}$$

Let Y be an oriented manifold of odd dimension $2\ell - 1$. Let $\widetilde{D}^Y$ be the operator acting on $C^\infty(Y, \Lambda(T^*Y))$ such that if $\alpha \in C^\infty(Y, \Lambda^p(T^*Y))$,

$$\widetilde{D}^Y \alpha = i^{\ell+p(p-1)} \left((-1)^p d *^{TY} - *^{TY} d\right) \alpha . \tag{1.38}$$

In view of (1.35), it is clear that $\widetilde{D}^Y$ is a Dirac operator acting on $C^\infty(Y, \Lambda(T^*Y))$. Moreover it preserves $C^\infty(Y, \Lambda^{\text{even}}(T^*Y))$ and $C^\infty(Y, \Lambda^{\text{odd}}(T^*Y))$. By (1.36), it $\widetilde{D}^Y$ splits into two equivalent operators, $\widetilde{D}^Y = \widetilde{D}^{Y,\text{even}} \oplus \widetilde{D}^{Y,\text{odd}}$. Also one verifies easily that $\ker(\widetilde{D}^Y)$ consists of the harmonic forms, i.e.

$$\ker \widetilde{D}^Y \simeq H^{\cdot}(Y) . \tag{1.39}$$

Let $\eta^{\widetilde{D}^{Y,\text{even}}}(s)$ be the eta function of $\widetilde{D}^{Y,\text{even}}$.

Let $\mathcal{L}$ be the multiplicative genus associated to

$$\mathcal{L}(x) = \frac{x}{\tanh(x)} . \tag{1.40}$$

We then have the signature formula of [**APS1,** Theorem 4.14] for $\operatorname{sign}(Z)$.

Theorem 1.6. *The following identity holds,*

$$\operatorname{sign}(Z) = \int_Z \mathcal{L}(TZ, \nabla^{TZ}) - \eta^{\widetilde{D}^{\partial Z,\text{even}}}(0) . \tag{1.41}$$

Proof : The proof of [**APS1**] consists in first using Theorem 1.4 with $\xi = S^{TZ}$. In a second step, the kernel of the corresponding Dirac operator D^Z is related to the kernel of the Dirac operator on $Z \bigcup_{\partial Z} \partial Z \times] - \infty, 0]$, which is Z extended by an infinite cylinder. □

d) Manifolds with conical singularities : the index theorem of Cheeger and Chou.

Let Y be a smooth manifold equipped with a Riemannian metric g^{TY} on TY. Let $C(Y) = Y\times]0,+\infty[$ be the metric cone equipped with the metric $dr^2+r^2g^{TY}$. Then $C(Y)$ can be compactified into a honest cone with vertex δ. Let $C^{[0,\ell]}(Y)$ be the truncated cone $\{x=(r,y)\in C(Y), 0\leq r\leq \ell\}$.

Let Z be a smooth manifold with boundary $Y=\partial Z$, taken as in Section 1 c). Let $\widehat{Z}_\ell = Z\bigcup_{Y\times\{\ell\}} C^{[0,\ell]}(\partial Z)$ be the manifold Z with the cone $C^{[0,\ell]}(\partial Z)$ attached. Let $g^{T\widehat{Z}_\ell}$ be a metric on $\widehat{Z}_\ell$, which coincides with the conical metric $dr^2+\frac{r^2}{\ell^2}g^{T\partial Z}$ on $C^{[0,\ell]}(\partial Z)$.

Assume that Z is oriented, even dimensional and spin. Put $n=\dim Z$. Let $S^{T\widehat{Z}_\ell} = S_+^{T\widehat{Z}_\ell}\oplus S_-^{T\widehat{Z}_\ell}$ be the Hermitian vector bundle of $(T\widehat{Z}_\ell, g^{T\widehat{Z}_\ell})$ spinors. Let $S^{T\partial Z}$ be the vector bundle of $(T\partial Z, g^{T\partial Z})$ spinors. Then over $C^{[0,\ell]}(\partial Z)$, $S^{T\widehat{Z}_\ell}\simeq S^{T\partial Z}\oplus S^{T\partial Z}$.

Let (ξ, g^ξ, ∇^ξ) be a Hermitian vector bundle with unitary connection on Z, which is product near ∂Z. Then it extends to $\widehat{Z}_\ell$.

Let $D^{\widehat{Z}_\ell}$ be the formally self-adjoint Dirac operator on $\widehat{Z}^\ell$ associated to $(g^{T\widehat{Z}_\ell}, \nabla^\xi)$.

To simplify the exposition we will assume that $D^{\partial Z}$ is invertible. Let $\ell>0$ be large enough so that for any $\lambda\in \mathrm{Sp}(D^{\partial Z}), \ell|\lambda|>\frac{1}{2}$.

Then by following ideas of Cheeger **[C1, 2]**, Chou showed in **[Ch]** that D_ℓ extends to a self-adjoint operator with domain the first Sobolev space $H^1(\widehat{Z}_\ell, S^{T\widehat{Z}_\ell}\otimes\xi)$. Moreover for $p\geq 1$, $D_+^{\widehat{Z}_\ell}$ is a Fredholm operator $H^p(\widehat{Z}_\ell, S_+^{\widehat{Z}_\ell}\otimes\xi)\to H^{p-1}(\widehat{Z}_\ell, S_-^{\widehat{Z}_\ell}\otimes\xi)$ and

$$\mathrm{Ind}\,(D_+^{\widehat{Z}_\ell}) = \dim\ker D_+^{\widehat{Z}_\ell} - \dim\ker D_-^{\widehat{Z}_\ell}. \tag{1.42}$$

Let D_+^Z be the Atiyah-Patodi-Singer operator considered in Section 1 e).

Proposition 1.7. *The following identity holds*

$$\mathrm{Ind}\,(D_+^Z) = \mathrm{Ind}\,(D_+^{\widehat{Z}_\ell}). \tag{1.43}$$

PROOF : If s is a H^1 section of $S_+^{T\widehat{Z}_\ell}\otimes\xi$ such that $D_+^{\widehat{Z}_\ell}s=0$, then on $C^{[0,\ell]}(\partial Z)$,

$$\left(\frac{\partial}{\partial r}+\frac{n-1}{2r}\right)s+\frac{\ell D^{\partial Z}s}{r}=0. \tag{1.44}$$

Write s in the form

$$s=\sum_{\lambda\in \mathrm{Sp}(D^{\partial Z})} s_\lambda(r),\ D^{\partial Z}s_\lambda=\lambda s_\lambda. \tag{1.45}$$

From (1.44), one deduces easily that since $\int_{[0,\ell]\times Z} |h|^2 r^{n-1} dr dx < +\infty$, if $\ell\lambda \geq \frac{1}{2}$, then $s_\lambda = 0$. Recall that if $\lambda \in \mathrm{Sp}(D^{\partial Z}), \ell|\lambda| > \frac{1}{2}$. So if $D^{\widehat{Z}_\ell} s = 0$, then $P_{\geq 0} s_{|\partial Z} = 0$.

We then find that $\ker D_+^Z \simeq \ker D_+^{\widehat{Z}_\ell}$. Similarly, one proves that $\ker D_-^Z \simeq \ker D_-^{\widehat{Z}_\ell}$. Our Proposition follows. □

REMARK 1.8. Proposition 1.7 asserts the remarkable fact that while Atiyah-Patodi-Singer conditions are global on ∂Z, the apparently "local" L_2 conditions of Cheeger and Chou imposes on the kernels and cokernels of $D^{\widehat{Z}_\ell}$ the global boundary conditions of Atiyah-Patodi-Singer. This is because any neighborhood of the vertex δ encodes the global geometry of the cross section.

Cheeger[**C1,2**] and Chou [**Cho**] apply the heat equation method to the index problem considered above. In particular the functional calculus over cones introduced by Cheeger allows an explicit evaluation of the heat kernel on $C(\partial Z)$. A direct computation shows that

$$\operatorname{Ind} D_+^{\widehat{Z}_\ell} = \int_A \widehat{A}(TZ, \nabla^{TZ}) \operatorname{ch}(\xi, \nabla^\xi) - \tfrac{1}{2}\eta^{D^{\partial Z}}(0) . \tag{1.46}$$

In (1.46), $\frac{1}{2}\eta(0)$ appears as the contribution of the vertex δ to the index. Of course given Proposition 1.7, this result fits with Theorem 1.4.

e) The L_2 cohomology of manifolds with conical singularities.

Let $\widehat{Z} = \widehat{Z}_1$ be taken as before. Let d be the de Rham operator on $\widehat{Z}$, let d^* be the formal adjoint of d with respect to the metric $g^{T\widehat{Z}}$.

For simplicity, assume that $\widehat{Z}$ is even dimensional. In [**C3**], Cheeger calculated the L_2 cohomology of a class of manifolds with singularities, the riemannian pseudo-manifolds, whose simplest version is the above manifold $\widehat{Z}$. In this case the strong closures of d and d^* are adjoint to each other, and so the L^2 cohomology exhibits Poincaré duality. This remarkable fact raised the question of the connection between the L_2 cohomology and the intersection cohomology of the corresponding spaces of Goresky and Mc Pherson. For a history of the subject, we refer to [**Kl**].

For our $\widehat{Z}$ considered above, if $H_{(2)}(\widehat{Z})$ denote the L_2 cohomology of $\widehat{Z}$, by [**C1**], [**C2,** p. 132, 133],

$$\begin{aligned} H^i_{(2)}(Z) &= H^i(Z) \text{ for } i < \frac{\dim Z}{2}, \\ &= \text{the image of } H^i(Z,\partial Z) \text{ in } H^i(Z) \text{ for } i = \frac{\dim Z}{2}, \\ &= H^i(Z,\partial Z) \text{ for } i > \frac{\dim Z}{2}. \end{aligned} \tag{1.47}$$

In particular if $\operatorname{sign}_2(Z)$ is the L_2 signature of $\widehat{Z}$,

$$\operatorname{sign}_2(Z) = \operatorname{sign}(Z). \tag{1.48}$$

Put

$$D = d + d^* . \tag{1.49}$$

In this case, Cheeger showed that the operator D is essentially self-adjoint on its obvious domain, and that $\ker D \simeq H_{(2)}(\widehat{Z})$.

The L_2 signature $\operatorname{sign}_2(X)$ can be shown to be equal to $\operatorname{Ind}(D_+)$. By using a heat equation formula for the signature, Cheeger **[C2]** obtains the formula

$$\operatorname{sign}_2(\widehat{Z}) = \int_Z \mathcal{L}(TZ, \nabla^{TZ}) - \eta^{\widetilde{D}^{\partial Z, \mathrm{even}}}(0). \tag{1.50}$$

In view of (1.48), it is natural that formula (1.50) coincides with the Atiyah-Patodi-Singer formula for $\operatorname{sign}(Z)$. In **[C2]**, Cheeger used (1.50) as a starting point for the construction of the $\mathcal{L}$-classes of pseudo-manifolds in terms of the eta invariants of the links.

The explicit computation of the L_2 signature for more general spaces is of considerable interest **[Mü2]**, **[St]**. Formula (1.50) for the L_2 signature of $\widehat{Z}$ appears as the prototype of such formulas.

When $\widehat{Z}$ is odd dimensional, there is an obstruction to Poincaré duality, which lies in the middle dimensional cohomology of ∂Z. When $\operatorname{sign}(\partial Z) = 0$, Cheeger **[C1]** has shown how to restore Poincaré duality by imposing $*$-invariant boundary conditions.

f) The *b*-calculus of Melrose.

To attack the index problem of Atiyah-Patodi-Singer from a different point of view, Melrose has developed a new machinery, the *b*-calculus. In **[Me]**, Melrose introduces the idea of a *b*-metric $\frac{dr^2}{r^2} + g^{TY}$ on the cylinder $]0, +\infty[\times Y$ (which differs from the conical metric $dr^2 + r^2 g^{TY}$ by the factor $\frac{1}{r^2}$).

Let us assume for simplicity that $D^{\partial Z}$ is invertible. If $\check{Z}$ is the manifold Z with the cylinder of $Y = \partial Z$ attached, Melrose considers the index problem for the Dirac operator associated to a corresponding *b*-metric. The operator $D^{\check{Z}}$ is still Fredholm, but the corresponding heat kernels are no longer trace class. Still they have a *b*-trace, i.e. a renormalized trace.

In **[Me]**, Melrose shows that in the appropriate context, if,

$$\alpha_t = b\text{-}\mathrm{Tr}\left[\exp(-tD_-^{\check{Z}} D_+^{\check{Z}})\right] - b\text{-}\mathrm{Tr}\left[\exp(-tD_+^{\check{Z}} D_-^{\check{Z}})\right], \tag{1.51}$$

then

$$\begin{aligned} \lim_{t\to+\infty} \alpha_t &= \operatorname{Ind}(D_+^{\check{Z}}), \\ \lim_{t\to 0} \alpha_t &= \int_Z \widehat{A}(TZ, \nabla^{TZ}) \operatorname{ch}(\xi, \nabla^\xi). \end{aligned} \tag{1.52}$$

The fundamental fact is that in (1.52), α_t is non constant. The eta invariant $\frac{\eta(0)}{2}$ then appears as a formula for $-\int_0^{+\infty} \frac{d\alpha}{dt} dt$.

II. The local families index theorem, adiabatic limits and the $\tilde{\eta}$ form.

In this Section, we review various results connected with a refinement of the families index theorem of Atiyah-Singer [**AS2**], the local families index theorem [**B2**].

In a) and b), we briefly state the families index theorem of Atiyah-Singer [**AS2**], [**APS2**] in its cohomological form. In c) and d), we introduce one essential technical tool, the superconnections of Quillen [**Q1**], which provide a refinement of Chern-Weil theory. In e), when the fibres Z of the fibration $\pi : M \to B$ are even dimensional, we construct the corresponding Levi-Civita superconnection, and we state the local families index theorem of [**B2**]. Also we obtain an associated odd form $\tilde{\eta}$ on B, which transgresses the families index theorem [**AS2**] at the level of differential forms. In f), we review the results of [**Q2**], [**BF1,2**] on determinant bundles. In g), we give the local families index theorem of [**BF2**] when the fibres Z are odd dimensional, and we construct an associated even $\tilde{\eta}$ form on B. In h), we relate the eta invariant to the component of degree 0 of $\tilde{\eta}$. In i), we state the holonomy theorem [**BF2**], [**C4**] in the form suggested by Witten [**W**]. In j), we give results of [**BC1**], [**D**] on the adiabatic limit of eta invariants. Finally in k), we state the families index theorem for families of manifolds with boundary [**BC2,3**], [**MeP1,2**].

a) The case where Z is even dimensional.

Let $\pi : X \to B$ be a submersion with compact fibre Z. Assume that the fibres are oriented and spin. Let g^{TZ} be a metric on the relative tangent bundle TZ, and let S^{TZ} be the vector bundle on Z of the (TZ, g^{TZ}) spinors.

Let $(\xi, g^{\xi}, \nabla^{\xi})$ be a Hermitian vector bundle on X, equipped with a unitary connection.

For $b \in B$, let D_b^Z be the Dirac operator acting on $C^\infty(Z_b, (S^{TZ} \otimes \xi)_{Z_b})$. Then $(D_b^Z)_{b \in S}$ is a family of elliptic self-adjoint operators.

Assume first that the fibres Z are even dimensional. Then $S^{TZ} = S_+^{TZ} \oplus S_-^{TZ}$. Also D_b^Z interchanges $C^\infty(Z_b, (S_+^{TZ} \otimes \xi)_{|Z_b})$ and $C^\infty(Z_b, (S_-^{TZ} \otimes \xi)_{|Z_b})$. Let $D_{+,b}^Z$ be the restriction of D_b^Z to $C^\infty(Z_b, (S_+^{TZ} \otimes \xi)_{|Z_b})$. Then $(D_{+,b}^Z)_{b \in B}$ is a family of Fredholm operators over B.

By Atiyah-Singer [**AS2**], the family $(D_{+,b})$ defines an element $\operatorname{Ind} D_+ \in K^0(B)$. When $\ker D_{\pm,b}^Z$ is of locally constant dimension,

$$\operatorname{Ind}(D_+^Z) = \left[\ker D_+^Z - \ker D_-^Z\right] \text{ in } K^0(B). \tag{2.1}$$

In the general case [**AS2**], one can perturb the family (D_+^Z) by a family of fibrewise regularizing operator so as to represent $\operatorname{Ind}(D_+^Z)$ by an explicit difference bundle on B.

In [**AS2**], Atiyah and Singer have calculated the index of a general family of elliptic operators in terms of the principal symbol of these operators. Recall that the Chern character map $K^0(B) \otimes_{\mathbf{Z}} \mathbf{Q} \to H^{\text{even}}(B, \mathbf{Q})$ is an identification of $\mathbf{Q}$-modules. In [**AS2**], Atiyah-Singer obtain the formula

$$\operatorname{ch}(\operatorname{Ind} D_+^Z) = \pi_* \left[\widehat{A}(TZ) \operatorname{ch}(\xi)\right] \text{ in } H^{\text{even}}(B, \mathbf{Q}). \tag{2.2}$$

When B is reduced to a point, Ind $D_+ \in \mathbf{Z}$, and (2.2) reduces to the Atiyah-Singer index theorem **[AS1]**,

$$\text{(2.3)} \qquad \operatorname{Ind}(D^Z_+) = \int_Z \widehat{A}(TZ)\operatorname{ch}(\xi) \text{ in } \mathbf{Z}\,.$$

b) The case where Z is odd dimensional.

If Z is odd dimensional, $(D^Z)_{s\in S}$ is a family of self-adjoint operators. By **[AS3]**, **[APS2,** Section 3], it defines an element $\operatorname{Ind}(D) \in K^1(B)$.

Again there is a Chern character map ch : $K^1(B)\otimes_{\mathbf{Z}}\mathbf{Q} \to H^{\text{odd}}(B,\mathbf{Q})$ which is compatible with Bott periodicity. Then the obvious analogue of (2.2) still holds.

c) Superconnections : the $\mathbf{Z}_2$-graded case.

Let A be a $\mathbf{Z}_2$-graded algebra. If $a, b \in A$, we define the supercommutator $[a,b]$ by

$$\text{(2.4)} \qquad [a,b] = ab - (-1)^{\deg a \deg b} ba\,.$$

Let $\pi : E = E_+ \oplus E_- \to B$ be a complex $\mathbf{Z}_2$-graded vector bundle. As we saw in Section 1 a), the bundle of algebras End(E) is $\mathbf{Z}_2$-graded.

Consider the $\mathbf{Z}_2$-graded bundle of algebras $\Lambda(T^*B)\widehat{\otimes}\operatorname{End}(E)$. We extend the supertrace $\operatorname{Tr_s}$: $\operatorname{End}(E) \to \mathbf{C}$ defined in (1.7) to a map : $\Lambda(T^*B)\widehat{\otimes}\operatorname{End}(E) \to \Lambda(T^*B)$, by the formula

$$\text{(2.5)} \qquad \operatorname{Tr_s}[\omega A] = \omega \operatorname{Tr_s}(A)\,,\ \omega \in \Lambda(T^*B), A \in \operatorname{End}(E)\,.$$

Using (1.9), one verifies that $\operatorname{Tr_s}$ still vanishes on supercommutators in $\Lambda(T^*_{\mathbf{R}}B)\widehat{\otimes}\operatorname{End}(E)$.

The vector bundle $\Lambda(T^*B)\widehat{\otimes}E$ is naturally $\mathbf{Z}_2$-graded.

DEFINITION 2.1. A superconnection A is an odd differential operator acting on $C^\infty(B,\Lambda(T^*B)\widehat{\otimes}E)$ such that if $\omega \in C^\infty(B,\Lambda(T^*B))$, $s \in C^\infty(B,E)$,

$$\text{(2.6)} \qquad A(\omega s) = d\omega s + (-1)^{\deg\omega}\omega A s\,.$$

Let $\nabla^E = \nabla^{E_+} \oplus \nabla^{E_-}$ be a connection on E preserving the splitting $E = E_+ \oplus E_-$. Then if $S = A - \nabla^E$, S is a smooth section of $(\Lambda(T^*S)\widehat{\otimes}\operatorname{End}(E))^{\text{odd}}$. Conversely, any superconnection can be written in the form $A = \nabla^E + S$.

DEFINITION 2.2. The curvature of the superconnection A is the operator A^2.

Since A is odd, one verifies easily that A^2 is a tensor, so that $A^2 \in C^\infty(B,(\Lambda(T^*B)\widehat{\otimes}\operatorname{End}(E))^{\text{even}})$.

One has the Bianchi identity

$$\text{(2.7)} \qquad [A, A^2] = 0\,.$$

Let φ be the endomorphism of $\Lambda(T^*S)^{\text{even}} : \omega \to (2i\pi)^{-\deg\omega/2}\omega$.

DEFINITION 2.3. Let $\mathrm{ch}(E,A)$ be the even form

$$\mathrm{ch}(E,A) = \varphi\,\mathrm{Tr_s}\left[\exp(-A^2)\right]. \tag{2.8}$$

Now we have the result of Quillen **[Q1]**.

Theorem 2.4. *The even form* $\mathrm{ch}(E,A)$ *is closed and its cohomology class* $[\mathrm{ch}(E,A)]$ *is given by*

$$[\mathrm{ch}(E,A)] = \mathrm{ch}(E). \tag{2.9}$$

PROOF : Using the Bianchi identity (2.7), and the vanishing of $\mathrm{Tr_s}$ on supercommutators, we get

$$d\,\mathrm{Tr_s}\left[\exp(-A^2)\right] = \mathrm{Tr_s}\left[\left[A,\exp(-A^2)\right]\right] = 0. \tag{2.10}$$

Therefore the form $\mathrm{ch}(E,A)$ is closed. By universality, we find that $[\mathrm{ch}(E,A)]$ does not depend on A. Taking $A = \nabla^E$ as before, we get (2.9). □

EXAMPLE. Let $\pi : E \to B$ be a real oriented even dimensional spin vector bundle. Let g^E be an Euclidean metric on V. Let $S^E = S^E_+ \oplus S^E_-$ be the vector bundle of (E, g^E) spinors.

If $Y \in E$, let $c(Y)$ denote the Clifford action of Y on S^E. Then $\sqrt{-1}c(Y)$ is a self-adjoint odd endomorphism of S^E.

Let ∇^E be an Euclidean connection on E, let ∇^{S^E} be its lift to S^E. Put

$$A = \pi^*\nabla^{S^E} + \sqrt{-1}c(Y). \tag{2.11}$$

Then A is a superconnection on $\pi^* S^E$.

Clearly

$$A^2 = \nabla^2 + \sqrt{-1}\left[\nabla^{S^E}, c(Y)\right] + |Y|^2. \tag{2.12}$$

The form $\mathrm{ch}(E,A)$ on the total space of E is gaussian-shaped along the fibres of E. Therefore $\mathrm{ch}(E,A)$ represents a cohomology class in $H_c^{\mathrm{even}}(E,\mathbf{Q})$, the cohomology with compact support in E. By Mathai-Quillen [**MQ,** Theorem 4.5 and 4.10],

$$\mathrm{ch}(E,A) = (-1)^{\dim E/2}\pi^*\widehat{A}^{-1}(E,\nabla^E)\omega, \tag{2.13}$$

where ω is a closed form on E representing the Thom class of E.

Note that in (2.13), one cannot make $c(Y) = 0$, if we want to produce a class in $H_c^{\mathrm{even}}(V,\mathbf{Q})$.

Let $\pi : E = E_+ \oplus E_- \to B$ be a $\mathbf{Z}_2$-graded complex vector bundle. Let $g^E = g^{E_+} \oplus g^{E_-}$ be a Hermitian metric on E such that E_+ and E_- are orthogonal in E. Let $\nabla^E = \nabla^{E_+} \oplus \nabla^{E_-}$ be a split unitary connection on $E = E_+ \oplus E_-$.

Let $V \in C^\infty(S, \text{End}^{\text{odd}}(E))$ be self-adjoint. For $t > 0$, let A_t be the superconnection

$$A_t = \nabla^E + \sqrt{t}V\,. \tag{2.14}$$

We extend φ to a map $\Lambda(T^*B) \otimes_{\mathbf{R}} \mathbf{C} \to \Lambda(T^*B) \otimes_{\mathbf{R}} \mathbf{C}$.

DEFINITION 2.5. Put

$$\begin{aligned} \alpha_t &= \varphi \operatorname{Tr_s}\left[\exp(-A_t^2)\right]\,, \\ \beta_t &= \frac{1}{(2i\pi)^{1/2}} \varphi \operatorname{Tr_s}\left[\frac{\partial A_t}{\partial t}\exp(-A_t^2)\right]\,. \end{aligned} \tag{2.15}$$

One verifies easily that, the forms α_t, β_t are real. Also by Theorem 2.4, α_t is closed, and $[\alpha_t] = \text{ch}(E)$. By an obvious extension of Chern-Simons theory to superconnections,

$$\frac{\partial}{\partial t}\alpha_t = -d\beta_t\,. \tag{2.16}$$

Clearly

$$\alpha_0 = \text{ch}(E, \nabla^E)\,. \tag{2.17}$$

Assume that $\ker V$ has locally constant dimension, i.e. $\ker V$ is a $\mathbf{Z}_2$-graded smooth vector subbundle of E. Then

$$[E] = [\ker V] \text{ in } K^0(B)\,. \tag{2.18}$$

Let $\nabla^{\ker V}$ be the orthogonal projection of ∇^E on $\ker V$.

Now we state a result by Berline-Vergne [**BeV,** Theorem 1.9], [**BeGeV,** Theorems 9.2 and 9.7].

Theorem 2.6. *As $t \to +\infty$,*

$$\begin{aligned} \alpha_t &= \text{ch}(\ker V, \nabla^{\ker V}) + \mathcal{O}\left(\frac{1}{\sqrt{t}}\right)\,, \\ \beta_t &= \mathcal{O}\left(\frac{1}{t^{3/2}}\right)\,. \end{aligned} \tag{2.19}$$

DEFINITION 2.7. Put

$$\widehat{\eta} = \int_0^{+\infty} \beta_t dt\,. \tag{2.20}$$

Theorem 2.8. *The form $\widehat{\eta}$ is real and odd, and moreover*

$$d\widehat{\eta} = \mathrm{ch}(E, \nabla^E) - \mathrm{ch}(\ker V, \nabla^{\ker V}). \tag{2.21}$$

PROOF : This follows from (2.16), (2.17), (2.19). □

DEFINITION 2.9. We will say that (E, g^E, ∇^E, V) splits if $E = \ker V \oplus \mathrm{Im}\ V$, if the connection ∇^E preserves $\ker D$ and $\mathrm{Im}\ V$, and V is a unitary odd section of $\mathrm{End}\,(\mathrm{Im}\ V)$ preserving $\nabla^{\mathrm{Im}\ V}$.

Now we state a result characterizing the form $\widehat{\eta}$ uniquely. This is an obvious analogue of corresponding results for Bott-Chern classes in **[BGS1]**.

Theorem 2.10. *There exists a unique way to associate to (E, g^E, ∇^E, V) an odd form $\widehat{\eta}$ in $C^\infty(B, \Lambda^{\mathrm{odd}}(T^*B))/dC^\infty(B, \Lambda^{\mathrm{even}}(T^*B))$ having the following three properties*

a) *$\widehat{\eta}$ is functorial.*

b) *If (E, g^E, ∇^E, V) splits, then $\widehat{\eta} = 0$.*

c) *The following identity holds*

$$d\widehat{\eta} = \mathrm{ch}(E, \nabla^E) - \mathrm{ch}(\ker V, \nabla^{\ker V}). \tag{2.22}$$

PROOF : If (E, g^E, ∇^E, V) splits, then

$$A_t^2 = \nabla^{E,2} + tP^{\mathrm{Im}\ V} \tag{2.23}$$

and so

$$\beta_t = 0. \tag{2.24}$$

It follows that if $\widehat{\eta}$ is taken as in (2.20),

$$\widehat{\eta} = 0. \tag{2.25}$$

So the form $\widehat{\eta}$ of Definition 2.9 has properties a), b), c). To establish uniqueness, one verifies easily that over $S \times [0,1]$, one can deform (E, g^E, ∇^E, V) at $s = 0$ into a split object at $s = 1$. Let $(\widetilde{E}, g^{\widetilde{E}}, \nabla^{\widetilde{E}}, \widetilde{V})$ be the corresponding object on $S \times [0,1]$. Then if $\widehat{\eta}$ is taken as in our Theorem,

$$\widehat{\eta} = \int_{[0,1]} \left(\mathrm{ch}\, \widetilde{E}, \nabla^{\widetilde{E}}) - \mathrm{ch}(\ker \widetilde{V}, \nabla^{\ker \widetilde{V}})\right) \text{ modulo coboundaries}, \tag{2.26}$$

which characterizes the class of $\widehat{\eta}$ uniquely. □

d) Superconnections : the odd case.

Let now $\pi : E \to B$ be a Hermitian vector bundle on B.

Let σ be an odd variable such that $\sigma^2 = 1$. Then $E \otimes \mathbf{C}(\sigma)$ is a $\mathbf{Z}_2$-graded vector bundle, and $\mathrm{End}\,(E) \otimes \mathbf{C}(\sigma)$ is a $\mathbf{Z}_2$-graded algebra. Let Tr^σ be the functional from $\Lambda(T^*B)\widehat{\otimes}(\mathrm{End}\,(E) \otimes \mathbf{C}(\sigma))$ into $\Lambda(T^*B)$ such that if $\omega \in A(T^*B)$, $A \in \mathrm{End}\,(E)$,

$$\begin{aligned} &\mathrm{Tr}^\sigma(\omega A) = 0 . \\ &\mathrm{Tr}^\sigma(\omega A \sigma) = \omega\ \mathrm{Tr}(A) . \end{aligned} \tag{2.27}$$

Again Tr^σ vanishes on supercommutators.

DEFINITION 2.11. A superconnection is an odd differential operator acting on $C^\infty(B, \Lambda(T^*B))\widehat{\otimes}(E \otimes \mathbf{C}(\sigma))$ such that (2.6) still holds

Let A^2 be the curvature of A. Then $A^2 \in C^\infty(B, (\Lambda(T^*B)\widehat{\otimes}(\mathrm{End}\,(E)\otimes \mathbf{C}(\sigma))^{\mathrm{even}})$.

Now we have the result of Quillen [**Q1,** Section 5].

Proposition 2.12. *The odd form*

$$\alpha = \mathrm{Tr}^\sigma\left[\exp(-A^2)\right] \tag{2.28}$$

is closed and exact.

PROOF : By proceeding as in the proof of Theorem 2.4, we see that α is closed. Moreover, we find that $[\alpha]$ does not depend on A. Also if $A = \nabla^E$, $\alpha = 0$. The proof of our Proposition is complete. □

Let $\pi : (E, g^E) \to B$ be a complex Hermitian vector bundle. Let ∇^E be a unitary connection on (E, g^E).

Let $V \in C^\infty(B, \mathrm{End}\,(E))$ be self-adjoint. For $t > 0$, let A_t be the superconnection

$$A_t = \nabla^E + \sqrt{t} V \sigma . \tag{2.29}$$

DEFINITION 2.13. Put

$$\begin{aligned} \alpha_t &= (2i)^{1/2} \varphi\, \mathrm{Tr}^\sigma\left[\exp(-A_t^2)\right] , \\ \beta_t &= \frac{1}{\sqrt{\pi}} \varphi\, \mathrm{Tr}^\sigma\left[\frac{\partial A_t}{\partial t} \exp(-A_t^2)\right] , \end{aligned} \tag{2.30}$$

As before, the forms α_t, β_t are real and by Proposition 2.12, the form α_t is closed and exact. Again as in (2.16),

$$\frac{\partial \alpha_t}{\partial t} = -d\beta_t . \tag{2.31}$$

Clearly

$$\alpha_0 = 0 . \tag{2.32}$$

Assume that $\ker(V)$ is of locally constant dimension, so that $\ker V$ is a vector subbundle of E. Let $\nabla^{\ker(V)}$ be the orthogonal projection of ∇^E on $\ker V$.

By [**BeV,** Theorem 1.9], [**BeGeV,** Theorems 9.2 and 9.7], we have

Theorem 2.14. *As $t \to +\infty$,*

$$\alpha_t = \mathcal{O}\left(\frac{1}{\sqrt{t}}\right),$$
$$\beta_t = \mathcal{O}\left(\frac{1}{t^{3/2}}\right). \tag{2.33}$$

DEFINITION 2.15. Put

$$\widehat{\eta} = \int_0^{+\infty} \beta_t dt. \tag{2.34}$$

Theorem 2.16. *The even form $\widehat{\eta}$ is real and closed.*

PROOF : By (2.32), (2.33)

$$\alpha_0 = 0, \alpha_\infty = 0. \tag{2.35}$$

Using (2.31), we find that

$$d\widehat{\eta} = 0. \tag{2.36}$$

□

Let $E_{>0}$ ($E_{<0}$) be the direct sum of the eigenspaces of E associated to positive (resp. negative) eigenvalues of V. Put $E_0 = \ker V$. Then E splits orthogonally as

$$E = E_{>0} \oplus E_{<0} \oplus E_0. \tag{2.37}$$

The following result was proved in [**BC**, Theorem 2.43].

Theorem 2.17. *The cohomology class $[\widehat{\eta}]$ of $\widehat{\eta}$ does not depend on ∇^E. More precisely*

$$[\widehat{\eta}] = \tfrac{1}{2}\left(\mathrm{ch}(E_{>0}) - \mathrm{ch}(E_{<0})\right). \tag{2.38}$$

PROOF : Using Theorem 2.16 and the universality of $\widehat{\eta}$, it is clear that $[\widehat{\eta}]$ does not depend on the metric g^E, the connection ∇^E or V as long as the splitting is kept fixed and orthogonal.

By deforming V into $\frac{V}{|V|}$ (which acts like 0 on E_0), we may and will assume that V is $+1$ on $E_{>0}$, -1 on $E_{<0}$ and 0 on E_0, and suppose that ∇^E is a split connection $\nabla^E = \nabla^{E_{>0}} \oplus \nabla^{E_{<0}} \oplus \nabla^{E_0}$. Then we get

$$\beta_t = \frac{1}{\sqrt{\pi}}\varphi\,\mathrm{Tr}^\sigma\left[\frac{V\sigma}{2\sqrt{t}}\exp\left(-\left(\nabla^{E,2} + t1_{E_{>0}\oplus E_{<0}}\right)\right)\right] \tag{2.39}$$

so that
(2.40)
$$\widehat{\eta} = \tfrac{1}{2}\varphi\,\mathrm{Tr}^\sigma\left[V\sigma\exp(-\nabla^{E,2})\right] = \tfrac{1}{2}\left(\mathrm{ch}(E_{>0}, \nabla^{E_{>0}}) - \mathrm{ch}(E_{<0}, \nabla^{E_{<0}})\right)$$

from which we get (2.38). □

REMARK 2.18. Observe that

$$\widehat{\eta} = \frac{1}{2\sqrt{\pi}} \int_0^{+\infty} t^{-1/2} \varphi \operatorname{Tr}^{\sigma} \left[V\sigma \exp\left(- \left(\nabla^E + \sqrt{t} V\sigma \right)^2 \right) \right] dt. \tag{2.41}$$

In particular

$$\widehat{\eta}^{(0)} = \frac{1}{2\sqrt{\pi}} \int_0^{+\infty} t^{-1/2} \operatorname{Tr}\left[V \exp\left(-tV^2 \right) \right] dt. \tag{2.42}$$

We discover that the expression for $\widehat{\eta}^{(0)}$ is formally exactly the same as the one in (1.25) for $\frac{1}{2}\eta^A(0)$. By (2.38) or by (2.42),

$$\widehat{\eta}^{(0)} = \tfrac{1}{2} \left(\operatorname{rk}(E_{>0}) - \operatorname{rk}(E_{<0}) \right). \tag{2.43}$$

Of course in (1.25), $\frac{1}{2}\eta^A(0)$ is a renormalized version of (2.42). This formal analogy will be very important in the sequel.

e) The local families index theorem : the even dimensional case.
We now make the same assumptions as in Section 2a). For $b \in B$, put

$$H_{\pm,b} = C^{\infty}(Z_b, (S^{TZ} \otimes \xi)_{|Z_b}). \tag{2.44}$$

Then $H = H_+ \oplus H_-$ is an infinite dimensional $\mathbf{Z}_2$-graded vector bundle on B.

Let $T^H X$ be a smooth vector bundle of TX such that $TX = T^H X \oplus TZ$.

We claim that $(T^H X, g^{TZ})$ determines a canonical Euclidean connection ∇^{TZ} on (TZ, g^{TZ}) [**B2**, Theorem 1.9]. We give two descriptions of ∇^{TZ}. In fact, if g^{TX} is a metric on TX such that $T^H X$ is orthogonal to TZ and that g^{TZ} is the restriction of g^{TX} to TZ, if ∇^{TX} is the Levi-Civita connection on (TX, g^{TX}), then

$$\nabla^{TZ} = P^{TZ} \nabla^{TX}. \tag{2.45}$$

Another description of ∇^{TZ} is as follows [**BC**, Section 4] :

- Fibrewise, ∇^{TZ} is the Levi-Civita connection of (Z, g^{TZ}).
- If $U \in TB$, if $U^H \in T^H X$ lifts U, if V is a smooth section of TZ,

$$\nabla^{TZ}_{U^H} V = \left[U^H, V \right] + \tfrac{1}{2} \left(L_{U^H} g^{TZ} \right) V. \tag{2.46}$$

Let g^{TB} be a Riemannian metric on g^{TB}. Let ∇^{TB} be the Levi-Civita connection on (TB, g^{TB}). Then ∇^{TB} lifts to a connection $\nabla^{T^H X}$ on $T^H X$.

Let T be the torsion of the connection of the connection $\nabla^{T^H X} \oplus \nabla^{TZ}$ on $TX = T^H X \oplus TZ$. Then by [**B2**, Theorem 1.9], T does not depend on g^{TB}. More precisely

- T vanishes on $TZ \times TZ$.

- If $U, V \in TB$,

$$T(U^H, V^H) = -P^{TZ}\left[U^H, V^H\right]. \tag{2.47}$$

- If $U \in TB$, $V \in TZ$

$$T(U^H, V) = \tfrac{1}{2} L_{U^H} g^{TZ} V. \tag{2.48}$$

Let ∇'^{TX} be the Levi-Civita connection on $(TX, \pi^* g^{TB} \oplus g^{TZ})$. Put

$$S = \nabla'^{TX} - \nabla^{TX}. \tag{2.49}$$

Since ∇'^{TX} is torsion free, if $U, V, W \in TX$,

$$2\langle S(U)V, W\rangle + \langle T(U,V), W\rangle + \langle T(W,U), V\rangle - \langle T(V,W), U\rangle. \tag{2.50}$$

Then by (2.50), the tensor $\langle S(.).,.\rangle$ does not depend on g^{TB}.

The connection ∇^{TZ} lifts to a unitary connection $\nabla^{S^{TZ}} = \nabla^{S_+^{TZ}} \oplus \nabla^{S_-^{TZ}}$ on $S^{TZ} = S_+^{TZ} \oplus S_-^{TZ}$. Let $\nabla^{S^{TZ}\otimes\xi}$ be the obvious connection on $S^{TZ} \otimes \xi$.

Let dv_Z be the Riemannian volume form along the fibres Z. If U is a smooth section of TB, the Lie derivative operator L_{U^H} acts on tensors along the fibre Z. Put

$$\mathrm{div}_Z(U) = \frac{L_{U^H} dv_Z(x)}{dv_Z(x)}. \tag{2.51}$$

Then one verifies easily that $\mathrm{div}_Z(U)$ is a tensor.

In the sequel, we identify smooth sections of H on B to smooth sections of $S^{TZ} \otimes \xi$ on X.

DEFINITION 2.19. Let ∇^H be the connection on H such that if $U \in TS$, if h is a smooth section of H on S,

$$\nabla^H_U s = \nabla^{S^{TZ}\otimes\xi}_{U^H} s. \tag{2.52}$$

Put

$$\nabla^{H,u}_U = \nabla^H_U + \tfrac{1}{2}\mathrm{div}_Z(U). \tag{2.53}$$

Then one verifies easily that $\nabla^{H,u}$ is a unitary connection on H, preserving H_+ and H_-. Moreover the curvatures of ∇^H and $\nabla^{H,u}$ take their values in first order differential operators acting along the fibres Z.

Let $f_1, \ldots, f_m$ be a basis of TB, let $f^1, \ldots, f^m$ be the dual basis of T^*B. Set

$$c(T) = \tfrac{1}{2} f^\alpha f^\beta c(T(f_\alpha^H, f_\beta^H)). \tag{2.54}$$

Then $c(T) \in (\Lambda(T^*B)\widehat{\otimes}\,\mathrm{End}\,(H))^{\mathrm{odd}}$.

DEFINITION 2.20. For $t > 0$, let A_t be the superconnection on H,

$$A_t = \nabla^{H,u} + \sqrt{t}D^Z - \frac{c(T)}{4\sqrt{t}}. \tag{2.55}$$

Observe that A_t^2 is fibrewise elliptic, so that $\exp(-A_t^2)$ is fibrewise trace class.

DEFINITION 2.21. Put

$$\begin{aligned} \alpha_t &= \varphi \operatorname{Tr_s}\left[\exp(-A_t^2)\right], \\ \beta_t &= \frac{1}{(2i\pi)^{1/2}} \varphi \operatorname{Tr_s}\left[\frac{\partial A}{\partial t} \exp(-A_t^2)\right]. \end{aligned} \tag{2.56}$$

Now we have the local families index theorem of [**B2**, Theorems 3.4, 4.12 and 4.16].

Theorem 2.22. *The forms α_t and β_t are real and the form α_t is closed. The cohomology class $[\alpha_t]$ of α_t is constant and*

$$[\alpha_t] = \operatorname{ch}(\operatorname{Ind} D^Z) \text{ in } H^{\text{even}}(B, \mathbf{Q}). \tag{2.57}$$

Also

$$\frac{\partial \alpha}{\partial t} = -d\beta_t. \tag{2.58}$$

Finally as $t \to 0$,

$$\begin{aligned} \alpha_t &= \pi_* \left(\widehat{A}(TZ, \nabla^{TZ}) \operatorname{ch}(\xi, \nabla^{\xi})\right) + \mathcal{O}(t), \\ \beta_t &= \mathcal{O}(1). \end{aligned} \tag{2.59}$$

REMARK 2.23. The fact that α_t is closed follows from the arguments in Theorem 2.4. The most difficult result is (2.59). In fact the main point of [**B2**] was to produce the "right" superconnection A_t such that a result like (2.59) would hold.

Observe that, in general, (2.59) does not hold for the "simpler" superconnection $\nabla^{H,u} + \sqrt{t}D^Z$.

Needless to say, from [**B2**, Theorem 2.22], we recover the cohomological form of the families index theorem of Atiyah-Singer [**AS2**] given in (2.2). Still our proof [**B2**] is completely local on the basis B, hence the fact Theorem 2.22 is called a local families index theorem. Still to prove (2.59), we use the fibrewise heat kernel for $\exp(-A_t^2)$, and prove a corresponding convergence result which is local on the fibre Z.

Let g^{TB} be a metric on TB. For $\varepsilon > 0$, let g_ε^{TX} be the metric on $TX = T^H X \oplus TZ$,

$$g_\varepsilon^{TX} = \frac{1}{\varepsilon}\pi^* g^{TB} \oplus g^{TZ}. \tag{2.60}$$

Assume that X and B are compact and that B is even dimensional, oriented and spin. Then X is also even dimensional, oriented and spin. Let S^{TB} be the vector bundle of (TB, g^{TB}) spinors. For $\varepsilon > 0$, let D_ε^X be the Dirac operator acting on $C^\infty(X, S^{TX} \otimes \xi)$, associated to $(g_\varepsilon^{TX}, \nabla^\xi)$.

Let $P_t^\varepsilon(x, x')$ $(x, x' \in X)$ be the smooth heat kernel for $\exp(-tD_\varepsilon^{X,2})$ associated to the volume $dv(x')$. In view of Theorems 1.2 and 2.22, it is natural to ask whether, given $t > 0$, as $\varepsilon \to 0$, $\operatorname{Tr_s}[P_t^\varepsilon(x, x)]$ has a limit.

Let $Q_t(x,x')$ be the smooth fibrewise kernel of $\exp(-A_t^2)$. Following a terminology introduced by physicists [**W**], the idea of studying the limit of certain quantities as $\varepsilon \to 0$ is called passing to the adiabatic limit.

A first result in that direction is as follows [**B2,** Theorem 5.3].

Theorem 2.24. *For* $t > 0$,

$$\lim_{\varepsilon\to 0} \operatorname{Tr_s}\left[P_t^\varepsilon(x,x)\right] = \left\{\varphi \operatorname{Tr_s}\left[Q_t(x,x)\right] \widehat{A}(TB,\nabla^{TB})\right\}^{\max(B)} . \tag{2.61}$$

PROOF : In [**B2**], the idea is to view (2.61) as a consequence of the local index theorem over B with coefficients in the infinite dimensional $\mathbf{Z}_2$-graded vector bundle H.

Put

$$D^H = \sum c(f_\alpha)\left(\nabla^{\pi^* S^{TB}\widehat{\otimes} S^{TZ}\otimes\xi}_{f_\alpha^H} + \tfrac{1}{2}\operatorname{div}(f_\alpha^H)\right) . \tag{2.62}$$

Then a simple computation [**BC,** eq. (4.26)] shows that

$$D_\varepsilon^X = D^Z + \sqrt{\varepsilon} D^H - \varepsilon \tfrac{1}{8} c(f_\alpha)c(f_\beta)c(T(f_\alpha^H, f_\beta^H)) . \tag{2.63}$$

At least formally, (2.61) follows easily from local index theoretic techniques over B. □

Let ∇'^{TX}_ε be the Levi-Civita connection on (TX, g_ε^{TX}). Put

$$S_\varepsilon = \nabla'^{TX}_\varepsilon - \nabla^{TX} . \tag{2.64}$$

Then by (2.50), we find that

$$\begin{aligned} P^{TZ} S_\varepsilon &= P^{TZ} S , \\ P^{T^H X} S_\varepsilon &= \varepsilon P^{T^H X} S . \end{aligned} \tag{2.65}$$

From (2.64), (2.65), we find that as $\varepsilon \to 0$, the connection ∇_ε^{TX} has a limit. More precisely,

$$\nabla'^{TX}_\varepsilon \to \nabla^{TX} + P^{TZ} S . \tag{2.66}$$

From (2.66), we find [**BF2,** eq. (3.196)] that as $\varepsilon \to 0$,

$$\widehat{A}(TX, \nabla'^{TX}_\varepsilon) \to \widehat{A}(TZ, \nabla^{TZ}) \pi^* \widehat{A}(TB, \nabla^{TB}) . \tag{2.67}$$

Now while $\widehat{A}(TX, \nabla'^{TX}_\varepsilon)\operatorname{ch}(\xi, \nabla^\xi)$ appears naturally when applying the local index Theorem 1.2 to the operator D_ε^X over X, $\widehat{A}(TZ, \nabla^{TZ})\operatorname{ch}(\xi, \nabla^\xi)$ appears naturally in the local families index theorem stated in Theorem 2.22.

Given Theorem 2.24, the "local in the fibre" version of the local families index Theorem of **[B2]** stated in Theorem 2.22 is just the assertion that the following diagram commutes

$$\begin{array}{ccc} \operatorname{Tr_s}[P_t^\varepsilon(x,x)] & \xrightarrow[t\to 0]{} & \begin{array}{c}\{\widehat{A}(TX,\nabla'^{TX}_\varepsilon)\\ \operatorname{ch}(\xi,\nabla^\xi\}^{\max(X)}\end{array} \\ \Big\downarrow{\scriptstyle \varepsilon\to 0} & & \Big\downarrow{\scriptstyle \varepsilon\to 0} \\ \begin{array}{c}\{\varphi\operatorname{Tr_s}[Q_t(x,x)]\\ \widehat{A}(TB,\nabla^{TB})\}^{\max(B)}\end{array} & \xrightarrow[t\to 0]{} & \begin{array}{c}\{\{\widehat{A}(TZ,\nabla^{TZ})\operatorname{ch}(\xi,\nabla^\xi)\}^{\max(Z)}\\ \widehat{A}(TB,\nabla^{TB})\}^{\max(B)}\end{array}\end{array} \tag{2.68}$$

Needless to say, the explicit form for A_t in (2.55) was found by trying to make the above diagram commute by brute force. The comparison of formulas (2.55) and (2.63) for A_t and D_ε^X provides overwhelming evidence that A_t is the "right" superconnection.

Finally observe that given Theorems 1.2 and 2.24, and also (2.66), a proof of Theorem 2.22 can be given, which makes the commutativity of the above diagram a tautology, by showing that the convergence as $t \to 0$ in the upper row is uniform in $\varepsilon \in]0,1]$.

Assume that $\ker D^Z$ is of locally constant dimension. Then $\ker(D^Z)$ is a smooth subbundle of H. Let $\nabla^{\ker D^Z,u}$ be the orthogonal projection of $\nabla^{H,u}$ on $\ker D^Z$.

The we have the following result of Berline-Getzler-Vergne [**BeGeV**, Theorems 9.19 and 9.23], which extends Theorem 2.6 to an infinite dimensional situation.

Theorem 2.25. *As $t \to +\infty$,*

$$\begin{aligned}\alpha_t &= \operatorname{ch}(\ker D^Z, \nabla^{\ker D^Z,u}) + \mathcal{O}\left(\frac{1}{\sqrt{t}}\right), \\ \beta_t &= \mathcal{O}\left(\frac{1}{t^{3/2}}\right).\end{aligned} \tag{2.69}$$

DEFINITION 2.26. Let $\widetilde{\eta}$ be the odd form on B

$$\widetilde{\eta} = \int_0^{+\infty} \beta_t dt. \tag{2.70}$$

By (2.58), (2.59), (2.69) we get the following result.

Theorem 2.27. *The odd smooth form $\widetilde{\eta}$ on B is such that*

$$d\widetilde{\eta} = \pi_*\left[\widehat{A}(TZ,\nabla^{TZ})\operatorname{ch}(\xi,\nabla^\xi)\right] - \operatorname{ch}(\ker D^Z, \nabla^{\ker D^Z,u}). \tag{2.71}$$

REMARK 2.28. From equation (2.71), one deduces easily how, modulo coboundaries, $\widetilde{\eta}$ depends on $(T^H X, g^{TZ}, g^\xi)$. This is because any two sets of such data can be deformed into each other.

f) The determinant bundle.

We make the same assumptions as in Section 2 e).

Complex lines form a group under the $\otimes$ operation. In particular, if λ is a complex line, let λ^{-1} be the dual line, so that $\lambda \otimes \lambda^{-1} = \mathbf{C}$, the canonical complex line.

If E is a complex vector space, put

$$\det E = \Lambda^{\max} E \,. \tag{2.72}$$

If $E = E_+ \oplus E_-$ is a $\mathbf{Z}_2$-graded vector space, set

$$\det E = \det E_+ \otimes (\det E_-)^{-1} \,. \tag{2.73}$$

DEFINITION 2.29. For $b \in B$, set

$$\lambda_b = \left(\det \ker D_b^Z\right)^{-1} \,. \tag{2.74}$$

Then in **[Q2]**, Quillen has shown how to glue the λ_b's into a honest line bundle λ, even though, in general, the dimension of $\ker D_{\pm,b}^Z$ is not locally constant. The idea is as follows. For $a > 0$, let U_a be the open set

$$U_a = \left\{ b \in B, a \notin \mathrm{Sp}(D_b^{Z,2}) \right\} \,. \tag{2.75}$$

Let $H^{[0,a]}$ be the direct sum of the eigenspaces of $D_b^{Z,2}$ for eigenvalues $\mu \leq a$. Put

$$\lambda^{[0,a]} = \left(\det H^{[0,a]}\right)^{-1} \,. \tag{2.76}$$

Then $\lambda^{[0,a]}$ is a smooth line bundle on U_a.

Given $0 < a < a'$, let $H^{[a,a']}$ be the direct sum of eigenspace of $D_b^{Z,2}$ for eigenvalues $\mu \in [a, a']$. Set

$$\lambda^{[a,a']} = (\det H^{[a,a']})^{-1} \,. \tag{2.77}$$

Then $\lambda^{[a,a']}$ has a canonical nonzero section $\det D_+^{Z,[a,a']}$, which is smooth on $U^a \cap U^{a'}$. Also on $U^a \cap U^{a'}$,

$$\lambda^{[0,a']} \simeq \lambda^{[0,a]} \otimes \lambda^{[a,a']} \,. \tag{2.78}$$

Since $\lambda^{[a,a']}$ is canonically trivialized, on $U^a \cap U^{a'}$

$$\lambda^{[0,a']} \simeq \lambda^{[0,a]} \,. \tag{2.79}$$

DEFINITION 2.30. The inverse determinant bundle λ is the line bundle which restricts to $\lambda^{[0,a]}$ on U^a.

By (2.79), we find that for any $b \in B$,

$$\lambda_b \simeq \left(\det \ker D_b^Z\right)^{-1} . \tag{2.80}$$

By the Atiyah-Singer family index theorem **[AS2]** (see eq. (2.2)),

$$c_1(\lambda) = -\pi_* \left[\widehat{A}(TZ)\,\mathrm{ch}(\xi)\right]^{(2)} . \tag{2.81}$$

In **[Q2]**, **[BF1,2]**, Quillen and Bismut and Freed have shown how to equip λ with a smooth metric $\| \ \|_\lambda$ and a unitary connection ∇^λ such that

$$c_1(\lambda, \nabla^\lambda) = -\pi_* \left[\widehat{A}(TZ, \nabla^{TZ})\,\mathrm{ch}(\xi, \nabla^\xi)\right]^{(2)} . \tag{2.82}$$

Here we will concentrate on the construction of the imaginary part of the connection ∇^λ **[BF1,2]**. Assume first that $\ker D^Z$ is of locally constant dimension. Then $(\det \ker D^Z)^{-1}$ is a smooth line bundle and $\lambda \simeq (\det \ker D^Z)^{-1}$.

The connection $\nabla^{\ker D^Z,u}$ induces a connection ${}^1\nabla^\lambda$ on λ. Put

$${}^2\nabla = {}^1\nabla + +2i\pi\widetilde{\eta}^{(1)} . \tag{2.83}$$

Clearly

$$c_1(\lambda, {}^1\nabla^\lambda) = -\left[\mathrm{ch}(\ker D^Z, \nabla^{\ker D^Z,u})\right]^{(2)} . \tag{2.84}$$

From (2.84), we deduce that

$$c_1(\lambda, {}^2\nabla^\lambda) = -\left[\mathrm{ch}(\ker D^Z, \nabla^{\ker D^Z,u})\right]^{(2)} - d\widetilde{\eta}^{(1)} . \tag{2.85}$$

By (2.71), (2.85), we find that

$$c_1(\lambda, {}^2\nabla^\lambda) = -\pi_* \left[\widehat{A}(TZ, \nabla^{TZ})\,\mathrm{ch}(\xi, \nabla^\xi)\right]^{(2)} . \tag{2.86}$$

The connection ∇^λ in **[BF1,2]** of differs from ${}^1\nabla^\lambda$ by an exact real form, so that (2.82) follows from (2.86).

The remarkable fact is that even if $\ker D_\pm^Z$ is not of locally constant dimension, in **[BF1,2]**, it is possible to define the connection ${}^2\nabla^\lambda$ by formulas similar to (2.83). The idea is to construct over U^a a connection $\nabla^{\lambda^{[0,a]}}$ by a suitable modification of (2.83), and to establish that the connections $\nabla^{\lambda^{[0,a]}}$, suitably modified, define a connection ∇^λ on λ, for which (2.82) still holds.

g) The local families index theorem : the odd case.

Now we assume that the fibres Z are odd dimensional. Put

$$H_b = C^\infty\left(Z_b, (S^{TZ} \otimes \xi)_{|Z_b}\right) . \tag{2.87}$$

Then (D^Z) is a family of self-adjoint operators acting on H.

Take σ as in Section 2 d).

DEFINITION 2.31. For $t > 0$, put

$$A_t = \nabla^{H,u} + \sqrt{T} D^Z \sigma - \frac{c(T)\sigma}{4\sqrt{t}}. \tag{2.88}$$

Again, $\exp(-A_t^2)$ is fibrewise trace class.

DEFINITION 2.32. Put

$$\begin{aligned} \alpha_t &= (2i)^{1/2} \varphi \operatorname{Tr}^\sigma \left[\exp(-A_t^2)\right], \\ \beta_t &= \frac{1}{\sqrt{\pi}} \varphi \operatorname{Tr}^\sigma \left[\frac{\partial A_t}{\partial t} \exp(-A_t^2)\right]. \end{aligned} \tag{2.89}$$

Now we state a result of [**BF2,** Theorem 2.10].

Theorem 2.33. *The forms α_t and β_t are real, and the form α_t is closed. The cohomology class $[\alpha_t]$ of α_t is constant, and*

$$[\alpha_t] = \operatorname{ch}(D^Z) \text{ in } H^{\mathrm{odd}}(B, \mathbf{Q}). \tag{2.90}$$

Also

$$\frac{\partial \alpha}{\partial t} = -d\beta_t. \tag{2.91}$$

Finally as $t \to 0$

$$\begin{aligned} \alpha_t &= \pi_* \left[\widehat{A}(TZ, \nabla^{TZ}) \operatorname{ch}(\xi, \nabla^\xi)\right] + \mathcal{O}(t), \\ \beta_t &= \mathcal{O}(1). \end{aligned} \tag{2.92}$$

Assume now that $\ker D^Z$ is of locally constant dimension. Then $\ker D^Z$ is a vector bundle on B.

We have the obvious analogue of Theorems 2.14 and 1.25.

Theorem 2.34. *As $t \to +\infty$,*

$$\begin{aligned} \alpha_t &= \mathcal{O}\left(\frac{1}{\sqrt{t}}\right), \\ \beta_t &= \mathcal{O}\left(\frac{1}{t^{3/2}}\right). \end{aligned} \tag{2.93}$$

DEFINITION 2.35. Let $\widetilde{\eta}$ be the even form

$$\widetilde{\eta} = \int_0^{+\infty} \beta_t dt. \tag{2.94}$$

Theorem 2.36. *The even form $\widetilde{\eta}$ is such that*

$$d\widetilde{\eta} = \pi_* \left[\widehat{A}(TZ, \nabla^{TZ}) \operatorname{ch}(\xi, \nabla^{\xi}) \right] . \tag{2.95}$$

PROOF : This follows from (2.91)-(2.93). □

h) The odd local families index theorem and the eta invariant.
Observe that

$$\beta_t^{(0)} = \frac{1}{2\sqrt{\pi}} \operatorname{Tr} \left[\frac{D^Z}{\sqrt{t}} \exp(-tD^{Z,2}) \right] . \tag{2.96}$$

By (2.92), as $t \to 0$,

$$\beta_t^{(0)} = \mathcal{O}(1) . \tag{2.97}$$

Now by (1.25), (2.97) guarantees that the eta function $\eta^{D^Z}(s)$ is holomorphic at 0. Note that this result extends Theorem 1.4 to the case where Z does not necessarily bound. This result is also a consequence of [**APS3,** Theorem 4.5].

By (2.94), (2.96),

$$\widetilde{\eta}^{(0)} = \frac{1}{2\sqrt{\pi}} \int_0^{+\infty} t^{-1/2} \operatorname{Tr} \left[D^Z \exp(-tD^{Z,2}) \right] dt . \tag{2.98}$$

Using (1.25), (2.97), (2.98), we get

$$\widetilde{\eta}^{(0)} = \tfrac{1}{2} \eta^{D^Z}(0) . \tag{2.99}$$

Moreover in degree 1, (2.95) specializes to

$$d\widetilde{\eta}^{(0)} = \pi_* \left[\widehat{A}(TZ, \nabla^{TZ}) \operatorname{ch}(\xi, \nabla^{\xi}) \right]^{(1)} . \tag{2.100}$$

In view of (2.99), formula (2.100) gives a local expression for $d\frac{1}{2}\eta(0)$, which, when the fibres Z bound, can also be derived from the index theorem of Atiyah-Patodi-Singer [**APS1**].

When B is a point (i.e. in the case of a single fibre), the condition that $\ker(D^Z)$ is of locally constant dimension is empty. However in general, this condition is non empty, since it implies that the family (D^Z) is trivial in $K^1(B)$.

Set

$$\overline{\eta}^{D^Z}(s) = \tfrac{1}{2} \left(\eta^{D^Z}(s) + \dim \ker D^Z \right) . \tag{2.101}$$

Then $\overline{\eta}^{D^Z}(0)$ is called the reduced eta invariant of D^Z. In [**APS3,** Section 2], $\overline{\eta}^{D^Z}(0)$ is shown to define a smooth function with values in $\mathbf{R}/\mathbf{Z}$, and the general form of (2.100) is

$$d\overline{\eta}^{D^Z}(0) = \pi_* \left[\widehat{A}(TZ, \nabla^{TZ}) \operatorname{ch}(\xi, \nabla^{\xi}) \right]^{(1)} . \tag{2.102}$$

i) The holonomy Theorem.

Now we use the assumptions and notation of Sections 2 e) and 2 f).

Let $s \in S_1 \to c_s \in B$ be an oriented smooth curve to B. In **[W]**, Witten raised the question of calculating the holonomy of a connection ∇^λ on λ in terms of the eta invariant of the odd dimensional oriented compact spin manifold $M = \pi^{-1}(C)$.

Let g^{TX} be a Riemannian metric on TX, let g^{TB} be a metric on TB. Put

$$g_\varepsilon^{TX} = g^{TX} + \frac{1}{\varepsilon}\pi^* g^{TB}. \tag{2.103}$$

Let g_ε^{TM} be the metric on TM induced by g_ε^{TX} on TM.

We equip S_1 with the non trivial spin structure. Then since TZ is spin, TM inherits an obvious spin structure. Let D_ε^M be the Dirac operator on M associated to g_ε^{TM}, ∇^ξ.

Let $\tau_1^0 \in S_1$ be the parallel transport with respect to the connection ∇^λ along $s \in S_1 \to c_s$.

Then we have the following result by Bismut-Freed [**BF2,** Theorem 3.16] and Cheeger [**C4**].

Theorem 2.37. *The limit as $\varepsilon \to 0$ of $\overline{\eta}^{D_\varepsilon^M}(0) \in \mathbf{R}/\mathbf{Z}$ exists, and moreover*

$$\tau_1^0 = \exp(-2i\pi \lim_{\varepsilon \to 0} \overline{\eta}^{D_\varepsilon^M}(0)). \tag{2.104}$$

PROOF : Assume first that c bounds Δ in B. Then by (2.82)

$$\tau_1^0 = \exp\left(-2i\pi \int_{\pi^{-1}(\Delta)} \widehat{A}(TZ, \nabla^{TZ})\,\mathrm{ch}(\xi, \nabla^\xi)\right). \tag{2.105}$$

On the other hand, by the index Theorem of Atiyah-Patodi-Singer [**APS1**] (see Theorem 1.3),

$$\overline{\eta}^{D_\varepsilon^M}(0) = \int_{\pi^{-1}(\Delta)} \widehat{A}(TM, g_\varepsilon^{TM})\,\mathrm{ch}(\xi, \nabla^\xi) \text{ in } \mathbf{R}/\mathbf{Z}. \tag{2.106}$$

As in (2.67), one verifies easily that as $\varepsilon \to 0$,

$$\widehat{A}(TM, g_\varepsilon^{TM}) \to \widehat{A}(TZ, \nabla^{TZ}). \tag{2.107}$$

From (2.106), (2.107), we find that as $\varepsilon \to 0$,

$$\overline{\eta}^{D_\varepsilon^M}(0) \to \int_{\pi^{-1}(\Delta)} \widehat{A}(TZ, \nabla^{TZ})\,\mathrm{ch}(\xi, \nabla^\xi) \text{ in } \mathbf{R}/\mathbf{Z}. \tag{2.108}$$

By (2.105)-(2.108), we get (2.104) in this case. In general, using (2.102) and (2.107), one finds easily that $\lim_{\varepsilon \to 0} \overline{\eta}^{D_\varepsilon^M}(0) \in \mathbf{R}/\mathbf{Z}$ exists. The main point of

[BF2], [C4] is to extend (2.104) when c does not bound in B. Then if for $b \in C$, D_b^Z is invertible, a direct study of the formula (2.98) for $\overline{\eta}^{D_\varepsilon^M}(0)$ by the methods used in the proof of (2.68) shows that as $\varepsilon \to 0$,

$$\overline{\eta}^{D_\varepsilon^M}(0) \to \int_c \widetilde{\eta}, \tag{2.109}$$

from which (2.104) follows easily. When $\ker D^Z$ is nonzero and not even a vector bundle over c, a non trivial perturbation argument shows that (2.104) still holds. □

REMARK 2.38. Theorem 2.37 is one of the motivations for studying adiabatic limits of eta invariants, when instead of the circle c, the base of the fibration is arbitrary.

j) Adiabatic limits of eta invariants.

Assume first that B is an odd dimensional compact oriented spin Riemannian manifold. Let $E = E_+ \oplus E_-$ be a $\mathbf{Z}_2$-graded Hermitian vector bundle as in Section 2 c), and let $V \in \operatorname{End}(E)$ be a self-adjoint section of $\operatorname{End}^{\text{odd}}(E)$, such that $\ker V$ is of locally constant dimension. Let $V_\pm$ be the restriction of V to $E_\pm$. In the sequel, the assumptions of Section 2 c) will be in force. In particular $\nabla^E = \nabla^{E_+} \oplus \nabla^{E_-}$ is a split unitary connection on $E = E_+ \oplus E_-$, and $\nabla^{\ker V}$ is the orthogonal projection of ∇^E on $\ker V$. Also the odd form $\widehat{\eta}$ was defined in Definition 2.7.

Let $D^{B,E_\pm}$, $D^{B,\ker V_\pm}$ be the Dirac operators associated to the above data, acting on smooth actions over B of $S^{TB} \otimes E_\pm$, $S^{TB} \otimes \ker V_\pm$. Let $\overline{\eta}^{D^{B,E_\pm}}(0)$, $\overline{\eta}^{D^{B,\ker V_\pm}}(0)$ be the corresponding reduced eta invariants.

Theorem 2.39. *The following identity holds*

$$\begin{aligned}\overline{\eta}^{D^{B,E_+}}(0) - \overline{\eta}^{D^{B,E_-}}(0) \\ = \overline{\eta}^{D^{B,\ker V_+}}(0) - \overline{\eta}^{D^{B,\ker V_-}}(0) + \int_B \widehat{A}(TB, \nabla^{TB})\widehat{\eta} \ \textit{in}\ \mathbf{R}/\mathbf{Z}.\end{aligned} \tag{2.110}$$

PROOF : In view of (2.21), (2.102), it is clear that both sides of (2.110) vary in the same way. By a simple deformation argument, we may as well assume that $E_+ = E_{0,+} \oplus F$, $E_- = E_{0,-} \oplus F$, V is the identity on F and vanishes on $E_{0,+} \oplus E_{0,-}$, $\nabla^{E_+} = \nabla^{E_{0,+}} \oplus \nabla^F$, $\nabla^{E_-} = \nabla^{E_{0,-}} \oplus \nabla^F$. In this situation, by Theorem 2.10, $\widehat{\eta}$ vanishes, and (2.110) is a trivial identity. □

Assume now that B is instead even dimensional, that E is a Hermitian vector bundle, that V is a self-adjoint section of $\operatorname{End}(E)$ such that $\ker V$ is a vector bundle, and ∇^E is a unitary connection on E. We use the notation of Section 2 d).

Let $D^{B,E}$, $D^{B,\ker V}$ be the Dirac operators acting on smooth sections of $C^\infty(B, S^{TB} \otimes E)$, $C^\infty(B, S^{TB} \otimes \ker V)$. Clearly

$$\begin{aligned}\overline{\eta}^{D^{B,E}}(0) &= \tfrac{1}{2} \operatorname{Ind}(D^{B,E}) \text{ in } \mathbf{R}/\mathbf{Z} \\ \overline{\eta}^{D^{B,\ker V}}(0) &= \tfrac{1}{2} \operatorname{Ind}(D^{B,\ker V}) \text{ in } \mathbf{R}/\mathbf{Z}.\end{aligned} \tag{2.111}$$

Theorem 2.40. *The following identities holds*

$$\overline{\eta}^{D^{B,E}}(0) = \overline{\eta}^{D^{B,\ker V}}(0) + \int_B \widehat{A}(TB, \nabla^{TB})\widehat{\eta} \text{ in } \mathbf{R}/\mathbf{Z}. \tag{2.112}$$

PROOF : By Theorem 2.17 and the Atiyah-Singer index theorem [**AS1**],

$$\int_B \widehat{A}(TB, \nabla^{TB})\widehat{\eta} = \tfrac{1}{2}\left(\operatorname{Ind}(D_+^{B,E_{>0}}) - \operatorname{Ind}(D_+^{B,E_{<0}})\right), \tag{2.113}$$

and so

$$\int_B \widehat{A}(TB, \nabla^{TB})\widehat{\eta} = \tfrac{1}{2}\left(\operatorname{Ind}(D_+^{B,E_{>0}}) + \operatorname{Ind}(D_+^{B,E_{<0}})\right) \text{ in } \mathbf{R}/\mathbf{Z}. \tag{2.114}$$

By (2.111), (2.114), we get (2.112). □

Now we make again the same assumptions and use the same notation as in Section 2 c). Also we suppose that B is compact, oriented, spin and odd dimensional.

Let $\tau = \pm 1$ on $E_\pm$. Then $D'^B = \tau D^B + V$ is a self-adjoint elliptic operator acting on $C^\infty(B, S^{TB} \otimes E)$. Let $\overline{\eta}^{D'^B}(0)$ be the reduced eta invariant associated to D'^B. By a simple deformation argument, one finds that

$$\overline{\eta}^{D'^B}(0) = \overline{\eta}^{D^{B,E_+}}(0) - \overline{\eta}^{D^{B,E_-}}(0) \text{ in } \mathbf{R}/\mathbf{Z}. \tag{2.115}$$

Now for $\varepsilon > 0$, we replace g^{TB} by $\frac{g^{TB}}{\varepsilon}$. Let D'^B_ε be the corresponding Dirac operator. Then by (2.102), $\overline{\eta}^{D'^B_\varepsilon}(0)$ remains constant in $\mathbf{R}/\mathbf{Z}$. Now, we give a refinement of Theorem 2.39, established in [**BC1**, Theorem 2.28].

Theorem 2.41. *If* $\ker V = \{0\}$, *then the limit as* $\varepsilon \to 0$ *of* $\overline{\eta}^{D'^B_\varepsilon}(0)$ *exists in* $\mathbf{R}$, *and moreover*

$$\lim_{\varepsilon \to 0} \overline{\eta}^{D'^B_\varepsilon}(0) = \int_B \widehat{A}(TB, \nabla^{TB})\widehat{\eta}. \tag{2.116}$$

PROOF : The main point in the proof of [**BC1**] is to show that

$$\lim_{\varepsilon \to 0} \frac{1}{2\sqrt{\pi t}} \operatorname{Tr}\left[D'^B_\varepsilon \exp(-tD'^{B,2}_\varepsilon)\right] = \int_B \widehat{A}(TB, \nabla^{TB})\beta_t, \tag{2.117}$$

which in turn follows from local index theory techniques. □

If B is even dimensional, an obvious analogue of Theorem 2.41 holds.

Now we make the same assumptions as in Sections 2 a) and 2 e). Suppose that X and B are compact, that X is odd dimensional and that B is oriented and spin. Then X is also oriented and spin. Let g^{TZ}, g^{TB} be metrics on TZ, TB. Put

$$g_\varepsilon^{TX} = \frac{1}{\varepsilon}\pi^* g^{TB} \oplus g^{TZ}. \tag{2.118}$$

Let D_ε^X be the Dirac operator on (X, g_ε^{TX}) as in Section 2 e). Using the variation formula for eta invariants, one verifies easily that as $\varepsilon \to 0$, $\overline{\eta}^{D_\varepsilon^X}(0)$ converges in $\mathbf{R}/\mathbf{Z}$.

Now we state the main result of [**BC1**, Theorems 4.35 and 4.95].

Theorem 2.42. *If* $\ker D^Z = \{0\}$, *as* $\varepsilon \to 0$, $\overline{\eta}^{D_\varepsilon^X}(0)$ *converges in* $\mathbf{R}$, *and moreover*

$$\lim_{\varepsilon\to 0} \overline{\eta}^{D_\varepsilon^X}(0) = \int_B \widehat{A}(TB, \nabla^{TB})\widetilde{\eta}. \tag{2.119}$$

PROOF : Formally, the proof of Theorem 2.42 is closely related to the proof of Theorem 2.41. In fact, (2.119) is an infinite dimensional version of (2.116), as should be clear from formula (2.63). The proof has three main steps:

- One proves that as $\varepsilon \to 0$,

$$\frac{1}{\sqrt{\pi}} \operatorname{Tr}\left[\frac{D_\varepsilon^X}{2\sqrt{t}} \exp(-tD_\varepsilon^{X,2})\right] \to \int_B \widehat{A}(TB, \nabla^{TB})\beta_t. \tag{2.120}$$

- One controls the lowest eigenvalue of D_ε^X as $\varepsilon \to 0$.
- One uses a version of finite propagation speed to control the integrand in (2.98) uniformly in ε as $t \to +\infty$. □

REMARK 2.43. In [**D**], Dai has given a very interesting extension of Theorem 2.42 to the case where $\ker D^Z$ is not necessarily zero.

Dai's result apply in particular to the case where B is odd dimensional, and D_ε^X is the signature operator of [**APS1**] associated to the metric g_ε^{TX}. In this case, there is no spectral flow, so that $\lim_{\varepsilon\to 0} \overline{\eta}^{D_\varepsilon^X}(0)$ exists in $\mathbf{R}$. Using results of Mazzeo and Melrose [**MazMe**] relating small eigenvalues of $D_\varepsilon^{X,2}$ to the Leray spectral of the fibration, Dai obtains a formula extending (2.119) by adding to the right-hand side of (2.119) the reduced eta invariant of a signature operator on B (twisted by the cohomology of the fibres) and a sum of half integers. These integers are the "signatures" of the Leray spectral sequence (E_r, d_r), for $r \geq 3$.

Adiabatic limits of eta invariants appeared naturally in the context of the solution by Atiyah-Donnelly-Singer [**ADS**] of the Hirzebruch conjecture [**Hir**] on the signature of Hilbert modular varieties. The fibrations which appear in this context are fibrations by tori over a torus basis. The calculation of [**ADS**] was recovered in the context of $\widetilde{\eta}$ forms in [**BC5**]. For a L_2 approach to the same problem, we refer to [**Mül**].

k) The families index theorem for manifolds with boundary.

Let now X be a manifold with boundary, let B be a manifold, and let $\pi : X \to B$ be a fibration, whose fibres Z are smooth compact manifolds with boundary.

We assume the fibres Z to be even dimensional, oriented and spin. Let g^{TZ} be a metric on TZ, which is fibrewise product near ∂Z. Let $T^H X$ be a horizontal vector subbundle of TX, such that $T^H X_{|\partial X} \subset T\partial X$. Put $T^H \partial X = T^H X_{|\partial X}$. Then $T^H \partial X$ is a horizontal subbundle of $T\partial X$.

Let (ξ, g^ξ, ∇^ξ) be a Hermitian vector bundle on X with unitary connection.

Assume first that Z is even dimensional. For $b \in B$, let D_b^Z be the Dirac operator with the Atiyah-Patodi Singer boundary conditions on ∂Z.

In order that the index bundle $\operatorname{Ind}(D_+^Z)$ to be well-defined, it is crucial that the family of boundary Dirac operators $D^{\partial Z}$ does not have spectral flow. So we first assume that

$$\ker D^{\partial Z} = 0 . \tag{2.121}$$

In this case, (D_+^Z) is a family of Fredholm operators and its index $\operatorname{Ind}(D_+^Z) \in K^0(B)$ is well-defined.

Let $\widetilde{\eta}$ be the even form on B constructed in Definition 2.35, which is attached to the family $D^{\partial Z}$.

The following result is proved in Bismut-Cheeger [**BC3,** Theorem 6.11].

Theorem 2.44. *The following identity holds*

$$\operatorname{ch}(\operatorname{Ind} D_+^Z) = \pi_* \left[\widehat{A}(TZ, \nabla^{TZ}) \operatorname{ch}(\xi, \nabla^\xi)\right] - \widetilde{\eta} \text{ in } H^{\text{even}}(B, \mathbf{Q}) . \tag{2.122}$$

PROOF : The basic idea in [**BC2,3**] is to replace the manifolds with boundary Z by the manifolds with conical singularity $\widehat{Z}_\ell = Z \cup_{\partial Z} C^{[0,\ell]}(\partial Z)$. Then we equip the fibres $\widehat{Z}$ with a family of metric of conical type. By proceeding as in Proposition 1.7, for ℓ large enough, the Atiyah-Patodi-Singer family (D_+^Z) and the family of L_2 Dirac operators $(D_+^{\widehat{Z}_\ell})$ have the same index. To the family $\widehat{Z}_\ell$, one attaches a natural Levi-Civita superconnection A_t^ℓ, to which the techniques of [**B2**] are formally applied. Note here that the advantage of using $\widehat{Z}_\ell$ is that the Atiyah-Patodi-Singer boundary conditions only appear in implicit form.

Observe that equation (2.95) explains why the right-hand side of (2.122) is closed. □

In [**MeP1**], Melrose and Piazza have extended Theorem 2.44 in a fundamental way. In fact, they observe that even if $\ker D^{\partial Z}$ is non trivial, by the family index Theorem of Atiyah-Singer [**AS2**], the family $(D^{\partial Z}) \in K^1(B)$ is trivial. They show that if B is compact, the triviality of the family $(D^{\partial Z})$ is equivalent to the existence of a spectral section P, i.e. a smooth family of self-adjoint projections $P_b : C^\infty(\partial Z_b, (S^{T\partial Z} \otimes \xi)_{|Z_b}) \to C^\infty(\partial Z_b, (S^{T\partial Z} \otimes \xi)_{|Z_b})$, such that there is $R > 0$ for which for any $b \in B$,

$$
\begin{aligned}
D^{\partial Z} u = \lambda u\,, \lambda > R\,,\ & \text{then } Pu = u \\
\lambda < -R\,,\ & \text{then } Pu = 0\,.
\end{aligned} \tag{2.123}
$$

Then Melrose and Piazza [**MeP1**] prove that if for every $b \in B$, the Atiyah-Patodi-Singer projection $P_{\geq 0, b}$ is replaced by P_b, the family of Dirac operators $D^{Z,P}_{+,b}$ associated to the boundary conditions attached to P has a honest index bundle $\operatorname{Ind}(D^{Z,P}_{+})$. They construct an even form $\tilde{\eta}^P$ on B, formally similar to the form $\tilde{\eta}$ in (2.94). However in Melrose-Piazza's construction, the term $\sqrt{t}D^{\partial Z}$ is replaced by a more complicate expression $\sqrt{t}D'^{\partial Z}_t$, where $D'^{\partial Z}_t$ interpolates between $D^{\partial Z}$ for $t \ll 1$ and a suitable perturbation $D^{\partial Z} + A_P$ (with A_P depending on P) for $t \gg 1$. The family A_P is smoothing and such that $D^{\partial Z} + A_P$ is invertible. It can be seen as providing an explicit trivialization of the zero class $(D^{\partial Z}) \in K^1(B)$. Modulo exact forms, the form $\tilde{\eta}^P$ only depends on P and not on the particular choice of A_P.

Then Melrose and Piazza [**MeP1**] prove :

Theorem 2.45. *The following identity holds*

(2.124)

$$
\operatorname{ch}(\operatorname{Ind}(D^{Z,P}_{+})) = \pi_* \left[\widehat{A}(TZ, \nabla^{TZ}) \operatorname{ch}(\xi, \nabla^{\xi})\right] - \tilde{\eta}^P \quad in \quad H^{\text{even}}(B, \mathbf{Q})\,.
$$

Besides in [**MeP1**], Melrose and Piazza compare the forms $\tilde{\eta}^P$ for different choices of P.

When the fibres Z are odd dimensional and the family $D^{\partial Z}$ is invertible, Bismut-Cheeger [**BC4,** Section 6] conjectured a formula like (2.124) for a family of self-adjoint Dirac operators D^Z. This conjecture has been proved and extended by Melrose-Piazza [**MeP2**]. They adapted the idea of a spectral section in this new context, produced a superconnection whose Chern character forms are shown to represent the index, and established the corresponding index formula.

III. Analytic torsion forms and analytic torsion currents.

The purpose of this Section is to review the properties of the analytic torsion forms of **[BGS2]** and **[BK]** , and of the analytic torsion currents of **[B2]**, **[BGS4,5]**.

As explained in the introduction, analytic torsion forms are naturally associated to a family of Hermitian Dolbeault complexes. Analytic torsion currents are associated to an embedding $i : Y \to X$ and a resolution of a holomorphic Hermitian vector bundle η on Y by a holomorphic complex of Hermitian vector bundles on X. Analytic torsion forms and analytic torsion currents are secondary objects which refine the Riemann-Roch-Grothendieck theorem for submersions and immersions at the level of differential forms or currents.

This Section in organized as follows. In a), we construct the torsion forms associated to a holomorphic Hermitian complex of vector bundles **[BGS1]**, and we relate them to the secondary classes of Bott-Chern **[BoCh]**. In b), we consider a holomorphic submersion $\pi : X \to S$, and a holomorphic Hermitian vector bundle ξ on X. When this fibration is Kähler (in a sense to be described), we show that the Levi-Civita superconnection of Definition 2.20 "respects" the holomorphic structure of the problem. When $R\pi_*\xi$ is locally free, we construct analytic torsion forms on S, which refine on the $\widetilde{\eta}$ forms of Definition 2.26.

In c), we introduce the Quillen metrics on the inverse of the determinant of the cohomology $\lambda(\xi) = (\det R\pi_*\xi)^{-1}$. The construction of the Quillen metric only involves the component of degree 0 of the above analytic torsion forms. Then we state the curvature theorem of **[BGS1,3]** for the Quillen metric on $\lambda(\xi)$.

In d), we describe the results of **[BerB]** and **[Ma]** on the compatibility of the analytic torsion forms to the composition of submersions.

In e), we construct the analytic torsion currents of **[B2]**, **[BGS4,5]**. In f), we show that these currents are compatible to the composition of immersions.

In g) and h), we describe the results of **[BL]** and **[B5,6]** on the compatibility of the analytic torsion forms and analytic torsion currents to the composition of an immersion and a submersion.

In i), we give a short introduction to the proof of the main result in **[BL]**. In j), we develop a simple but crucial technical tool in **[BL]**, the Hodge theory of the resolution of a point.

In k), we explain the construction in **[B4]** of the analytic torsion forms associated to a short exact sequence of holomorphic vector bundles, which plays an important role in the proof of the main result in **[BL]** and **[B5]**. In particular the evaluation of **[B4]** produces the genus R of Gillet and Soulé **[GS3]** in the final formula.

a) The torsion forms of a holomorphic Hermitian complex.

Let S be a complex manifold. Let

$$(E, v) : 0 \to E_m \underset{v}{\to} \dots \underset{v}{\to} E_0 \to 0 \tag{3.1}$$

be a holomorphic complex of vector bundles on S . Put

$$E_+ = \bigoplus_{i \text{ even}} E_i \quad , \quad E_- = \bigoplus_{i \text{ odd}} E_i \,. \tag{3.2}$$

Then $E = E_+ \oplus E_-$ is $\mathbf{Z}_2$-graded.

Let $g^E = \bigoplus_{i=0}^{m} g^{E_i}$ be a Hermitian metric on E. Let $\nabla^E = \bigoplus_{i=0}^{m} \nabla^{E_i}$ be the holomorphic Hermitian connection on $E = \bigoplus_{i=0}^{m} E_i$. Let v^* be the adjoint of v. Put

$$V = v + v^* . \tag{3.3}$$

Then V is a self-adjoint section of End ${}^{\text{odd}}(E)$. For $t > 0$, put

$$\begin{aligned} A''_t &= \nabla^{E''} + \sqrt{t}v\,, \\ A'_t &= \nabla^{E'} + \sqrt{t}v^*\,, \\ A_t &= \nabla^{E} + \sqrt{t}V\,. \end{aligned} \tag{3.4}$$

Clearly

$$A_t = A''_t + A'_t\,. \tag{3.5}$$

Also A_t is a superconnection of the kind we already met in (2.14). Let N be the number operator on E, i.e. N acts by multiplication by k on E_k.

The following result is established in [**BGS1,** Proposition 1.6].

Proposition 3.1. *The following identities hold*

$$\begin{aligned} &A''^2_t = 0\ ,\quad A'^2_t = 0\,, \\ &A^2_t = [A''_t, A'_t]\ , \\ &[A''_t, A^2_t] = 0\ ,\quad [A'_t, A^2_t] = 0\,, \\ &\frac{\partial A''_t}{\partial t} = \frac{1}{2t}[A''_t, N]\ , \\ &\frac{\partial A'_t}{\partial t} = \frac{-1}{2t}[A'_t, N]\ . \end{aligned} \tag{3.6}$$

Proof : Since (E, v) is a holomorphic complex,

$$A''^2_t = 0\,. \tag{3.7}$$

Also

$$\frac{\partial A''_t}{\partial t} = \frac{v}{2\sqrt{t}} = \frac{1}{2t}[A''_t, N]\ . \tag{3.8}$$

The other identities in (3.6) follow easily from analogues of (3.7), (3.8). □

DEFINITION 3.2. For $t > 0$, put

$$
\begin{aligned}
\alpha_t &= \varphi \operatorname{Tr_s}\left[\exp(-A_t^2)\right], \\
\beta_t &= \frac{1}{(2i\pi)^{1/2}} \varphi \operatorname{Tr_s}\left[\frac{\partial A_t}{\partial t}\exp(-A_t^2)\right], \\
\gamma_t &= \varphi \operatorname{Tr_s}\left[N\exp(-A_t^2)\right].
\end{aligned}
\tag{3.9}
$$

Observe that in our context, the forms α_t and β_t were already introduced in Definition 2.5.

Let P^S be the set of smooth real forms on S which are sums of forms of type (p,p). Let $P^{S,0}$ be the subspace of the $\alpha \in P^S$ such that $\alpha = \overline{\partial}\beta + \partial\gamma$, with β and γ smooth.

Now we have the result of [**BGS1**, Theorem 1.15].

Theorem 3.3. *The forms α_t and γ_t lie in P^S. Moreover*

$$
\begin{aligned}
\frac{\partial \alpha_t}{\partial t} &= -d\beta_t, \\
\beta_t &= \frac{1}{2i\pi}\frac{(\overline{\partial}-\partial)\gamma_t}{2t}.
\end{aligned}
\tag{3.10}
$$

In particular

$$
\frac{\partial \alpha_t}{\partial t} = \frac{\overline{\partial}\partial}{2i\pi}\frac{\gamma_t}{t}. \tag{3.11}
$$

PROOF : We only prove part of Theorem 3.3. The first identity in (3.10) was already established in (2.16). Using (3.6), we obtain the second identity in (3.10). □

Assume now that the homology $H(E,v)$ is of locally constant dimension. Then $H(E,v)$ is a holomorphic $\mathbf{Z}$-graded holomorphic vector bundle on S. Clearly

$$
H(E,v) \simeq \ker V. \tag{3.12}
$$

Let $g^{H(E,v)}$ be the metric on $H(E,v)$ induces by g^E via (3.12).

One verifies easily that $\nabla^{\ker V} = P^{\ker V}\nabla^E$ is the holomorphic Hermitian connection on $(H(E,v), g^{H(E,v)})$. Put

$$
\operatorname{ch}'(E,g^E) = \sum_{i=0}^{m}(-1)^i i \operatorname{ch}(E_i, g^{E_i}). \tag{3.13}
$$

Then by an analogue of Theorem 2.6, as $t \to +\infty$,

$$
\gamma_t = \operatorname{ch}'(H(E,v), g^{H(E,v)}) + \mathcal{O}\left(\frac{1}{\sqrt{t}}\right). \tag{3.14}
$$

DEFINITION 3.4. For $s \in \mathbf{C}$, $0 < \mathrm{Re}(s) < \frac{1}{2}$, put

$$R(E, g^E)(s) = \frac{1}{\Gamma(s)} \int_0^{+\infty} t^{s-1}(\gamma_t - \gamma_\infty)dt. \tag{3.15}$$

By (3.14), $R(E, g^E)(s)$ is a holomorphic function of s, which extends holomorphically to $s = 0$.

DEFINITION 3.5. Set

$$T(E, g^E) = \frac{\partial}{\partial s} R(E, g^E)(s)|_{s=0}. \tag{3.16}$$

Recall that the odd form $\widehat{\eta}$ was defined in Definition 2.7. Now we have the result of [**BGS1,** Theorem 1.17].

Theorem 3.6. *The form $T(E, g^E)$ lies in P^S. Moreover*

$$\begin{aligned} &\frac{\overline{\partial}\partial}{2i\pi} T(E, g^E) = \mathrm{ch}(H(E, v), g^{H(E,v)}) - \mathrm{ch}(E, g^E), \\ &\frac{1}{2i\pi}(\overline{\partial} - \partial)\frac{T}{2}(E, g^E) = \widehat{\eta}. \end{aligned} \tag{3.17}$$

PROOF : These identities follow easily from Theorems 2.6 and 3.3. □

DEFINITION 3.7. We will say that $((E, v), g^E)$ is split, if $E_i = F_i \oplus F_{i-1} \oplus H_i$, $v|_{E_i}$ vanishes on $F_i \oplus H_i$ and is the identity on F_{i-1}, and the above splitting is orthogonal with respect to g^{E_i}.

Now we state a result of [**BGS1,** Corollary 1.30].

Theorem 3.7. *There is a unique way to associate to $((E, v), g^E)$, with $H(E, v)$ of locally constant dimension, a class $T(E, v) \in P^S/P^{S,0}$ such that*

a) *$T(E, g^E)$ is functorial.*

b) *If (E, g^E) is split, $T(E, g^E) = 0$.*

c) *The following identity holds,*

$$\frac{\overline{\partial}\partial}{2i\pi} T(E, g^E) = \mathrm{ch}(H(E, v), g^{H(E,v)}) - \mathrm{ch}(E, g^E). \tag{3.18}$$

PROOF : Existence is almost obvious by the above construction. As to uniqueness, observe that over $S \times \mathbf{P}^1$, one constructs easily a complex $(\widetilde{E}, \widetilde{v})$ and a metric $g^{\widetilde{E}}$ such that

$$((\widetilde{E}, v), g^{\widetilde{E}})_{S\times\{0\}} = ((E,v), g^E)\,, \quad ((\widetilde{E}, v), g^{\widetilde{E}})_{S\times\{0\}} \text{ is split}. \tag{3.19}$$

Using the obvious equation

$$\frac{\overline{\partial}\partial}{2i\pi} \log|z|^2 = \delta_{\{0\}} - \delta_{\{\infty\}}\,, \tag{3.20}$$

we get

(3.21)

$$T(E, g^E) = \int_{\mathbf{P}^1} \log|z|^2 \left(\mathrm{ch}\left(H(\widetilde{E}, \widetilde{v}), g^{H(\widetilde{E},\widetilde{v})}\right) - \mathrm{ch}(\widetilde{E}, g^{\widetilde{E}})\right) \quad \text{in} \quad P^S/P^{S,0}\,,$$

which guarantees uniqueness. □

REMARK 3.8. Observe that

$$R(E, g^E)^{(0)}(s) = \mathrm{Tr_s}\left[N[V^2]^{-s}\right], \tag{3.22}$$

so that

$$T(E, g^E)^{(0)} = -\ \mathrm{Tr_s}\left[N \log(V^2)\right]. \tag{3.23}$$

By (3.18), we get

$$c_1(\det E, g^E) = c_1(\det H(E,v), g^{\det H(E,v)}) - \frac{\overline{\partial}\partial}{2i\pi} T(E, g^E)^{(0)}. \tag{3.24}$$

The interpretation of (3.24) is easy. In fact there is a canonical holomorphic isomorphism **[KMu]**

$$\det E \simeq \det H(E,v)\,. \tag{3.25}$$

Let us briefly describe this isomorphism. First assume that $H(E,v) = \{0\}$, i.e. (E,v) is acyclic. Then (3.25) says that $\det E$ has a canonical non zero section $\tau(E,v)$. To construct $\tau(E,v)$, we choose $\omega_m \in \det E_m$, $\omega_m \neq 0$, $\omega_{m-1} \in \Lambda^{\dim E_{m-1} - \dim E_m} E_{m-1}$ such that $v\omega_m \wedge \omega_{m-1} \in \det E_{m-1}$ is non zero, $\omega_{m-2} \in \Lambda^{\dim E_{m-2} - \dim E_{m-1} + \dim E_m} E_{m-2}$ such that $v\omega_{m-1} \wedge \omega_{m-2} \in \det E_{m-2}$ is non zero... . These choices are possible because (E,v) is acyclic. Then

$$\tau(E,v) = \left(\omega_m \otimes (v\omega_m \wedge \omega_{m-1})^{-1} \otimes (v\omega_{m-1} \wedge \omega_{m-2}) \otimes \ldots\right)^{(-1)^m} \tag{3.26}$$

does not depend on the above choices.

When $H(E,v)$ is non zero, the construction of the canonical isomorphism (3.25) is similar.

Then one verifies easily that

$$g^{\det E} = g^{\det H(E,v)} \exp\{T(E, g^E)^{(0)}\}, \tag{3.27}$$

from which (3.24) follows immediately.

The class of forms $T(E,g^E) \in P^S/P^{S,0}$ appears as a prototype of Bott-Chern classes [**BoCh**], [**BGS1**]. Let us give a construction of these classes in the simplest case.

Let $\rho : F \to S$ be a holomorphic vector bundle. Let g^F, g'^F be two Hermitian metrics on F. Let Q be a characteristic polynomial. The following result in established in [**BGS1,** Theorem 1.29].

Theorem 3.8. *There exists a unique way to assign to (F, g^F, g'^F) a class $\widetilde{Q}(F, g^F, g'^F) \in P^S/P^{S,0}$ such that*

a) *$\widetilde{Q}(F, g^F, g'^F)$ is functorial.*

b) *If $g^F = g'^F$, $\widetilde{Q}(F, g^F, g'^F) = 0$.*

c) *The following identity holds*

$$\frac{\bar{\partial}\partial}{2i\pi}\widetilde{Q}(F, g^F, g'^F) = Q(F, g'^F) - Q(F, g^F). \tag{3.28}$$

PROOF : We just outline a construction of $\widetilde{Q}(F, g^F, g'^F)$ [**BGS1**]. Extend F to a vector bundle $\widetilde{F}$ on $S \times \mathbf{P}^1$. Let $g^{\widetilde{F}}$ be a metric on $\widetilde{F}$ such that $g^{\widetilde{F}}_{S\times\{0\}} = g^F$, $g^{\widetilde{F}}_{S\times\{\infty\}} = g'^F$. Put

$$\widetilde{Q}(F, g^F, g'^F) = -\int_{\mathbf{P}^1} \log(|z|^2) Q(\widetilde{F}, g^{\widetilde{F}}). \tag{3.29}$$

Then by (3.20), (3.28) holds. □

Let $g^E = \bigoplus_{i=0}^{m} g^{E_i}$, $g'^E = \bigoplus_{i=0}^{m} g'^{E_i}$ be two set of Hermitian metrics on E. Let $g^{H(E,v)}$, $g'^{H(E,v)}$ be the corresponding metrics on $H(E,v)$.

Theorem 3.9. *The following identity holds*

$$\begin{aligned} T(E, g'^E) - T(E, g^E) = \widetilde{\mathrm{ch}}(H(E,v), g^{H(E,v)}, g'^{H(E,v)}) \\ - \widetilde{\mathrm{ch}}(E, g^E, g'^E) \quad \text{in} \quad P^S/P^{S,0}. \end{aligned} \tag{3.30}$$

PROOF : Using Theorem 3.7, our Theorem is a straightforward consequence of the uniqueness of Bott-Chern classes stated in Theorem 3.8. □

b) The Levi-Civita superconnection of a Kähler fibration and the analytic torsion forms.

Let $\pi : X \to S$ be a holomorphic submersion with compact fibre Z. Let ξ be a holomorphic vector bundle on X. Let $R\pi_*\xi$ be the direct image of ξ.

Let ω^X be a real closed $(1,1)$-form on X, such that the restriction of ω^X to TZ is the Kähler form ω^{TZ} of a Hermitian metric g^{TZ} on $TZ = TX/S$. If $J^{T_\mathbf{R}Z}$ is the complex structure of $T_\mathbf{R}Z$, if $U, V \in T_\mathbf{R}Z$, $\omega^{TZ}(U,V) =$

$\langle U, J^{T_{\mathbf{R}}Z}V\rangle$. Let g^ξ be a Hermitian metric on ξ, let ∇^ξ be the holomorphic Hermitian connection on (ξ, g^ξ). Let T^HX be the orthogonal bundle to TZ with respect to ω^X.

Let $(\Omega(Z, \xi_{|Z}), \overline{\partial}^Z)$ be the family of relative Dolbeault complex along the fibres Z. We equip $\Omega(Z, \xi_{|Z})$ with the L_2 metric attached to g^{TZ}, g^ξ,

$$\langle s, s'\rangle = \int_Z \langle s, s'\rangle \frac{dv_Z}{(2\pi)^{\dim Z}} . \tag{3.31}$$

Let $\overline{\partial}^{*Z}$ be the formal adjoint of $\overline{\partial}^Z$ with respect to (3.31). Set

$$D^Z = \overline{\partial}^Z + \overline{\partial}^{Z*} . \tag{3.32}$$

Then by **[Hi]**, $\sqrt{2}D^Z$ is a family of standard Dirac operators along the fibre Z. The only minor difference is that the fibres Z only have spin^c structure.

To the data (g^{TZ}, T^HX) we can associate the objects constructed in Section 2 e).

The following result is proved in [**BGS2,** Theorem 1.7].

Theorem 3.10. *The connection $\nabla^{T_{\mathbf{R}}Z}$ on $T_{\mathbf{R}}Z$ preserves the complex structure of $T_{\mathbf{R}}Z$. It induces the holomorphic Hermitian connection on (TZ, g^{TZ}).*

As a 2-form, T is of complex type $(1,1)$.

Let $\nabla^{\Lambda(T^{*(0,1)}Z)\otimes\xi}$ be the connection induced by ∇^{TZ}, ∇^ξ on $\Lambda(T^{*(0,1)}Z)\otimes \xi$.

If $U \in TZ$, let $U^H \in T^HX$ be the horizontal lift of U.

Definition 3.11. If $U \in T_{\mathbf{R}}S$, if s is a smooth section of $\Omega(Z, \xi_{|Z})$ over S, put

$$\nabla_U^{\Omega(Z,\xi_{|Z})} s = \nabla_{U^H}^{\Lambda(T^{*(0,1)}Z)\otimes\xi} s . \tag{3.33}$$

The following result is established in [**BGS2,** Theorem 1.14].

Theorem 3.12. *The connection $\nabla^{\Omega(Z,\xi_{|Z})}$ on $\Omega(Z, \xi_{|Z})$ preserves the Hermitian product (3.31) on $\Omega(Z, \xi_{|Z})$. Its curvature is of complex type* $(1,1)$. *Also*

$$\left[\nabla^{\Omega(Z,\xi_{|Z})''}, \overline{\partial}^Z\right] = 0 \ , \quad \left[\nabla^{\Omega(Z,\xi_{|Z})'}, \overline{\partial}^{Z*}\right] = 0 . \tag{3.34}$$

Amazingly enough, $(\Omega(Z, \xi_{|Z}), \overline{\partial}^Z)$ appears to be a "holomorphic" Hermitian vector bundle compact over S. By (3.34), we find that

$$\left(\nabla^{\Omega(Z,\xi_{|Z})''} + \overline{\partial}^Z\right)^2 = 0 \ , \quad \left(\nabla^{\Omega(Z,\xi_{|Z})'} + \overline{\partial}^{Z*}\right)^2 = 0 . \tag{3.35}$$

The explanation for (3.35) given in [**BGS2,** Theorem 2.8] is that using the smooth identification $\Lambda(T^{*(0,1)}X) \simeq \Lambda(T^{*(0,1)}Z)\widehat{\otimes}\pi^*\Lambda(T^{*(0,1)}S)$, $\nabla^{\Omega(Z,\xi_{|Z})''} + \overline{\partial}^Z$ is exactly the full Dolbeault operator $\overline{\partial}^X$ acting on $\Omega(X, \xi)$.

Recall that $\Lambda(T^{*(0,1)}X) \otimes \xi$ is a $c(T_{\mathbf{R}}Z)$ Clifford module. Namely if $X \in TZ$, let $X^* \in T^{*(0,1)}Z$ correspond to X by the metric g^{TZ}. Then if $X \in TZ$, $Y \in \overline{TZ}$, put

$$c(X) = \sqrt{2}X^* \wedge \ , \quad c(Y) = -\sqrt{2}i_Y \,. \tag{3.36}$$

Extend c to a linear map $T_{\mathbf{R}}Z \otimes_{\mathbf{R}} \mathbf{C} \to \operatorname{End}((\Lambda(T^{*(0,1)}Z) \otimes \xi)$. Then if $X, X' \in T_{\mathbf{R}}Z \otimes_{\mathbf{R}} C$,

$$c(X)c(X') + c(X')c(X) = -2\,\langle X, X'\rangle_{g^{TZ}} \,. \tag{3.37}$$

Let (f_α) be a basis of $T_{\mathbf{R}}S$, let (f^α) be the dual basis of $T^*_{\mathbf{R}}S$. Put

$$\begin{aligned} c(T^{(1,0)}) &= \tfrac{1}{2} f^\alpha f^\beta c(T^{(1,0)}(f^H_\alpha, f^H_\beta))\,, \\ c(T^{0,1)}) &= \tfrac{1}{2} f^\alpha f^\beta c(T^{(0,1)}(f^H_\alpha, f^H_\beta))\,. \end{aligned} \tag{3.38}$$

With the notation in (2.54),

$$c(T) = c(T^{(1,0)}) + c(T^{(0,1)})\,. \tag{3.39}$$

DEFINITION 3.13. For $t > 0$, put

$$\begin{aligned} B''_t &= \sqrt{t}\ \overline{\partial}^Z + \nabla^{\Omega(Z,\xi_{|Z})''} - \frac{c(T^{(1,0)})}{2\sqrt{2t}}\,, \\ B'_t &= \sqrt{t}\ \overline{\partial}^{Z*} + \nabla^{\Omega(Z,\xi_{|Z})'} - \frac{c(T^{(0,1)})}{2\sqrt{2t}}\,, \\ B_t &= B''_t + B'_t\,. \end{aligned} \tag{3.40}$$

Then by (2.55) and Theorem 3.12, for $t > 0$, B_t is exactly the superconnection $A_{\frac{t}{2}}$ in the sense of **[B1]**, i.e. B_t is a Levi-Civita superconnection.

Put

$$\omega^X = \omega^{TZ} + \omega^H\,. \tag{3.41}$$

In particular $\omega^H \in \pi^*\Lambda^2(T^*_{\mathbf{R}}S)$ is the restriction of ω^X to $T^H_{\mathbf{R}}X = \pi^*T_{\mathbf{R}}S$.

Let N_V be the number operator of $\Omega(Z, \xi_{|Z})$, i.e. N_V acts by multiplication by k on $\Omega^k(Z, \xi_{|Z})$.

DEFINITION 3.14. For $t > 0$, put

$$N_t = N_V + \frac{i\omega^H}{t}\,. \tag{3.42}$$

Then $N_t \in \left(\Lambda(T^*_{\mathbf{R}}S)\widehat{\otimes} \operatorname{End}(\Omega(Z, \xi_{|Z}))\right)^{\text{even}}$.

The following result is proved in [**BGS2**, Theorem 2.6].

Theorem 3.15. *The following identities hold,*

$$
\begin{aligned}
&B''^2_t = 0 \ , \quad B'^2_t = 0 \, , \\
&B^2_t = [B''_t, B'_t] \, , \\
&[B''_t, B^2_t] = 0 \ , \quad [B'_t, B^2_t] = 0 \, , \\
&\frac{\partial B''_t}{\partial t} = \frac{-1}{2t} [B''_t, N_t] \, , \\
&\frac{\partial B'_t}{\partial t} = \frac{1}{2t} [B'_t, N_t] \, .
\end{aligned}
\tag{3.43}
$$

REMARK 3.16. The identities in (3.43) are remarkable. They guarantee that the Levi-Civita superconnection B_t also has natural holomorphic properties, i.e. it splits as $B_t = B''_t + B'_t$. Besides, by comparing (3.43) with (3.6), N_t appears as the right "number operator" associated to B_t.

DEFINITION 3.17. For $t > 0$, set

$$
\begin{aligned}
&\alpha_t = \varphi \ \mathrm{Tr_s} \left[\exp(-B^2_t)\right] \, , \\
&\beta_t = \left(\frac{1}{2i\pi}\right)^{1/2} \varphi \ \mathrm{Tr_s} \left[\frac{\partial B_t}{\partial t} \exp(-B^2_t)\right] \, , \\
&\gamma_t = \varphi \ \mathrm{Tr_s} \left[N_t \exp(-B^2_t)\right] \, .
\end{aligned}
\tag{3.44}
$$

Now we state a result taken from [**BGS3**, Theorems 2.9 and 2.16].

Theorem 3.18. *The forms $\alpha_t, \beta_t, \gamma_t$ are real. The forms α_t and γ_t lie in P^S. The cohomology class of α_t is constant, and*

$$
[\alpha_t] = \mathrm{ch}(R\pi_* \xi) \quad in \quad H^{\mathrm{even}}(S, \mathbf{R}) \, .
\tag{3.45}
$$

Also,

$$
\begin{aligned}
&\frac{\partial \alpha_t}{\partial t} = -d\beta_t, \\
&\beta_t = -\frac{1}{2i\pi} (\overline{\partial} - \partial) \frac{\gamma_t}{2t} \, .
\end{aligned}
\tag{3.46}
$$

In particular,

$$
\frac{\partial \alpha_t}{\partial t} = -\frac{\overline{\partial}\partial}{2i\pi} \frac{\gamma_t}{t} \, .
\tag{3.47}
$$

Finally as $t \to 0$,

$$
\begin{aligned}
&\alpha_t = \pi_* \left[\mathrm{Td}(TZ, g^{TZ}) \, \mathrm{ch}(\xi, g^\xi)\right] + \mathcal{O}(t), \\
&\gamma_t = \frac{C_{-1}}{t} + C_0 + \mathcal{O}(t) \ , \quad C_{-1}, C_0 \in P_S \, .
\end{aligned}
\tag{3.48}
$$

PROOF : We just sketch the proof of part of Theorem 3.18. Equation (3.45) follows from Theorem 2.22. Equation (3.46) follows from (3.43) as in (3.10). The first equation in (3.48) follows from Theorem 2.22. The second equation in (3.48) is proved in [**BGS2**] by local index theoretic techniques. □

Now we assume that $R\pi_*\xi$ is locally free. So $R\pi_*\xi$ is a holomorphic **Z**-graded vector bundle on S, and moreover $(R\pi_*\xi)_s \simeq H(Z_s, \xi_{|Z_s})$. Since $H(Z, \xi_{|Z}) \simeq \ker D^Z$, $R\pi_*\xi$ inherits a smooth metric $g^{R\pi_*\xi}$.

Theorem 3.19. *As $t \to +\infty$,*

$$\begin{aligned} \alpha_t &= \mathrm{ch}(R\pi_*\xi, g^{R\pi_*\xi}) + \mathcal{O}\left(\frac{1}{\sqrt{t}}\right), \\ \gamma_t &= \mathrm{ch}'(R\pi_*\xi, g^{T\pi_*\xi}) + \mathcal{O}\left(\frac{1}{\sqrt{t}}\right). \end{aligned} \tag{3.49}$$

PROOF : With the notation of Theorem 2.25, using [**BGS3,** Theorem 3.11] (which relies on (3.35)), one shows easily that $\nabla^{\ker D^Z, u}$ is just the holomorphic Hermitian connection on $(R\pi_*\xi, g^{R\pi_*\xi})$. Theorem 3.19 is then an obvious modification of Theorem 2.25. □

DEFINITION 3.20. For $s \in \mathbf{C}$, $0 < \mathrm{Re}(s) < \frac{1}{2}$, put

$$R(\omega^X, g^\xi) = -\frac{1}{\Gamma(s)} \int_0^{+\infty} t^{s-1}(\gamma_t - \gamma_\infty)dt. \tag{3.50}$$

By (3.48), (3.49), $R(\omega^X, g^\xi)$ is a holomorphic function of s, which extend holomorphically near $s = 0$.

DEFINITION 3.21. Set

$$T(\omega^X, g^\xi) = \frac{\partial}{\partial s} R(\omega^X, g^\xi)(0). \tag{3.51}$$

Recall that the form $\widetilde{\eta}$ was defined in Definition 2.26. Then we have the result of [**BGS2,** Theorem 2.20], [**BK,** Theorem 3.9].

Theorem 3.22. *The form $T(\omega^X, g^\xi)$ lies in P^S. Moreover*

$$\begin{aligned} \frac{\overline{\partial}\partial}{2i\pi} T(\omega^X, g^\xi) &= \mathrm{ch}(R\pi_*\xi, g^{R\pi_*\xi}) - \pi_* \left[\mathrm{Td}(TZ, g^{TZ})\,\mathrm{ch}(\xi, g^\xi)\right], \\ \frac{1}{2i\pi}(\overline{\partial} - \partial)\frac{T}{2}(E, g^E) &= \widetilde{\eta}. \end{aligned} \tag{3.52}$$

PROOF : Equation (3.52) follows from Theorems 3.18 and 3.19. □

In Remark 2.28, we observed that the dependence of $\widetilde{\eta}$ (modulo coboundaries) on the various geometric data is quite explicit. Also in Theorem 3.9, we found that the dependence of $T(E, g^E) \in P^S/P^{S,0}$ on the metric g^E can be explicitly given in terms of Bott-Chern classes.

It is then natural to ask how $T(\omega^X, g^\xi)$ depends on (ω^X, g^ξ). In fact if only g^ξ is made to vary, equation (3.52) and the methods of **[BGS1]** used in the proof of Theorem 3.7 provide the answer immediately. However if ω^X also varies, the answer certainly does not rely on the methods of Theorem 3.7. In fact, for Theorem 3.22 to hold, it is crucial for ω^X to be closed. So in order to calculate $T(\omega'^X, g^\xi) - T(\omega^X, g^\xi)$ using (3.52), a necessary condition would be, for example, that the fibres Z have the same volume for ω^X and ω'^X.

Let (ω'^X, g'^ξ) be taken as before. The following "anomaly formulas" were established in [**BK**, Theorem 3.10], extending earlier work in degree 0 [**BGS3**, Theorem 1.23].

Theorem 3.23. *The following identity holds*

$$(3.53)\quad T(\omega'^X, g'^\xi) - T(\omega^X, g^\xi) = \widetilde{\mathrm{ch}}\left(R\pi_*\xi, g^{R\pi_*\xi}, g'^{R\pi_*\xi}\right)$$
$$-\pi_*\left[\widetilde{\mathrm{Td}}(TZ, g^{TZ}, g'^{TZ})\,\mathrm{ch}(\xi, g^\xi) + \mathrm{Td}(TZ, g'^{TZ})\widetilde{\mathrm{ch}}(\xi, g^\xi, g'^\xi)\right] \quad \text{in } P^S/P^{S,0}.$$

In particular, $T(\omega^X, g^\xi) \in P^S/P^{S,0}$ only depends on (g^{TZ}, g^ξ).

The last statement in Theorem 3.23 is if particular importance. It says that, as should be the case, the class of $T(\omega^X, g^\xi)$ in $P^S/P^{S,0}$ only depends on the geometric data which appear in the right-hand side of the first equation in (3.52).

c) Quillen metrics.

Assume first that S is a point. Let g^{TZ}, g^ξ be the Hermitian metrics on TZ, ξ.

DEFINITION 3.24. Put

$$(3.54)\qquad \theta(s) = -\,\mathrm{Tr_s}\left[N_V\left[D^{Z,2}\right]^{-s}\right].$$

Then $\theta(s)$ is a linear combination of the zeta functions of the Laplacian $D^{Z,2}$ acting on forms in $\Omega(Z, \xi_{|Z})$ of degree $0, 1 \ldots, \dim Z$.

Put

$$(3.55)\qquad \lambda(\xi) = (\det H(Z, \xi_{|Z}))^{-1}.$$

Then $\lambda(\xi)$ is a complex line. The metric $g^{H(Z,\xi_{|Z})}$ induces a metric $|\ \ |_{\lambda(\xi)}$ on $\lambda(\xi)$.

In [**Q2**], Quillen introduced the following metrics on $\lambda(\xi)$.

DEFINITION 3.25. The Quillen metric $\| \quad \|_{\lambda(\xi)}$ on $\lambda(\xi)$ is given by

$$\| \quad \|_{\lambda(\xi)} = | \quad |_{\lambda(\xi)} \exp\left\{ -\tfrac{1}{2} \frac{\partial \theta}{\partial s}(0) \right\}. \tag{3.56}$$

The underlying motivation for formula (3.56) is equation (3.27). In fact (3.56) is a way of making sense of the metric $g^{(\det \Omega(Z,\xi_{|Z}))^{-1}}$, which $(\det \Omega(Z, \xi_{|Z}))^{-1}$ does not exist. The quantity $\frac{\partial \theta}{\partial s}(0)$ is called the Ray-Singer analytic torsion [**RS**].

Let now $\pi : X \to S$ be a holomorphic sumersion with compact fibre Z. Let ξ be a holomorphic vector bundle on X.

By a construction due to Grothendieck-Knudsen-Munford [**KMu**], there is a canonically defined holomorphic line bundle $\lambda(\xi)$ on S, called the inverse of the determinant of the direct image $R\pi_*\xi$. In particular if $s \in S$, there is a canonical isomorphism

$$\lambda(\xi)_s \simeq \left(\det H(Z_s, \xi_{|Z_s})\right)^{-1}. \tag{3.57}$$

Needless to say, if $R\pi_*\xi$ is locally free,

$$\lambda(\xi) = \left(\det R\pi_*\xi\right)^{-1}. \tag{3.58}$$

In the general case, we will still use the notation $\lambda(\xi) = (\det R\pi_*\xi)^{-1}$.

Let g^{TZ}, g^{ξ} be arbitrary Hermitian metrics on TZ, ξ. Then by the construction in Definition 3.25 and using (3.57), the fibre $\lambda(\xi)_s$ can be equipped with the Quillen metric $\| \cdot \|_{\lambda(\xi)_s}$.

A first result on Quillen metrics is as follows [**BGS3,** Theorem 3.14].

Theorem 3.26. *The Quillen metric is a smooth metric on* $\lambda(\xi)$.

PROOF : If $R\pi_*\xi$ is locally free, $| \quad |_{\lambda(\xi)}$ and $\| \quad \|_{\lambda(\xi)}$ are smooth. The remarkable fact is that in the general case, $\| \quad \|_{\lambda(\xi)}$ is still smooth. However formula (3.27) partly explains the smoothness of $\| \quad \|_{\lambda(\xi)}$. □

DEFINITION 3.27. We will say that $\pi : X \to S$ is locally Kähler if there is a covering of S by open sets U such that $\pi^{-1}(U)$ is Kähler.

We now state the result of [**BGS3,** Theorem 1.27].

Theorem 3.28. *Assume that* $\pi : X \to S$ *is locally Kähler and that* g^{TZ} *is fibrewise Kähler. Then*

$$c_1(\lambda(\xi), \| \quad \|_{\lambda(\xi)}) = -\pi_* \left[\mathrm{Td}(TZ, g^{TZ}) \,\mathrm{ch}(\xi, g^{\xi})\right]^{(2)}. \tag{3.59}$$

PROOF : Clearly, we can assume that X is Kähler. Let g^{TX} be a Kähler metric on TX, with Kähler form ω^X, and assume first that g^{TZ} is the metric on TZ induced by g^{TX}. By (3.50), (3.54),

$$R(\omega^X, g^\xi)^{(0)}(s) = \theta(s)\,, \tag{3.60}$$

and so

$$T(\omega^X, g^\xi)^{(0)} = \frac{\partial\theta}{\partial s}(0)\,. \tag{3.61}$$

Suppose that $R\pi_*\xi$ is locally free. By (3.52), we get

$$\frac{\overline{\partial}\partial}{2i\pi} T(\omega^X, g^\xi)^{(0)} = -c_1(\lambda(\xi), \| \ \|_{\lambda(\xi)}) - \pi_*\left[\mathrm{Td}(TZ, g^{TZ})\,\mathrm{ch}(\xi, g^\xi)\right]^{(2)}\,. \tag{3.62}$$

From (3.62), we get (3.59). If $R\pi_*\xi$ is not locally free, more work is needed to establish (3.59) **[BGS3]**.

Suppose now that g'^{TZ} is a metric on TZ which is only fibrewise Kähler. Then by (3.53),

$$\log\left(\frac{\| \ \|'_{\lambda(\xi)}}{\| \ \|_{\lambda(\xi)}}\right)^2 = \int_Z \widetilde{\mathrm{Td}}(TZ, g^{TZ}, g'^{TZ})\,\mathrm{ch}(\xi, g^\xi)\,. \tag{3.63}$$

From (3.59) (established for g^{TZ}) and (3.63), we get (3.59) for g'^{TZ}. □

d) Adiabatic limits of Quillen metrics, analytic torsion forms, and composition of submersions.

Let $\pi : X \to S$ be a submersion of compact complex manifolds. Let ξ be holomorphic vector bundle on X. We assume that $R\pi_*\xi$ is locally free.

Let g^{TX} be a Kähler metric on X, let ω^X be the corresponding Kähler form. Let g^{TS} be a Kähler metric on S. Let g^ξ be a Hermitian metric on ξ.

Put

$$\begin{aligned} \lambda &= \left(\det H(X, \xi_{|X})\right)^{-1}\,, \\ \lambda' &= \otimes\left(\det H(S, R^i\pi_*\xi)\right)^{(-1)^{i+1}}\,. \end{aligned} \tag{3.64}$$

Let

$$\Omega(X,\xi) = F^0\Omega(X,\xi) \supset F^1\Omega(X,\xi) \supset \ldots \supset F^{\dim S+1}\Omega(X,\xi) = 0 \tag{3.65}$$

be the obvious filtration by the partial degree in $\Lambda(T^{*(0,1)}S)$ of the Dolbeault complex $\Omega(X,\xi)$. Let (E_r, d_r) be the associated spectral sequence. Then

$$E_2^{(p,q)} = H^p(S, R^q\pi_*\xi)\,. \tag{3.66}$$

By (3.66), it follows that the lines λ and λ' are canonically isomorphic.

We can equip the line λ with the Quillen metric associated to g^{TX}, g^{ξ}, and the line λ' with the Quillen metric associated to $g^{TS}, g^{R\pi_*\xi}$.

Consider the exact sequence

$$0 \to TZ \to TX \to \pi^* TS \to 0\,. \tag{3.67}$$

Let $\widetilde{\mathrm{Td}}(TX, TS, g^{TX}, g^{TS}) \in P^X / P^{X,0}$ be the Bott-Chern class **[BoCh]**, **[BGS1]** such that

(3.68)

$$\frac{\bar\partial\partial}{2i\pi}\widetilde{\mathrm{Td}}(TX, TS, g^{TX}, g^{TS}) = \mathrm{Td}(TX, g^{TX}) - \mathrm{Td}(TZ, g^{TZ})\pi^*\,\mathrm{Td}(TS, g^{TS})\,.$$

The following result is established in [**BerB,** Theorem 3.1].

Theorem 3.29. *The following identity holds*

$$\log\left(\frac{\| \quad \|_\lambda}{\| \quad \|_{\lambda'}}\right)^2 = -\int_S \mathrm{Td}(TS, g^{TS}) T(\omega^X, g^{\xi}) + \int_X \widetilde{\mathrm{Td}}(TX, TS, g^{TX}, g^{TS})\,\mathrm{ch}(\xi, g^{\xi})\,. \tag{3.69}$$

We identify $\frac{\mathrm{Td}'}{\mathrm{Td}}(x)$ to the corresponding additive genus. By definition, the genus Td' is the product of the additive genus $\frac{\mathrm{Td}'}{\mathrm{Td}}$ and the multiplicative genus Td.

In [**BerB,** Theorem 3.2], it is shown that (3.69) is essentially equivalent to the following statement. For $\varepsilon > 0$, put

$$g_\varepsilon^{TX} = g^{TX} + \frac{1}{\varepsilon}\pi^* g^{TS}\,. \tag{3.70}$$

Let $\| \quad \|_{\lambda,\varepsilon}$ be the Quillen metric on λ associated to $(g_\varepsilon^{TX}, g^{\xi})$.

Theorem 3.30. *As* $\varepsilon \to 0$,

$$\log\left(\frac{\| \quad \|^2_{\lambda,\varepsilon}}{\| \quad \|^2_{\lambda}}\right) - \int_X \pi^*\,\mathrm{Td}'(TS)\,\mathrm{Td}(TZ)\,\mathrm{ch}(\xi)\log(\varepsilon) \to -\int_S \mathrm{Td}(TS, g^{TS}) T(\omega^X, g^{\xi}) + \log\left(\frac{\| \quad \|_{\lambda'}}{\| \quad \|_{\lambda}}\right)^2. \tag{3.71}$$

REMARK 3.31. The proof of Theorems 3.29 and 3.30 is a combination of the adiabatic limit techniques of Bismut-Cheeger **[BC1]**, and of the Leray spectral arguments of Mazzeo-Melrose **[MazMe]** and Dai **[D]**.

Recently, Ma **[Ma]** has established an extension of Theorem 3.29 for the higher analytic torsion forms $T(\omega^X, g^{\xi})$. Namely let

$$\begin{array}{ccccc} Z & \longrightarrow & W & & \\ \downarrow{\scriptstyle \pi_{Z/Y}} & & \downarrow{\scriptstyle \pi_{W/V}} & \searrow^{\pi_{W/S}} & \\ Y & \longrightarrow & V & \xrightarrow[\pi_{V/S}]{} & S \end{array} \tag{3.72}$$

be a commutative diagram of holomorphic submersions, with compact fibres Z and Y. Let ξ be a holomorphic vector bundle on W. Assume that $R\pi_{W/S*}\xi$, $R\pi_{W/V*}\xi$ and $R\pi_{V/S*}R\pi_{W/V*}\xi$ are locally free.

Let g^ξ be a Hermitian metric on ξ. Let ω^W, ω^V be $(1,1)$ closed forms on W, V having the properties described in Section 3 b). Then by proceeding as in Section 3 b), the three direct images vector bundles described above inherit Hermitian metrics.

Let $T_{W/V}(\omega^W, g^\xi), T_{W/S}(\omega^W, g^\xi), T_{V/S}(\omega^V, g^{R\pi_{W/V*}\xi})$ be the analytic torsion forms on V and S associated to $\pi_{W/V}, \pi_{W/S}, \pi_{V/S}$, and the given metrics.

A problem which arises in **[Ma]** is the adequate definition of

$$\alpha = \widetilde{\mathrm{ch}}\left(R\pi_{W/S*}\xi, R\pi_{V/S*}R\pi_{W/V*}\xi, g^{R\pi_{W/S*}\xi}, g^{R\pi_{W/S*}R\pi_{W/V*}\xi}\right), \tag{3.73}$$

so that

$$\frac{\bar\partial\partial}{2i\pi}\alpha = \mathrm{ch}\left(R\pi_{V/S*}R\pi_{W/V*}\xi, g^{R\pi_{V/S*}R\pi_{W/V*}\xi}\right) - \mathrm{ch}\left(R\pi_{W/S*}\xi, g^{R\pi_{W/S*}\xi}\right). \tag{3.74}$$

In fact there is a spectral sequence E_r of sheaves on S such that $E_2 = R\pi_{W/S*}R\pi_{W/V*}\xi$, which abuts to $R\pi_{W/S*}\xi$.

Under an adequate assumption of ampleness, this spectral sequence is trivial, so that the definition of α is easy. If the E_r are locally free, there is also a natural definition of α. In general, if W and V are projective, a definition of α is given in **[Ma]**.

Then Ma's result is as follows.

Theorem 3.32. *The following identity holds*

$$\begin{aligned} T_{W/S}(\omega^W, g^\xi) &= T_{V/S}(\omega^V, g^{R\pi_{W/V*}\xi}) \\ &\quad + \pi_{W/S*}\left[\mathrm{Td}(TY, g^{TY}) T_{W/V}(\omega^W, g^\xi)\right] \\ &\quad + \alpha - \pi_{W/S*}\left[\widetilde{\mathrm{Td}}(TZ, TY, g^{TZ}, g^{TY})\,\mathrm{ch}(\xi, g^\xi)\right] \quad \text{in } P^S/P^{S,0}. \end{aligned} \tag{3.75}$$

e) Analytic torsion currents.

Let $i : Y \to X$ be an embedding of complex manifolds. Let η be a holomorphic vector bundle on Y. Let

$$(\xi, v) : 0 \to \xi_m \underset{v}{\to} \xi_{m-1} \cdots \underset{v}{\to} \xi_0 \to 0 \tag{3.76}$$

be a holomorphic complex of vector bundles on X, which, together with a holomorphic restriction map $r : \xi_{0|Y} \to \eta$, provides a resolution of the sheaf $i_*\mathcal{O}_Y(\eta)$, i.e. we have an exact sequence of sheaves

$$0 \to \mathcal{O}_X(\xi_m) \underset{v}{\to} \cdots \to \mathcal{O}_X(\xi_0) \to i_*\mathcal{O}_Y(\eta) \to 0. \tag{3.77}$$

In particular the complex (ξ, v) is acyclic on $X \setminus Y$. If $y \in Y$, $U \in TX_y$, let $\partial_U v(y)$ be the derivative of v in any holomorphic trivialization of (ξ, v)

near y. Then by [**B3**, Theorem 1.2], $\partial_U v(y)$ acts on $H((\xi, v)_y)$, the action only depends on the image $z \in N_{Y/X,y}$ of U, and will be denoted by $\partial_z v(y)$. Let π be the projection $N_{Y/X} \to Y$. Then by [**B3**, Theorem 1.2], there is a canonical isomorphism of holomorphic complexes on $N_{Y/X}$,

$$(\pi^* H(\xi, v)_{|Y}), \partial_z v) \simeq \left(\pi^*(\Lambda N^*_{Y/X} \otimes \eta), i_z\right) , \tag{3.78}$$

where in the right-hand side of (3.78), appears a Koszul complex. Let $g^\xi = \bigoplus_{i=0}^{m} g^{\xi_i}$ be a Hermitian metric on $\xi = \bigoplus_{i=0}^{m} \xi_i$. Let $g^{N_{Y/X}}, g^\eta$ be Hermitian metrics on $N_{Y/X}, \eta$.

Put

$$V = v + v^* . \tag{3.79}$$

By finite dimensional Hodge theory,

$$H(\xi, v)_{|Y}) \simeq \ker V_{|Y} . \tag{3.80}$$

As a subbundle of $\xi_{|Y}$, $\ker V_{|Y}$ inherits a Hermitian metric. Let $g^{H(\xi,v)_{|Y}}$ be the corresponding metric on $H(\xi, v)_{|Y}$.

DEFINITION 3.33. We will say that $g^{\xi_0}, \ldots, g^{\xi_m}$ verify assumption (A) with respect to $g^{N_{Y/X}}, g^\eta$ if the identification (3.78) is an isometry.

By [**B3**, Proposition 1.6], given metrics $g^{N_{Y/X}}, g^\eta$, there exist metrics $g^{\xi_0}, \ldots, g^{\xi_m}$ such that (A) is verified.

In the sequel we assume that (A) holds.

Let ∇^ξ be the holomorphic Hermitian connection on ξ. For $t > 0$, put

$$A_t = \nabla^\xi + \sqrt{t} V . \tag{3.81}$$

Let N_H be the number operator of (ξ, v).

DEFINITION 3.34. For $t \geq 0$, put

$$\begin{aligned} \alpha_t &= \varphi \operatorname{Tr_s} \left[\exp(-A_t^2)\right] , \\ \gamma_t &= \varphi \operatorname{Tr_s} \left[N_H \exp(-A_t^2)\right] . \end{aligned} \tag{3.82}$$

Of course, equations (3.10), (3.11) are still valid. Now we give a result of [**B3**, Theorems 3.2 and 4.3], which replaces Theorem 2.6 and (3.14) in this new situation. Let δ_Y be the current of integration on Y.

Theorem 3.35. *As $t \to +\infty$,*

$$\begin{aligned} \alpha_t &= \operatorname{Td}^{-1}(N_{Y/X}, g^{N_{Y/X}}) \operatorname{ch}(\eta, g^\eta)\delta_Y + \mathcal{O}\left(\frac{1}{\sqrt{t}}\right), \\ \gamma_t &= -((\operatorname{Td})^{-1})'(N_{Y/X}, g^{N_{Y/X}}) \operatorname{ch}(\eta, g^\eta)\delta_Y + \mathcal{O}\left(\frac{1}{\sqrt{t}}\right) . \end{aligned} \tag{3.83}$$

REMARK 3.36. In(3.83), $\mathcal{O}\left(\frac{1}{\sqrt{t}}\right)$ is taken in the adequate Sobolev space of currents. Also, the convergence is shown to be microlocal in the set of currents whose wave front set is conormal to Y.

By Theorem 2.4 and by (3.83), we see that

$$\mathrm{ch}(\xi) = \mathrm{Td}^{-1}(N_{Y/X})\,\mathrm{ch}(\eta)\delta_Y \quad \text{in} \quad H^{\mathrm{even}}(X, \mathbf{Q})\,. \tag{3.84}$$

This is exactly the content of Riemann-Roch-Grothendieck for immersions, which says that

$$\mathrm{ch}(i_*\eta) = \mathrm{Td}^{-1}(N_{Y/X})\,\mathrm{ch}(\eta)\delta_Y \quad \text{in} \quad H^{\mathrm{even}}(X, \mathbf{Q})\,. \tag{3.85}$$

Using Theorem 3.35, we can now imitate Definition 3.5 and construct a current $T(\xi, g^\xi)$ on X by formulas (3.15), (3.16). The following result is proved in [**BGS4,** Theorem 2.5].

Theorem 3.37. *The current $T(\xi, g^\xi)$ is a sum of currents of type (p,p), and its wave front set lies in $N^*_{Y/X,\mathbf{R}}$. Moreover*

$$\frac{\bar{\partial}\partial}{2i\pi} T(\xi, g^\xi) = \mathrm{Td}^{-1}(N_{Y/X}, g^{N_{Y/X}})\,\mathrm{ch}(\eta, g^\eta)\delta_Y - \mathrm{ch}(\xi, g^\xi)\,. \tag{3.86}$$

Let P^X_Y be the set of currents on X, which are sums of currents of type (p,p), whose wave front set lies in $N^*_{Y/X,\mathbf{R}}$. We define $P^{X,0}_Y$ as in Section 3 a), with the adequate condition on the wave front set of β, γ.

The following extension of Theorem 3.9 is established in [**BGS5,** Theorem 2.5].

Theorem 3.38. *Let $(g^\xi, g^{N_{Y/X}}, g^\eta)$ and $(g'^\xi, g'^{N_{Y/X}}, g'^\eta)$ be triples of metrics verifying condition (A). Then*

$$\begin{aligned} T(\xi, g'^\xi) - T(\xi, g^\xi) = \Big(&\widetilde{\mathrm{Td}}^{-1}(N_{Y/X}, g^{N_{Y/X}}, g'^{N_{Y/X}})\,\mathrm{ch}(\eta, g^\eta) \Big) \\ &+ \mathrm{Td}^{-1}(N_{Y/X}, g'^{N_{Y/X}})\widetilde{\mathrm{ch}}(\eta, g^\eta, g'^\eta))\delta_Y \\ &- \widetilde{\mathrm{ch}}(\xi, g^\xi, g'^\xi)) \quad \text{in} \quad P^X_Y / P^{X,0}_Y\,. \end{aligned} \tag{3.87}$$

PROOF : The proof of Theorem 3.38 is essentially the same as the proof of Theorem 3.9. □

f) Compatibility of the currents $T(\xi, g^\xi)$ to the composition of immersions.

Let $i : Y \to X$, $i' : Y' \to X$ be two complex submanifolds of X intersecting transversally. In particular $\dim Y + \dim Y' \geq \dim X$. Let η, η' be holomorphic vector bundles on Y, Y', let (ξ, v), (ξ', v') be two holomorphic complexes of vector bundles on X which provide resolutions of $i_*\eta$, $i'_*\eta'$. Then one verifies easily that if $Y'' = Y \cap Y'$, if $i'' : Y'' \to X$ is the corresponding embedding, then $(\xi\widehat{\otimes}\xi', v+v')$ provides a resolution of $i''_*(\eta_{|Y''}\otimes\eta'_{|Y''})$.

Then we have the diagram

$$
\begin{array}{ccc}
Y'' & \longrightarrow & Y \\
\downarrow & \searrow & \downarrow \\
Y' & \longrightarrow & X
\end{array}
\tag{3.88}
$$

Let $(g^{N_{Y/X}}, g^{\eta}, g^{\xi})$ and $(g^{N_{Y'/X}}, g^{\eta'}, g^{\xi'})$ be Hermitian metrics verifying (A). Then we equip $N_{Y''/X} = N_{Y/X|Y''} \oplus N_{Y'/X|Y''}$ with the metric $g^{N_{Y''/X}} = g^{N_{Y/X|Y''}} \oplus g^{N_{Y'/X|Y''}}$. One verifies easily that $(g^{N_{Y''/Y}}, g^{\eta_{|Y''}} \otimes g^{\eta'_{|Y''}}, g^{\xi''})$ verifies (A).

Let $P^X_{Y\cup Y'}, P^{X,0}_{Y\cup Y'}$ be the obvious analogues of $P^X_Y, P^{X,0}_Y$ when replacing Y by $Y\cup Y'$.

The following result is established in [**BGS5**, Theorem 2.7].

Theorem 3.39. *The following identity holds*

$$
\begin{aligned}
T(\xi\widehat{\otimes}\xi', g^{\xi\otimes\xi'}) = {}& T(\xi, g^{\xi})\,\mathrm{ch}(\xi', g^{\xi'}) + \\
& \mathrm{Td}^{-1}(N_{Y/X}, g^{N_{Y/X}})\,\mathrm{ch}(\eta, g^{\eta}) T(\xi', g^{\xi'})\delta_Y \text{ in } P^X_{Y\cup Y'}/P^{X,0}_{Y\cup Y'}.
\end{aligned}
\tag{3.89}
$$

g) Complex immersions and Quillen metrics.

Assume now that X and Y are compact. Let g^{TX}, g^{TY} be Kähler metrics on TX, TY. Let $g^{N_{Y/X}}, g^{\eta}$ be Hermitian metrics on $N_{Y/X}, \eta$, let $g^{\xi} = \bigoplus_{i=0}^{m} g^{\xi_i}$ be a Hermitian metric on $\xi = \bigoplus_{i=0}^{m} \xi_i$ which verifies (A) with respect to $g^{N_{Y/X}}, g^{\eta}$.

Put

$$
\begin{aligned}
\lambda(\eta) &= (\det H(Y,\eta))^{-1}, \\
\lambda(\xi_i) &= (\det H(X,\xi_i))^{-1}, \\
\lambda(\xi) &= \bigotimes_{i=0}^{m} (\lambda(\xi_i))^{(-1)^i}.
\end{aligned}
\tag{3.90}
$$

We claim that there is a canonical isomorphism

$$\lambda(\xi) \simeq \lambda(\eta). \tag{3.91}$$

In fact let $H(X,\xi)$ be the hypercohomology of the sheaf $\mathcal{O}_X(\xi)$. Namely if δ is the natural Čech coboundary, we consider the complex $(\mathcal{O}_X(\xi), \delta+v)$. Needless to say, δ and v are graded so that

$$(\delta+v)^2 = \delta v + v\delta = 0. \tag{3.92}$$

Also if $N_{\check{\text{C}}\text{ech}}$ is the natural Čech number operator, we grade $(\mathcal{O}_X(\xi), \delta + v)$ by $N_{\check{\text{C}}\text{ech}} - N_H$ so that $\delta + v$ increase the total degree by 1. Then $r : (\mathcal{O}_X(\xi), \delta + v) \to (\mathcal{O}_Y(\eta), \delta)$ is a quasi-isomorphism, so that

$$H(X,\xi) \simeq H(Y,\eta)\,. \tag{3.93}$$

Now there is a spectral sequence whose E_1 is given by

$$E_1^{(p,q)} = H^q(X, \xi_{-p})\,. \tag{3.94}$$

By (3.94), we get

$$(\det H(X,\xi))^{-1} \simeq \bigotimes_p \left(\left(\det E_1^{(p,.)}\right)^{-1}\right)^{(-1)^p}, \tag{3.95}$$

which is equivalent to

$$(\det H(X,\xi))^{-1} \simeq \bigotimes (\lambda(\xi_i))^{(-1)^i}\,. \tag{3.96}$$

By (3.93), (3.96) we get (3.91).

Now $\lambda(\xi)$ and $\lambda(\eta)$ are equipped with Quillen metrics $\|\ \|_{\lambda(\xi)}$ and $\|\ \|_{\lambda(\eta)}$. It is natural to compare these metrics. This question was first varied by Gillet and Soulé [**GS3**] in their program to prove a Riemann-Roch-Grothendieck formula in Arakelov geometry.

In fact if A is the ring of integers of a number field k, if $X \to \operatorname{Spec}(A)$ is an arithmetic variety, if ξ is an algebraic vector bundle on X, then $H(X,\xi)$ is A-module. If $\lambda(\xi) = (\det H(X,\xi))^{-1}$, then if $\lambda(\xi)$ is equipped with a metric at places at infinity, $\lambda(\xi)$ has an Arakelov degree $\deg \lambda(\xi)$. The idea in [**GS3**] is to precisely equip $\lambda(\xi)$ with Quillen metrics at the places at infinity.

In [**GS3**], Gillet and Soulé gave a conjectural formula for $\deg \lambda(\xi)$ in terms of arithmetic characteristic classes. Still, when calculating the ratio of two Quillen metrics on the same algebraic object, only the contributions at infinity remain, so that [**GS3**] suggests a comparison formula which should be valid for any complex Kähler manifold.

Now we describe a result of Bismut-Lebeau [**BL,** Theorem 0.1] where the conjectured comparison formula was established.

First we introduce the Gillet-Soulé series R [**GS3**]. Let $\zeta(s) = \sum_1^{+\infty} \frac{1}{n^s}$ be the Riemann zeta function.

DEFINITION 3.40. Let $R(x)$ be the power series

$$R(x) = \sum_{\substack{n \geq 1 \\ n \text{ odd}}} \left(2\frac{\zeta'(-n)}{\zeta(-n)} + \sum_1^n \frac{1}{j} \right) \zeta(-n) \frac{x^n}{n!}\,. \tag{3.97}$$

We identify $R(x)$ with the additive genus $\sum_1^q R(x_i)$.

Let $\widetilde{\mathrm{Td}}(TX_{|Y}, g^{TY}, g^{TY_{|Y}}, g^{N_{Y/X}}) \in P^Y/P^{Y,0}$ be the Bott-Chern class, such

$$(3.98)\quad \frac{\bar\partial\partial}{2i\pi}\widetilde{\mathrm{Td}}(TX_{|Y}, g^{TY}, g^{TX_{|Y}}, g^{N_{Y/X}}) = \mathrm{Td}(TX_{|Y}, g^{TX_{|Y}}) - \mathrm{Td}(TY, g^{TY})\,\mathrm{Td}(N_{Y/X}, g^{N_{Y/X}}).$$

Then the result of Bismut-Lebeau [**BL**, Theorem 0.1] is as follows.

Theorem 3.41. *The following identity holds*

$$(3.99)\quad \log\left(\frac{\|\ \ \|_{\lambda(\xi)}}{\|\ \ \|_{\lambda(\eta)}}\right)^2 = -\int_X \mathrm{Td}(TX, g^{TX})T(\xi, g^\xi) + \int_Y \frac{\widetilde{\mathrm{Td}}(TX_{|Y}, g^{TY}, g^{TX_{|Y}}, g^{N_{Y/X}})}{\mathrm{Td}(N_{Y/X}, g^{N_{Y/X}})}\mathrm{ch}(\eta, g^\eta) - \int_X \mathrm{Td}(TX)R(TX)\,\mathrm{ch}(\xi) + \int_Y \mathrm{Td}(TY)R(TY)\,\mathrm{ch}(\eta).$$

PROOF : Some details on the proof of Theorem 3.41 are given in Section 3 i)- 3 k). □

h) Analytic torsion forms, analytic torsion currents and the composition of an immersion and a submersion.

Let now $i : W \to V$ be an embedding of complex manifolds, let $\pi_V : V \to S$ be a holomorphic submersion with compact fibre X, which restricts to a submersion $\pi_W : W \to S$ with compact fibre Y. Then we have the diagram

$$(3.100)\quad \begin{array}{ccccc} Y & \longrightarrow & W & & \\ \downarrow{\scriptstyle i} & & \downarrow{\scriptstyle i} & \searrow^{\pi_W} & \\ X & \longrightarrow & V & \xrightarrow{\pi_V} & S \end{array}$$

Let η be a holomorphic vector bundle on W, let (ξ, v) be a holomorphic complex of vector bundles of X which resolves $i_*\eta$. Of course, fibrewise, the situation is the same as in Section 3 g). Equivalently, the case where S is a point is just the case considered in Section 3 g).

Assume that $R\pi_{W*}\eta$ is locally free. Then $R\pi_{W*}\eta \simeq H(Y, \eta_{|Y})$. By (3.93), $R\pi_{V*}\xi$ is also locally free.

Let ω^V (resp. ω^W) be a $(1,1)$ closed form on V (resp. W). Let $g^\xi = \bigoplus_{i=0}^m g^{\xi_i}$, $g^{N_{Y/X}}$, g^η be Hermitian metrics on $\xi = \bigoplus_{i=0}^m \xi_i$, $N_{Y/X}$, η, which verify assumption (A) (keeping in mind that $N_{Y/X} = N_{W/V}$).

Let $T(\omega^W, g^\eta) \in P^S$ be the analytic torsion forms constructed in Section 3 b). They verify the equation

(3.101)
$$\frac{\overline{\partial}\partial}{2i\pi} T(\omega^W, g^\eta) = \mathrm{ch}\left(H(Y, \eta_{|Y}), g^{H(Y,\eta_{|Y})}\right) - \pi_{W*}\left[\mathrm{Td}(TY, g^{TY})\,\mathrm{ch}(\eta, g^\eta)\right].$$

Similarly, to the family of double complexes $(\Omega(X, \xi_{|X}), \overline{\partial}^X + v)$, we can associate analytic torsion forms $T(\omega^V, g^\xi) \in P^S$, which verify the equation

(3.102)
$$\frac{\overline{\partial}\partial}{2i\pi} T(\omega^V, g^\xi) = \mathrm{ch}(H(X, \xi_{|X}), g^{H(X,\xi_{|X})}) - \pi_{V*}\left[\mathrm{Td}(TX, g^{TX})\,\mathrm{ch}(\xi, g^\xi)\right].$$

Since $H(X, \xi_{|X}) \simeq H(Y, \eta_{|Y})$, the Bott-Chern class $\widetilde{\mathrm{ch}}(H(Y, \eta_{|Y}), g^{H(X,\xi_{|X})}, g^{H(Y,\eta_{|Y})})$ is well-defined.

The main result proved in [**B5,** Theorem 0.1], which extends Theorem 3.41 to the relative situation, is as follows.

Theorem 3.42. *The following identity holds*

$$\begin{aligned}(3.103)\quad & \widetilde{\mathrm{ch}}(H^{(Y,\eta|_Y)}, g^{H(X,\xi|_X)}, g^{H(Y,\eta|_Y)}) - T(\omega^W, g^\eta) + T(\omega^V, g^\xi)\\ & - \int_X \mathrm{Td}(TX, g^{TX}) T(\xi, g^\xi) + \int_Y \frac{\widetilde{\mathrm{Td}}(TX|_W, g^{TY}, g^{TX|_W}, g^{N_{Y/X}})}{\mathrm{Td}(N_{Y/X}, g^{N_{Y/X}})}\,\mathrm{ch}(\eta, g^\eta)\\ & - \int_X \mathrm{Td}(TX) R(TX)\,\mathrm{ch}(\xi) + \int_Y \mathrm{Td}(TY) R(TY)\,\mathrm{ch}(\eta) = 0 \quad \text{in} \quad P^S/P^{S,0}.\end{aligned}$$

Assume now that for $j > 0$, $R^j \pi_{V*} \xi_k = 0$, $(0 \le k \le m)$, $R^j \pi_{W*} \eta = 0$. Then we have a holomorphic complex of vector bundles K on S,

(3.104).
$$K : 0 \to H^0(X, \xi_m) \to H^0(X, \xi_{m-1}) \dots \to H^0(X, \xi_0) \to H^0(X, \xi_{|X}) \to 0$$

Let $\widetilde{\mathrm{ch}}(K, g^K) \in P^S/P^{S,0}$ be the Bott-Chern class such that

$$\begin{aligned}(3.105)\quad \frac{\overline{\partial}\partial}{2i\pi} \widetilde{\mathrm{ch}}(K, g^K) &= \mathrm{ch}\left(H^0(X, \xi_{|X}), g^{H^0(X,\xi_{|X})}\right)\\ & - \sum_{i=0}^{m} (-1)^i\,\mathrm{ch}\left(H^0(X, \xi_{i|X}), g^{H^0(X,\xi_{i|X})}\right).\end{aligned}$$

The following result is proved in [**B5,** Theorem 0.2].

Theorem 3.43. *The following identity holds*

$$(3.106)\quad T(\omega^V, g^\xi) - \sum_{i=0}^{m} (-1)^i T(\omega^V, g^{\xi_i}) - \widetilde{\mathrm{ch}}(K, g^K) = 0 \quad \text{in} \quad P^S/P^{S,0}.$$

REMARK 3.44. Needless to say, Theorems 3.39, 3.41-3.43 are compatible to each other.

i) A sketch of the proof of Theorem 3.41.

Using the anomaly formulas of Theorem 3.23 as in (3.63), and also Theorem 3.38, to establish Theorem 3.41, we may and will assume that $g^{TY}, g^{N_{Y/X}}$ are the metrics induced by g^{TX} on $TY, N_{Y/X}$.

Put

$$E = C^{\infty}(X, \Lambda(T^{*(0,1)}X)\widehat{\otimes}\xi)\,. \tag{3.107}$$

We define the total **Z**-grading on E by the operator $N_V - N_H$. Then $\overline{\partial}^X + v$ acts on E and

$$(\overline{\partial}^X + v)^2 = 0\,. \tag{3.108}$$

Therefore $(E, \overline{\partial}^X + v)$ is a **Z**-graded complex, whose hypercohomology $H(E, \overline{\partial}^X + v)$ is finite dimensional. Dolbeault's theory shows that

$$H(E, \overline{\partial}^X + v) \simeq H(X, \xi)\,. \tag{3.109}$$

Set

$$F = C^{\infty}(Y, \Lambda(T^{*(0,1)}Y)\widehat{\otimes}\eta)\,. \tag{3.110}$$

The restriction map $r : \xi_{0|Y} \to \eta$ extends to a map of complexes $r : (E, \overline{\partial}^X + v) \to (F, \overline{\partial}^Y)$. By [**BL,** Theorem 1.7], it induces the canonical identification $H(X, \xi) \simeq H(Y, \eta)$.

Put

$$\widetilde{\lambda}(\xi) = (\det H(X, \xi))^{-1}\,. \tag{3.111}$$

Then by (3.93), (3.96)

$$\widetilde{\lambda}(\xi) \simeq \lambda(\xi) \simeq \lambda(\eta)\,. \tag{3.112}$$

Moreover by imitating Definition 3.25, we can equip $\widetilde{\lambda}(\xi)$ with a Quillen metric $\|\ \|_{\widetilde{\lambda}(\xi)}$. A first step in the proof of Theorem 3.41 is the simple fact, established in [**BL,** Theorem 2.1], that

$$\|\ \|_{\widetilde{\lambda}(\xi)} = \|\ \|_{\lambda(\xi)}\,. \tag{3.113}$$

To establish Theorem 3.41, we must then compute $\operatorname{Log}\left(\dfrac{\|\ \|_{\widetilde{\lambda}(\xi)}}{\|\ \|_{\lambda(\eta)}}\right)^2$.

Put

$$D^X = \overline{\partial}^X + \overline{\partial}^{X*}, \quad V = v + v^*. \tag{3.114}$$

For $T \geq 0$, it is clear that

$$(\overline{\partial}^X + Tv)^2 = 0. \tag{3.115}$$

Also for $T > 0$

$$H^{\cdot}(E, \overline{\partial}^X + Tv) \simeq H^{\cdot}(E, \overline{\partial}^X + v). \tag{3.116}$$

For $T \geq 0$, put

$$A_T = D^X + TV. \tag{3.117}$$

Then by Hodge theory,

$$\ker A_T = H^{\cdot}(E, \overline{\partial}^X + Tv), \tag{3.118}$$

so that for $T > 0$,

$$\ker(A_T) \simeq H^{\cdot}(E, \overline{\partial}^X + Tv) \simeq H^{\cdot}(F, \overline{\partial}^Y). \tag{3.119}$$

Set

$$D^Y = \overline{\partial}^Y + \overline{\partial}^{Y*}. \tag{3.120}$$

Then

$$\ker D^Y \simeq H(Y, \eta). \tag{3.121}$$

Now we describe in some detail a few difficulties which appear in the evaluation of $\operatorname{Log}\left(\dfrac{\| \quad \|_{\widetilde{\lambda}(\xi)}}{\| \quad \|_{\lambda(\eta)}}\right)^2$ given in [**BL**]. We will here take a toy object to describe these difficulties.

Let $\chi(\xi), \chi(\eta)$ be the Euler characteristics of $\mathcal{O}_X(\xi), \mathcal{O}_Y(\eta)$. Then, by the Mc Kean-Singer formula [**McKS**], for $t > 0$, $T \geq 0$,

$$\begin{aligned} \chi(\xi) &= \operatorname{Tr_s}\left[\exp(-(tA_T)^2\right], \\ \chi(\eta) &= \operatorname{Tr_s}\left[\exp(-(tD_Y)^2\right]. \end{aligned} \tag{3.122}$$

Of course

$$\chi(\xi) = \chi(\eta). \tag{3.123}$$

Let $P_t^T(x, x')$ $(x, x' \in X)$, $Q_t(y, y')$ $(y, y' \in Y)$ be the smooth kernels for $\exp(-(tA_T)^2)$, $\exp(-(tD^Y)^2)$ with respect to $dv_X(x')$, $dv_Y(y')$. Here $\operatorname{Tr_s}\left[P_t^T(x, x)\right] dv_X(x)$, $\operatorname{Tr_s}\left[Q_t(y, y)\right] dv_Y(y)$ will be be considered as currents on X.

In [**BL,** Sections 9 and 13], the following result is proved.

Theorem 3.45. *For any* $t > 0$,

$$\lim_{T\to+\infty} \operatorname{Tr_s}\left[P_t^T(x,x)\right] dv_X(x) = \operatorname{Tr_s}\left[Q_t(y,y)\right] dv_Y(y)\,. \tag{3.124}$$

Also, by the local index theorem given in Theorem 1.2,

$$\begin{aligned} &\lim_{t\to 0} \operatorname{Tr_s}\left[P_t^T(x,x)\right] dv_X(x) = \left\{\operatorname{Td}(TX, g^{TX})\operatorname{ch}(\xi, g^{\xi})\right\}^{\max},\\ &\lim_{t\to 0} \operatorname{Tr_s}\left[Q_t(y,y)\right] dv_Y(y) = \left\{\operatorname{Td}(TY, g^{TY})\operatorname{ch}(\eta, g^{\eta})\right\}^{\max}. \end{aligned} \tag{3.125}$$

We then have the non commutative diagram

$$\begin{array}{ccc} \operatorname{Tr_s}[P_t^T(x,x)]dv_X(x) & \xrightarrow[t\to 0]{} & \{\operatorname{Td}(TX,g^{TX})\operatorname{ch}(\xi,g^{\xi})\}^{\max} \\ \Big\downarrow{\scriptstyle T\to+\infty} & & \\ \operatorname{Tr_s}[Q_t(y,y)]dv_Y(y)\delta_Y & \xrightarrow[t\to 0]{} & \{\operatorname{Td}(TY,g^{TY})\operatorname{ch}(\eta,g^{\eta})\}^{\max}\delta_Y \end{array} \tag{3.126}$$

Needless to say, (3.126) fits with (3.122), (3.123), because of (3.86).
Observe that

$$tA_{\frac{T}{t}} = tD^X + TV\,. \tag{3.127}$$

We use the notation in Section 3 e). A simple application of local index techniques shows that

$$\lim_{t\to 0} \operatorname{Tr_s}\left[P_t^{T/t}(x,x)\right] dv_X(x) = \left\{\operatorname{Td}(TX, g^{TX})\alpha_{T^2}\right\}^{\max}. \tag{3.128}$$

In view of (3.83), (3.124), (3.128), we have the new diagram

$$\begin{array}{ccc} \operatorname{Tr_s}[P_t^{T/t}(x,x)]dv_X(x) & \xrightarrow[t\to 0]{} & \{\operatorname{Td}(TX,g^{TX})\alpha_{T^2}\}^{\max} \\ & & \Big\downarrow{\scriptstyle T\to+\infty} \\ \Big\downarrow{\scriptstyle T\to+\infty} & & \left\{\frac{\operatorname{Td}(TX,g^{TX})}{\operatorname{Td}(N_{Y/X},g^{N_{Y/X}})}\operatorname{ch}(\eta,g^{\eta})\right\}^{\max}\delta_Y \\ & & \\ \operatorname{Tr_s}[Q_t(y,y)]dv_Y(y)\delta_Y & \xrightarrow[t\to 0]{} & \{\operatorname{Td}(TY,g^{TY})\operatorname{ch}(\eta,g^{\eta})\}^{\max}\delta_Y \end{array} \tag{3.129}$$

The whole point is now to find how to close the gaps in the diagrams (3.126), (3.129) at the level of currents (the gap is of course 0 in cohomology).
Clearly

$$tA_{T/t^2} = tD^X + \frac{T}{t}V\,. \tag{3.130}$$

Consider the exact sequence of holomorphic Hermitian vector bundles on Y

$$0 \to TY \to TX_{|Y} \to N_{Y/X} \to 0 \tag{3.131}$$

and more generally any short exact sequence as in (3.131). In [**B4**], for $T > 0$, a form $\delta_T \in P^Y$ is constructed, such that

$$\begin{aligned} \lim_{T\to 0} \delta_T &= \frac{\mathrm{Td}(TX_{|Y}, g^{TX_{|Y}})}{\mathrm{Td}(N_{Y/X}, g^{N_{Y/X}})}, \\ \lim_{T\to+\infty} \delta_T &= \mathrm{Td}(TY, g^{TY}). \end{aligned} \tag{3.132}$$

Then, in [**BL,** Section 12], it is shown that for $T > 0$,

$$\lim_{t\to 0} \mathrm{Tr_s}\left[P_t^{T/t^2}(x,x)\right] dv_X(x) = \{\delta_T \, \mathrm{ch}(\eta, g^\eta)\}^{\max} \delta_Y . \tag{3.133}$$

Then we have the new diagram

$$\begin{array}{ccc} \mathrm{Tr_s}\,[P_t^0(x,x)]dv_X(x) & & \frac{\mathrm{Td}(TX,g^{TX})}{\mathrm{Td}(N_{Y/X},g^{N_{Y/X}})}\,\mathrm{ch}(\eta,g^\eta)\delta_Y \\ \Big\uparrow {\scriptstyle T\to 0} & & \Big\uparrow {\scriptstyle T\to 0} \\ \mathrm{Tr_s}\,[P_t^{T/t^2}(x,x)]dv_X(x) & \xrightarrow[t\to 0]{} & \{\delta_T\,\mathrm{ch}(\eta,g^\eta)\}^{\max}\delta_Y \\ \Big\downarrow {\scriptstyle T\to+\infty} & & \Big\downarrow {\scriptstyle T\to+\infty} \\ \mathrm{Tr_s}\,[Q_t(y,y)]dv_Y(y)\delta_Y & \xrightarrow[t\to 0]{} & \{\mathrm{Td}(TY,g^{TY})\,\mathrm{ch}(\eta,g^\eta)\}^{\max}\delta_Y \end{array} \tag{3.134}$$

Let P_∞^T be the orthogonal projection operator on $\ker(A_T)$. Then P_∞^T is given by a smooth kernel $P_\infty^T(x,x')$ on X. Similarly let Q_∞ be the orthogonal projection operator on $\ker(D_Y)$, and let $Q_\infty(y,y')$ be the corresponding kernel on Y.

Then by [**BL,** Section 10], we have the diagram

$$\begin{array}{ccc} \mathrm{Tr_s}\,[P_t^T(x,x)]dv_X(x) & \xrightarrow[t\to+\infty]{} & \mathrm{Tr_s}\,[P_\infty^T(x,x)]dv_X(x) \\ \Big\downarrow {\scriptstyle T\to+\infty} & & \Big\downarrow {\scriptstyle T\to+\infty} \\ \mathrm{Tr_s}\,[Q_t(y,y)]dv_Y(y)\delta_Y & \xrightarrow[t\to+\infty]{} & \mathrm{Tr_s}\,[Q_\infty(y,y)]dv_Y(y)\delta_Y \end{array} \tag{3.135}$$

Part of the proof of Theorem 3.41 [**BL**] consists in putting together the diagrams (3.126), (3.129), (3.134) and (3.135).

j) The Hodge theory of the resolution of the point.

In the proof in [**BL**] of Theorem 3.41, or in the construction of the form δ_T in [**B4**], the following toy model of an embedding $i : Y \to X$ appears naturally.

Let $V_{\mathbf{R}}$ be a real even dimensional vector space, let J be a complex structure on $V_{\mathbf{R}}$. Let V be the corresponding complex vector space, so that $V_{\mathbf{R}} \otimes_{\mathbf{R}} \mathbf{C} = V \oplus \overline{V}$. Put $n = \dim V$. Recall that if $z \in V$, z represents $Z = v + \overline{v} \in V_{\mathbf{R}}$.

Let i be the embedding $\{0\} \to V$. If $z \in V$, let i_z be the interior multiplication by z. Let

(3.136)

$$(\Lambda V^*, \sqrt{-1} i_z) : 0 \to \Lambda^n V^* \xrightarrow[\sqrt{-1} i_z]{} \Lambda^{n-1} V^* \to \dots \xrightarrow[\sqrt{-1} i_z]{} \Lambda^0(V^*) = \mathbf{C} \to 0$$

be the obvious Koszul complex. Let r be the restriction map $\alpha \in \Lambda^0(V^*)_{|0} \to \alpha \in \mathbf{C}$. Then by [**GrH,** p. 688], the complex $(\Lambda V^*, \sqrt{-1} i_z)$ provides a resolution of the sheaf $i_* \mathbf{C}$.

Let $(C^\infty(V, \Lambda \overline{V}^* \widehat{\otimes} \Lambda V^*), \overline{\partial}^V)$ be the Dolbeault complex over V of smooth sections of $\Lambda \overline{V}^* \widehat{\otimes} \Lambda V^*$.

Let N_V, N_H be the number operator of $\Lambda \overline{V}^*$, ΛV^*. The $\mathbf{Z}$-grading of the complex $(C^\infty(V, \Lambda \overline{V}^* \otimes \Lambda V^*), \overline{\partial}^V + \sqrt{-1} i_y)$ is given by $N_V - N_H$.

Let $r : C^\infty(V, \Lambda \overline{V}^* \otimes \Lambda V^*) \to \mathbf{C}$ be such that if $\alpha \in C^\infty(V, \Lambda^p(\overline{V}^*))$, $\beta \in C^\infty(V, \Lambda^q(V^*))$, then

$$\begin{aligned} r(\alpha \widehat{\otimes} \beta) &= 0 \text{ if } p + q > 0\,, \\ &= \alpha_0 \beta_0 \text{ if } p = 0, q = 0\,. \end{aligned} \tag{3.137}$$

Observe that $\Lambda V_R^* \otimes_{\mathbf{R}} \mathbf{C} = \Lambda \overline{V}^* \otimes \Lambda V^*$, and that r is nothing else than the restriction of a smooth form on V to $\{0\}$.

Let $\mathbf{C}$ be the trivial complex, equipped with the chain map $\overline{\partial}^{\{0\}} = 0$. By the arguments of [**BL,** Theorem 1.7], the chain map

$$r : (C^\infty(V, \Lambda \overline{V}^* \widehat{\otimes} \Lambda V^*), \overline{\partial}^V + \sqrt{-1} i_z) \to (\mathbf{C}, \overline{\partial}^{\{0\}}) \tag{3.138}$$

is a quasi-isomorphism. In particular if $\mathcal{H}$ is the hypercohomology of $(C^\infty(V, \Lambda \overline{V}^* \widehat{\otimes} \Lambda V^*), \overline{\partial}^V + \sqrt{-1} i_z)$,

$$\begin{aligned} \mathcal{H}^p &= 0 \text{ if } p \neq 0\,, \\ &= \mathbf{C} \text{ if } p = 0\,, \end{aligned} \tag{3.139}$$

and $r = \mathcal{H} \to \mathbf{C}$ identify canonically $\mathcal{H}$ with $\mathbf{C} = H^0(\{0\}, \mathbf{C})$.

Let now g^V be a Hermitian metric on V. Let $\langle\, , \rangle_{\Lambda(\overline{V}^*) \widehat{\otimes} \Lambda(V^*)}$ be the corresponding Hermitian product on $\Lambda(\overline{V}^*) \widehat{\otimes} \Lambda(V^*)$ and let dv_V be the associated volume form on V. Then we equip $C_c^\infty(V, \Lambda \overline{V}^* \otimes \Lambda V^*)$ with the Hermitian product

$$\langle s, s' \rangle = \left(\frac{1}{2\pi} \right)^{\dim V} \int_{V_R} \langle \alpha, \beta \rangle_{\Lambda(\overline{V}^*) \widehat{\otimes} \Lambda(V^*)} \, dv_V \,. \tag{3.140}$$

Let $\overline{\partial}^{V^*}$ be the formal adjoint of $\overline{\partial}^V$. Then $\overline{\partial}^{V^*} - \sqrt{-1}i_z^*$ is the formal adjoint of $\overline{\partial}^V + \sqrt{-1}i_z$.

Let θ be the Kähler form of $V_{\mathbf{R}}$, i.e.

$$\theta(X,Y) = \langle X, JY\rangle_{V_{\mathbf{R}}} . \tag{3.141}$$

Let $L = \theta\wedge$, and let Λ be the adjoint of L. Put

$$S = -(L + \Lambda) . \tag{3.142}$$

By [**B4**, Proposition 1.4],

$$(\overline{\partial}^V + \sqrt{-1}i_z + \overline{\partial}^{V^*} - \sqrt{-1}i_z^*)^2 = -\frac{\Delta^V}{2} + \frac{|Z|^2}{2} + S . \tag{3.143}$$

Then the Laplacian (3.143) is an harmonic oscillator.

The following elementary result is proved in [**B4**, Theorem 1.6].

Theorem 3.46. *Let $\beta \in C^\infty(V, \Lambda(\overline{V}^*)\widehat{\otimes}\Lambda(V^*))$ be given by*

$$\beta = \exp\left(\theta - \frac{|Z|^2}{2}\right) . \tag{3.144}$$

Then β has total degree 0, and moreover

$$\|\beta\|_{L_2} = 1 . \tag{3.145}$$

Also

$$\begin{aligned}(\overline{\partial}^V + \sqrt{-1}i_z)\beta &= 0 ,\\ (\overline{\partial}^{V^*} - \sqrt{-1}i_z^*)\beta &= 0 .\end{aligned} \tag{3.146}$$

Moreover β spans the 1-dimensional kernel of $(\overline{\partial}^V + \sqrt{-1}i_z + \overline{\partial}^{V^*} - \sqrt{-1}i_z^*)^2$. Finally

$$r\beta = 1 \tag{3.147}$$

i.e. β represents canonically $1 \in H^0(\{0\}, \mathbf{C})$ in $C^\infty(V, \Lambda(\overline{V}^*)\widehat{\otimes}\Lambda(V^*))$.

REMARK 3.47. Several remarks are in order here. First note that θ, as a $(1,1)$ form, has total degree 0, so that indeed β is of total degree 0. Also observe that $\|\beta\|_{L_2} = 1$ and $r\beta = 1$, so that $1 \to \beta$ is an isometry.

Now we go back to the formalism of Section 3 h).

Let $\pi' : N_{Y/X} \to Y$ be the obvious projection. Consider the embedding $i' : Y \to N_{Y/X}$ (where Y is identified with the zero section of $N_{Y/X}$).

A fundamental fact, established in [**BL,** Section 10] is that for $T \to +\infty$, $\ker(A_T) \simeq H^\bullet(E, \overline{\partial}^X + Tv) \simeq H^\bullet(F, \overline{\partial}^Y)$ can be asymptotically described as follows. Put

$$D^Y = \overline{\partial}^Y + \overline{\partial}^{Y*}. \tag{3.148}$$

Then

$$\ker(D^Y) \simeq H(Y, \eta). \tag{3.149}$$

Take $\alpha \in \ker(D^Y) \simeq H(Y, \eta)$. Then by [**BL,** Section 10], the element $\gamma_T \in \ker(A_T) \simeq H(X, \xi)$ canonically identified with α can be described asymptotically in a tubular neighborhood of Y by

$$\gamma_T \simeq \pi'^* \beta \exp\left(\theta_{N_{Y/X}} - \frac{T|Z|^2}{2}\right). \tag{3.150}$$

Of course γ_T can be viewed locally as a smooth section of $\Lambda(T^{*(0,1)}X)\widehat{\otimes}\xi$ because of (3.80).

k) The forms δ_T : A toy model for the analytic torsion forms.

Let Y be a complex manifold. Let

$$0 \to L \underset{i}{\to} M \underset{j}{\to} N \to 0 \tag{3.151}$$

be a short exact sequence of holomorphic vector bundle on Y. Let g^M be a Hermitian metric on M, let g^L be the induced metric on L. By identifying N to $L^\perp$, let g^N be the metric induced by g^M on N.

Let $\mathcal{L}, \mathcal{M}$ be the total spaces of L, M. Then we have the diagram

$$\begin{array}{ccccc} L & \longrightarrow & \mathcal{L} & & \\ \downarrow i & & \downarrow i & \overset{\pi_{\mathcal{L}/S}}{\searrow} & \\ M & \longrightarrow & \mathcal{M} & \xrightarrow[\pi_{\mathcal{M}/S}]{} & S \end{array} \tag{3.152}$$

Also on $\mathcal{M}$, the Koszul complex $\left(\pi_{\mathcal{M}}^* \Lambda N^*, \sqrt{-1} i_{j(z)}\right)$ is a resolution of the constant sheaf $i_* \mathbf{C}$.

Let $\nabla^L, \nabla^M, \nabla^N$ be the holomorphic Hermitian connections on L, M, N, and let R^M, R^L, R^N be their curvatures. Then the connections ∇^L, ∇^M define horizontal subbundles $T^H\mathcal{L}$, $T^H\mathcal{M}$. Put

$$\begin{aligned} \omega^{\mathcal{L}} &= \frac{i\overline{\partial}\partial |z|_L^2}{2}, \\ \omega^{\mathcal{M}} &= \frac{i\overline{\partial}\partial |z|_M^2}{2}. \end{aligned} \tag{3.153}$$

Clearly

$$\omega^{\mathcal{L}} = i^* \omega^{\mathcal{M}} . \tag{3.154}$$

Also $\omega^{\mathcal{L}}, \omega^{\mathcal{M}}$ induce the tautological Kähler forms along the fibres L, M. Moreover one verifies easily that $T^H\mathcal{L}, T^H\mathcal{M}$ are exactly the orthogonal bundles to L, M with respect to $\omega^{\mathcal{L}}, \omega^{\mathcal{M}}$.

Then we are in a situation formally similar to the one we met in Section 3 g). We will then construct the associated Levi-Civita superconnection. In our context, if $U, V \in T_{\mathbf{R}}S$,

$$\begin{aligned} T^{\mathcal{L}}(U^H, V^H) &= R^L(U,V)Z , \\ T^{\mathcal{M}}(U^H, V^H) &= R^M(U,V)Z . \end{aligned} \tag{3.155}$$

Let E, F be the bundles on Y of smooth sections of $\Lambda(\overline{M}^*)\widehat{\otimes}\Lambda(N^*), \Lambda(\overline{N}^*)$ along the fibres M, L. Let ∇^E, ∇^F be the connections on E, F constructed as in Definition 3.11.

DEFINITION 3.48. For $T > 0$, let $\mathcal{B}_T^{\mathcal{M}}$ be the Levi-Civita superconnection on E,

(3.156)
$$\mathcal{B}_T^M = \left(\overline{\partial}^M + \overline{\partial}^{M*}\right) + \sqrt{T}\left(\sqrt{-1}i_{j(z)} + (\sqrt{-1}i_{j(z)})^*\right) + \nabla^E - \frac{c(R^M Z)}{2\sqrt{2}} ,$$

Similarly by making $T = 0$ in (3.156), we can construct the superconnection $\mathcal{B}^N$.

Recall that $\Lambda(\overline{M}^*), \Lambda(N^*)$ are $c(M_R), c(N_R)$ Clifford modules. Let $c, \widehat{c}$ denote the corresponding Clifford actions.

We have the C^∞ identification $M = L \oplus N$. Put

$$A = \nabla^M - \nabla^L \oplus \nabla^N . \tag{3.157}$$

Then A exchanges $L_{\mathbf{R}}$ and $N_{\mathbf{R}}$.

Let $P^L : M \to L, P^N : M \to N$ be the orthogonal projection operators. If (f_α) is a basis of $T_{\mathbf{R}}S$, put

$$\widehat{c}(AP^L Z) = -\Sigma f^\alpha \widehat{c}(A(f_\alpha)P^L Z) . \tag{3.158}$$

Let $e_1, \ldots, e_{2n}$ be an orthonornal basis of $N_{\mathbf{R}}$. Put

$$S = \frac{\sqrt{-1}}{2} \sum_1^{2n} c(e_i)\widehat{c}(e_i) . \tag{3.159}$$

Then by [**B4**, Theorem 3.10] (or by the more general curvature identity of [**B2**, Theorem 3.5], if $e_1, \ldots, e_{2m}$ is an orthonormal basis of $M_{\mathbf{R}}$,

$$\begin{aligned} \mathcal{B}_T^{\mathcal{M},2} = {} & -\tfrac{1}{2} \sum_1^{2m} \left(\nabla_{e_i} + \tfrac{1}{2}\langle R^M Z, e_i\rangle\right)^2 \\ & + \frac{T\left|P^N Z\right|^2}{2} + \sqrt{T} S + \frac{\sqrt{-T}}{\sqrt{2}} \widehat{c}(AP^L Z) \\ & + \tfrac{1}{2} \operatorname{Tr}\left[R^M\right] + R^{\Lambda(N^*)} . \end{aligned} \tag{3.160}$$

In particular,

$$(3.161)\qquad \mathcal{B}_0^{\mathcal{M},2} = -\tfrac{1}{2}\sum_1^m \left(\nabla_{e_i} + \tfrac{1}{2}\left\langle R^M Z, e_i\right\rangle\right)^2 + \tfrac{1}{2}\operatorname{Tr}\left[R^M\right] + R^{\Lambda(N^*)}.$$

With some surprise, we see that $\mathcal{B}_0^{\mathcal{M},2}$ is nothing else than the Getzler operator **[Ge1,2]** [**BeGeV,** Proposition 4.19] in local index theory. The fact that the Getzler operator (obtained by rescaling the square of the Dirac operator) is itself a square is a surprise. Another surprising feature of $\mathcal{B}_T^{\mathcal{M},2}$ is that its matrix part only acts on $\Lambda(\overline{N}^*)\widehat{\otimes}\Lambda(N^*)$ and not on $\Lambda(\overline{M}^*)\widehat{\otimes}\Lambda(N^*)$.

For $T > 0$, $\mathcal{B}_T^2$ is essentially a perturbation of the harmonic oscillator in (3.143). For $T > 0$, let $S_T(Z,Z')$ $(Z, Z' \in M_{\mathbf{R}})$ be the smooth kernel of the operator $\exp\left(-\mathcal{B}_T^{\mathcal{M},2}\right)$ with respect to $\dfrac{dv_M(Z')}{(2\pi)^{\dim M}}$. Then $S_T(Z,Z') \in \Lambda(T^*_{\mathbf{R}}Y)\widehat{\otimes}\operatorname{End}(\Lambda(\overline{N}^*)\widehat{\otimes}\Lambda(N^*))$. Then a simple fact proved in [**B4,** Theorem 4.2] is that $Z \to \operatorname{Tr_s}[S_T(Z,Z)]$ only depends on $j(Z) \in N_{\mathbf{R}}$. Also one verifies easily that if $Z \in N_{\mathbf{R}}$,

$$(3.162)\qquad |S_T(Z,Z)| \le c(T)\exp(-C(T)\,|Z|^2).$$

The operator $\exp(-\mathcal{B}_T^2)$ is in general not trace class. Still we can define a generalized supertrace as follows.

DEFINITION 3.49. Set

$$(3.163)\qquad \operatorname{Tr_s}\left[\exp\left(-\mathcal{B}_T^{\mathcal{M},2}\right)\right] = \int_{N_{\mathbf{R}}} \operatorname{Tr_s}\left[S_T(Z,Z)\right]\frac{dv_N(Z)}{(2\pi)^{\dim N}}.$$

Put

$$(3.164)\qquad \delta_T = \Phi\operatorname{Tr_s}\left[\exp\left(-\mathcal{B}_T^{\mathcal{M},2}\right)\right].$$

Then we have the result of [**B4,** Theorems 4.8 and 7.7].

Theorem 3.50. *The forms δ_T lie in P^B, they are closed, and their cohomology class does not depend on T. Moreover as $T \to 0$,*

$$(3.165)\qquad \delta_T = \frac{\operatorname{Td}(M,g^M)}{\operatorname{Td}(N,g^N)} + \mathcal{O}(T),$$

and as $T \to +\infty$

$$(3.166)\qquad \delta_T = \operatorname{Td}(L,g^L) + \mathcal{O}\left(\frac{1}{\sqrt{T}}\right).$$

Let N_H be the number operator of $\Lambda(N^*)$. Set

$$(3.167)\qquad \varepsilon_T = \Phi\operatorname{Tr_s}\left[N_H\exp(-\mathcal{B}_T^2)\right]$$

(where again, the right-hand side of (3.167) is a generalized supertrace).

By [**B4**, Theorem 4.6]

$$\frac{\partial}{\partial T}\delta_T = \frac{\bar{\partial}\partial}{2i\pi}\frac{\varepsilon_T}{T}. \tag{3.168}$$

Also by [**B4**, Theorem 7.7], as $T \to +\infty$,

$$\varepsilon_T = \frac{\dim N}{2}\operatorname{Td}(L, g^L) + \mathcal{O}\left(\frac{1}{\sqrt{T}}\right). \tag{3.169}$$

DEFINITION 3.51. For $s \in \mathbf{C}, 0 < \operatorname{Re}(s) < 1/2$, put

$$\mathbf{B}(s) = \frac{1}{\Gamma(s)}\int_0^{+\infty} T^{s-1}(\varepsilon_T - \varepsilon_\infty)dT. \tag{3.170}$$

Then $\mathbf{B}(s)$ extends holomorphically near $s = 0$.

DEFINITION 3.52. Put

$$\mathbf{B}(L, M, g^M) = \frac{\partial}{\partial s}\mathbf{B}(s)_{|s=0}. \tag{3.171}$$

Clearly $\mathbf{B}(L, M, g^M)$ is a generalized analytic torsion form on Y.

The following result is established in [**B4**, Theorem 8.3].

Theorem 3.53. *The form* $\mathbf{B}(L, M, g^M)$ *lies in* P^Y. *Moreover*

$$\frac{\bar{\partial}\partial}{2i\pi}\mathbf{B}(L, M, g^M) = \operatorname{Td}(L, g^L) - \frac{\operatorname{Td}(M, g^M)}{\operatorname{Td}(N, g^N)}. \tag{3.172}$$

The construction of $\mathbf{B}(L, M, g^M)$ is functorial. In view of (3.172), a natural question is to evaluate

$$\mathbf{B}(L, M, g^M) + \operatorname{Td}^{-1}(N, g^N)\widetilde{\operatorname{Td}}(L, M, g^M). \tag{3.173}$$

in $P^Y/P^{Y,0}$. Using equation (3.172), it is enough to calculate (3.173) in a split situation.

DEFINITION 3.54. Let $D(x)$ be the formal power series

$$D(x) = \sum_{\substack{n \geq 1 \\ n \text{ odd}}}\left(\Gamma'(1) + \frac{2\zeta'(-n)}{\zeta(-n)} + \sum_1^n \frac{1}{j}\right)\zeta(-n)\frac{x^n}{n!}. \tag{3.174}$$

We identify $D(x)$ to the corresponding additive genus. Let $\mathrm{Td}(L), D(N)$ be the classes of $\mathrm{Td}(L,g^L)$, $\mathrm{Td}(N,g^N)$ in $P^Y/P^{Y,0}$. Clearly they do not depend on g^L, g^N.

Then we have the result of [**B4**, Theorem 8.5].

Theorem 3.55. *The following identity holds*

(3.175)

$$\mathbf{B}(L,M,g^M) = -\mathrm{Td}^{-1}(N,g^N)\widetilde{\mathrm{Td}}(L,M,g^M) + \mathrm{Td}(L)D(N) \text{ in } P^Y/P^{Y,0}.$$

PROOF : As explained in (3.173), it is enough to evaluate $\mathbf{B}(L,M,g^M)$ in the case where the exact sequence splits holomorphically and metrically.

Let $\varphi(T,x)$ be the function

$$\varphi(T,x) = \frac{4}{T}\sinh\left(\frac{x+\sqrt{x^2+4T}}{4}\right)\sinh\left(\frac{-x+\sqrt{x^2+4T}}{4}\right). \tag{3.176}$$

□

We identify $\frac{\partial\varphi/\partial x}{\varphi}(T,x)$, as a function of x with the corresponding additive genus. Then in [**B4**, eq. (8.26)-(8.28)], it is shown by explicit computation that in the split case, for $T>0$

$$\varepsilon_T - \varepsilon_\infty = -\mathrm{Td}(L,g^L)\frac{\partial\varphi/\partial x}{\varphi}(T,N,g^N). \tag{3.177}$$

Set

$$\begin{aligned} C(s,x) &= -\frac{1}{\Gamma(s)}\int_0^{+\infty} T^{s-1}\frac{\partial\varphi/\partial x}{\varphi}(T,x)dT, \\ D(x) &= \frac{\partial C}{\partial s}(0,x). \end{aligned} \tag{3.178}$$

We identify $D(x)$ with the corresponding additive genus. Then by (3.170), (3.173), (3.178), when (3.151) splits,

$$\mathbf{B}(L,M,g^M) = \mathrm{Td}(L,g^L)D(N,g^N). \tag{3.179}$$

Now we have the easy expressions for $\varphi(T,x)$

$$\varphi(T,x) = \prod_{k=1}^{+\infty}\left(1+\frac{ix}{2k\pi}+\frac{T}{4k^2\pi^2}\right)\left(1-\frac{ix}{2k\pi}+\frac{T}{4k^2\pi^2}\right), \tag{3.180}$$

which makes the computation of $\frac{\partial\varphi/\partial x}{\varphi}(T,x)$ quite pleasing.

In [**B4**, Appendix], Bismut and Soulé obtain the expression of $D(x)$ given in (3.174) by using (3.180) and the functional equation for $\zeta(s)$.

By (3.97), (3.174),

$$D(x) = R(x) + \Gamma'(1)\frac{\widehat{A}'}{\widehat{A}}(x).$$

In [**BL**], the term related to $\Gamma'(1)\frac{\widehat{A}'}{\widehat{A}}(x)$ disappears in the final result, because it is killed by a corresponding term in $T(\xi, g^\xi)$.

REMARK 3.56. In **[GS3]**, Gillet and Soulé have obtained the genus R by evaluating the analytic torsion of $\mathbf{P}_n(\mathbf{C})$ equipped with the Fubini study metric, and by calculating the degree over $\mathbf{P}_n(\mathbf{Z})$ of their Todd genus $\widehat{\mathrm{Td}}$. They obtained the R-genus as a defect in their conjectured Riemann-Roch formula in Arakelov theory.

It has in fact been made clear in the work of Bost **[Bos]** and Roessler **[Ro]** that the evaluation of the analytic torsion of $\mathbf{P}_n(\mathbf{C})$ can be obtained as a consequence of **[BL]**. Using the results of **[BGS5]**, the formula of **[GS3]** for $\mathbf{P}_n(\mathbf{Z})$ can then be also obtained as a consequence of **[BL]**.

The amazing almost coincidence of the genera R of **[GS3]** and D of **[B4]** is then explained.

References

[A] Atiyah, M.F. : K-Theory. New-York-Amsterdam: Benjamin 1967.

[ADS] Atiyah, M.F., Donnelly, H., Singer, I.M. : Eta invariants, signature defects of cusps and values of L functions. Ann. of Math. **18**, 131-177 (1983).

[Al] Alvarez-Gaumé, L. : Supersymmetry and the Atiyah-Singer index theorem. Comm. Math. Phys. **90**, 161-173 (1983).

[ABoP] Atiyah, M.F., Bott, R., Patodi, V.K. : On the heat equation and the index theorem. Invent. Math. **19**, 279-330 (1973).

[APS1] Atiyah, M.F., Patodi, V.K., Singer, I.M. : Spectral asymmetry and Riemannian geometry, I. Proc. Cambridge Philos. Soc. **77**, 43-69 (1975).

[APS2] Atiyah, M.F., Patodi, V.K., Singer, I.M. : Spectral asymmetry and Riemannian geometry, III. Proc. Cambridge Philos. Soc. **79**, 71-99 (1976).

[AS1] Atiyah, M.F., Singer, I.M. : The index of elliptic operators. III, Ann. of Math. **87**, 546-604 (1968).

[AS2] Atiyah, M.F., Singer, I.M. : The index of elliptic operators. IV, Ann. of Math. **93**, 119-138 (1971).

[AS3] Atiyah, M.F., Singer, I.M. : Index Theory for skew-adjoint Fredholm operators. Publ. Math. IHES, **37**, 5-26 (1969).

[BeV] Berline, M., Vergne, M. : A proof of Bismut local index theorem for a family of Dirac operators. Topology **26**, 435-463 (1987).

[BeGeV] Berline, N., Getzler E., Vergne, M. : Heat kernels and the Dirac operator. Grundl. der Math. Wiss. Band 298. Berlin-Heidelberg-New-York: Springer 1992.

[BerB] Berthomieu A., Bismut, J.-M. : Quillen metrics and higher analytic torsion forms. J. reine angew. Math. **457**, 85-184 (1994).

[B1] Bismut, J.-M. : The Atiyah-Singer index theorem : A probabilistic approach. I. The index theorem. J. Funct. Anal. **57**, 56-99 (1984).

[B2] Bismut, J.-M. : The index Theorem for families of Dirac operators : two heat equation proofs. Invent. Math. **83**, 91-151 (1986).

[B3] Bismut, J.-M. : Superconnection currents and complex immersions. Invent. Math. **99**, 59-113 (1990).

[B4] Bismut, J.-M. : Koszul complexes, harmonic oscillators and the Todd class. J.A.M.S. **3**, 159-256 (1990).

[B5] Bismut, J.-M. : Holomorphic families of immersions and higher analytic torsion forms. Astérisque **244**. Paris: SMF 1997.

[B6] Bismut, J.-M. : Familles d'immersions et formes de torsion analytique en degré supérieur. C.R. Acad. Sci. Paris, Série 1, **320**, 969-974 (1995).

[B7] Bismut, J.-M. : From Quillen metrics to Ray-Singer metrics. Some aspects of the Ray-Singer analytic torsion. In Topological methods in modern mathematics. L. Goldberg and A. Phillips eds. p. 273-324. Houston : Publish or Perish 1993.

[BC1] Bismut, J.-M., Cheeger, J. : η-invariants and their adiabatic limits. J.A.M.S. **2**, 33-70 (1989).

[BC2] Bismut, J.-M., Cheeger, J. : Families index for manifolds with boundary, superconnections and cones, I. J. F. Anal. **89**, 313-363 (1990).

[BC3] Bismut, J.-M., Cheeger, J. : Families index for manifolds with boundary, superconnections and cones, II. J. F. Anal. **90**, 306-354 (1990).

[BC4] Bismut, J.-M., Cheeger, J. : Remarks on the index theorem for families of Dirac operators on manifolds with boundary. In Differential Geometry. F. Cheeger et al. eds, 59-83, Harlow Longman (1991).

[BC5] Bismut, J.-M., Cheeger, J. : Transgressed Euler classes of $SL(2n, \mathbf{Z})$ vector bundles, adiabatic limits of eta invariants and special values of L functions. Ann. Scient. Ec. Norm. Sup. **25**, 335-391 (1992).

[BF1] Bismut, J.-M., Freed D.S. : The analysis of elliptic families I, Metrics and connections on determinants bundles. Comm. Math. Phys. **106**, 159-176 (1986).

[BF2] Bismut, J.-M., Freed D.S. : The analysis of elliptic families II, Dirac operators, eta invariants and the holonomy theorem. Comm. Math. Phys. **107**, 103-163 (1986).

[BGS1] Bismut, J.-M., Gillet, H., Soulé, C. : Analytic torsion and holomorphic determinant bundles, I. Comm. Math. Phys. **115**, 49-78 (1988).

[BGS2] Bismut, J.-M., Gillet, H., Soulé, C. : Analytic torsion and holomorphic determinant bundles, II. Comm. Math. Phys. **115**, 79-126 (1988).

[BGS3] Bismut, J.-M., Gillet, H., Soulé, C. : Analytic torsion and holomorphic determinant bundles, III. Comm. Math. Phys. **115**, 301-351 (1988).

[BGS4] Bismut, J.-M., Gillet, H., Soulé, C. : Bott-Chern currents and complex immersions. Duke Math. Journal **60**, 255-284 (1990).

[BGS5] Bismut, J.-M., Gillet, H., Soulé, C. : Complex immersions and Arakelov geometry. The Grothendieck Festschrift, P. Cartier and al. ed. Vol. I, pp. 249-331. Progress in Math. n° 86. Boston : Birkhaüser 1990.

[BK] Bismut, J.-M., Köhler, K. : Higher analytic torsion forms for direct images and anomaly formulas, J. Alg. Geom. **1**, 647-684 (1992).

[BL] Bismut, J.-M., Lebeau, G. : Complex immersions and Quillen metrics, Publ. Math. IHES, **74**, 1-297 (1991).

[Bos] Bost, J.-M. : Analytic torsion of projective spaces and compatibility with immersions of Quillen metrics. I.M.R.N. **8**, 427-435 (1998).

[BoCh] Bott, R., Chern, S.S. : Hermitian vector bundles and the equidistribution of the zeroes of their holomorphic sections. Acta Math. **114**, 71-112 (1965).

[C1] Cheeger, J. : On the spectral geometry of spaces with cone-like singularities. Proc. Math. Acad. Sci. USA **76**, 2103-2166 (1979).

[C2] Cheeger, J. : Spectral geometry of singular Riemannian spaces. J. Diff. Geom. **18**, 575-657 (1983).

[C3] Cheeger, J. : On the Hodge theory of Riemannian pseudomanifolds. In Proceedings of Symposia in Pure Math., vol. 36, 91-146, Am. Math. Soc. Providence R.I. 1980.

[C4] Cheeger, J. : η-invariants, the adiabatic approximation and conical singularities. J. Diff. Geom. **26**, 175-221 (1987).

[CSi] Cheeger, J., Simons, J. : Differential characters and geometric invariants. In Geometry and Topology, J. Alexander and J. Harer ed., Lect. Notes in Math. n° 1167, p. 50-80. Berlin-Heidelberg-New-York: Springer 1985.

[ChS] Chern, S.S., Simons, J. : Characteristic forms and geometric invariants. Ann. of Math. **99**, 48-69 (1974).

[Cho] Chou, A.W. : The Dirac operator on spaces with conical singularities and positive scalar curvature. Trans. Am. Math. Soc. **289**, 1-40 (1985).

[D] Dai, X. : Adiabatic limits, non multiplicativity of signature and Leray spectral sequence, J.A.M.S. **4**, 265-321 (1991).

[Ge1] Getzler, E. : Pseudodifferential operator on supermanifolds and the Atiyah-Singer index theorem. Comm. Math. Phys. **92**, 163-178 (1983).

[Ge2] Getzler, E. : A short proof of the Atiyah-Singer Index Theorem, Topology **25**, 111-117 (1986).

[Gi1] Gilkey, P. : Curvature and the eigenvalues of the Laplacian for elliptic complexes. Adv. Math. **10**, 344-382 (1973).

[Gi2] Gilkey, P. : Invariance theory, the heat equation and the Atiyah-Singer index theorem. Washington : Publish or Perish 1984.

[GS1] Gillet, H., Soulé, C. : Arithmetic Intersection Theory. Publ. Math. IHES, **72**, 93-174 (1990).

[GS2] Gillet, H., Soulé, C. : Characteristic classes for algebraic vector bundles with Hermitian metrics. Ann. Math. I. **131**, 163-203 (1990). II. **131**, 205-238 (1990).

[GS3] Gillet, H., Soulé, C. : Analytic torsion and the arithmetic Todd genus. Topology, **30**, 21-54 (1991).

[GS4] Gillet, H., Soulé., C.: An arithmetic Riemann-Roch theorem. Inv. Math., **110**, 473-543 (1992).

[GrH] Griffiths, P., Harris, J. : Principles of algebraic geometry, Wiley, New York 1978.

[Hir] Hirzebruch, F. : Hilbert modular surfaces. Enseign. Math. 183-281 (1973).

[Hi] Hitchin, N.J. : Harmonic spinors. Adv. in Math. **14**, 1-55 (1974).

[Kl] Kleiman, S. L. : The development of intersection homology theory. A century of Mathematics in America. Part II. History of Mathematics 2, 543-585. Providence : A.M.S. 1989.

[KMu] Knudsen, F.F., Mumford, D. : The projectivity of the moduli space of stable curves, I : Preliminaries on "det" and "div". Math. Scand. **39**, 19-55 (1976).

[Ma] Ma, X. : Formes de torsion analytique et familles de submersion. C.R.A.S. **324** (Série I), 205-210 (1997).

[MazMe] Mazzeo, R., Melrose, R. : The adiabatic limit, Hodge cohomology and Leray's spectral sequence of a fibration. J. Diff. Geom. **31**, 185-213 (1990).

[McKS] McKean, H., Singer, I.M. : Curvature and the eigenforms of the Laplacian. J. Diff. Geom. **1**, 43-69 (1967).

[Me] Melrose, R. : The Atiyah-Patodi-singer index theorem. Wellesley: Peters 1993.

[MeP1] Melrose, R., Piazza, P. : Families of Dirac operators, boundaries and the *b*-calculus. To appear in J. of Diff. Geom.

[MeP2] Melrose, R., Piazza, P. : An index theorem for families of Dirac operators on odd dimensional manifolds with boundary. To appear in J. of Diff. Geom.

[MQ] Mathai, V., Quillen, D. : Superconnections, Thom classes, and equivariant differential forms. Topology **25**, 85-110 (1986).

[Mü1] Müller, W. : Signature defects of cusps of Hilbert modular varieties and values of L series at $s = 1$. J. Diff. Geom. **20**, 55-119 (1984).

[Mü2] Müller, W. : Manifolds with cusps of rank 1, spectral theory and L_2 index theorem. Lecture Notes in Math. n° 1244. Berlin-Heidelberg-New-York: Springer 1987.

[P1] Patodi, V.K. : Curvature and the eigenforms of the Laplace operator. J. Diff. Geom. **5**, 233-249 (1971).

[P2] Patodi, V.K. : An analytic proof of Riemann Roch Hirzebruch theorem for Kähler manifolds. J. Diff. Geom. **5**, 251-283 (1971).

[Q1] Quillen, D. : Superconnections and the Chern character. Topology **24**, 89-95 (1985).

[Q2] Quillen, D. : Determinants of Cauchy-Riemann operators over a Riemann surface. Funct. Anal. Appl. **14**, 31-34 (1985).

[RS] Ray, D.B., Singer, I.M. : Analytic torsion for complex manifolds. Ann. of Math. **98**, 154-177 (1973).

[Ro] Roessler, D. : An Adams-Riemann-Roch theorem in Arakelov geometry. To appear in Duke Math. J. (1999).

[St] Stern, M. : L^2 index theorems on locally symmetric spaces. Inventiones Math. **96**, 231-282 (1989).

[W] Witten, E. : Global anomalies in string theory. In Proceedings of the Symposium on anomalies, Geometry and Topology of Chicago (1985). W.A. Bardeen, A.R. White Eds, pp. 61-99. Singapore : World Scientific 1985.

Thurston's Hyperbolization of Haken Manifolds

by Jean-Pierre Otal

Juin 1997
CNRS-UMR 128, UMPA, ENS LYON
46, Allée. d'Italie
69364 Lyon, cedex 07, France
jpotal@umpa.ens-lyon.fr

À Paule Tilouine

in memoriam

INTRODUCTION

In the early 1970's, useful connections between 3-manifolds and Kleinian groups began to emerge and set the scene for Thurston's hyperbolization theorem.

— On the one hand, techniques from 3-dimensional topology improved the understanding of *Kleinian groups*, i.e. discrete torsion-free subgroups of $PSL_2(\mathbb{C})$, the group of orientation preserving isometries of the hyperbolic space $\mathbb{H}^3$. A. Marden recognized in **[Mard]** some important consequences of a theory of Waldhausen for the study of *geometrically finite groups* (cf. §1). A fundamental result of Waldhausen gives a necessary and sufficient condition under which a homotopy equivalence between *Haken manifolds* (see below) can be deformed to a diffeomorphism **[Wa2]**. Using this theorem, Marden obtained a necessary and sufficient condition under which an abstract isomorphism between two geometrically finite groups is induced by a quasi-conformal homeomorphism of $\mathbb{S}^2$, the formal boundary of $\mathbb{H}^3$. This condition was a step to fit the geometrically finite groups into Ahlfors-Bers theory of quasi-conformal deformations. It is also probably Marden who first posed in print the problem of giving conditions on a compact 3-manifold to be hyperbolic [**Mard**, p. 461]. We say that a compact orientable 3-manifold M is *hyperbolic* if its interior is diffeomorphic to the quotient of $\mathbb{H}^3$ by a geometrically finite group. Another equivalent definition is the following. Let M be a compact orientable 3-manifold and let P be the union of the tori contained in ∂M. We say that M is *hyperbolic* if $M - P$ carries *a hyperbolic metric* i.e. a complete metric of constant curvature -1 such that M is locally outwardly convex along $\partial M - P$ (cf. §1). Marden observed that *the irreducibility of* M (see below) and the triviality of the center of $\pi_1(M)$ are necessary conditions.

B. Maskit developed a construction for Kleinian groups from simpler ones. In particular, his *Combination theorems* provide sufficient conditions under which two Kleinian groups can be amalgamated in order to produce a new Kleinian group **[Mas1]**. The topological description of this amalgamation at the level of the quotient 3-manifolds is the gluing of the two corresponding 3-manifolds along a subsurface contained in their boundaries. This is parallel to the key construction used in the study of Haken manifolds, namely as the gluing of two simpler ones along incompressible parts of their boundaries.

R. Riley, exploiting different ideas, wrote a computer program to find discrete and faithful representations into $\mathrm{PSL}_2(\mathbb{C})$ of certain knot groups. Using this program, he gave explicitly the representation of the fundamental group of the figure 8 knot [**Ri**]. A fibering theorem of Stallings [**Stal**] then implied that the quotient of $\mathbb{H}^3$ by this Kleinian group was diffeomorphic to the complement of the figure 8 knot in $\mathbb{S}^3$.

Since the quotient of $\mathbb{H}^3$ by a Kleinian group is a complete hyperbolic 3-manifold which conversely determines the Kleinian group up to conjugacy, 3-manifolds were inevitable side products of Kleinian group theory; however, topologically interesting examples were slow to be discovered. F. Löbell provided in 1931 what was perhaps the first example of a *closed* hyperbolic manifold [**Lö**]. In 1970, E. Andreev succeeded in giving a complete combinatorial classification of 3-dimensional hyperbolic Coxeter groups with compact fundamental domains [**An**]: this result provided a huge family of closed hyperbolic 3-manifolds for it was known by a theorem of Selberg [**Sel**] that each such Coxeter group contains a finite index subgroup which is torsion-free (cf. [**Bo**] for examples of closed quotients of an arbitrary simply connected symmetric space). But the foremost example of a hyperbolic manifold to have been appraised as a topological manifold is probably *the hyperbolic dodecahedral space*, which appeared in 1933 [**WS**]. In this paper, H. Seifert and C. Weber describe this manifold as a 5-fold cover of $\mathbb{S}^3$ ramified over a link with two components and they compute its first homology group showing that it is a torsion group [**WS**, p. 252]. Furthermore, they observe that this manifold is not *a Seifert fibered space* [**Sei**], as a direct consequence of the triviality of the center of a cocompact Kleinian group [**WS**, p. 249].

— On the other hand, the progress made in understanding 3-manifolds, especially Haken manifolds, very gradually led to hyperbolic geometry. *An irreducible manifold* is a 3-manifold in which any embedded 2-sphere bounds a 3-ball. By a theorem of H. Kneser, any compact 3-manifold M without 2-spheres in the boundary, can be written as the connected sum of irreducible manifolds and copies of $\mathbb{S}^2 \times \mathbb{S}^1$ [**Kn**]. J. Milnor [**Mil**] proved that this decomposition is unique up to diffeomorphisms when M is closed and orientable; see [**He1**] for the generalization to the case with non-empty boundary. This justifies restricting the study of compact orientable 3-manifolds to irreducible orientable manifolds.

The Sphere theorem was proven by C. Papakyriakopoulos in the mid 1970's [**Pa**]: it says that if M is a 3-manifold with $\pi_2(M) \neq 0$, then M contains an embedded 2-sphere which represents a non-zero element of $\pi_2(M)$. Therefore, if M is irreducible, then $\pi_2(M) = 0$. It follows that the universal cover $\widetilde{M}$ of an irreducible manifold M with infinite fundamental group is contractible. It was then natural to ask what the topological type of $\widetilde{M}$ was. For instance, when M is a closed irreducible manifold with infinite (or torsion-free) fundamental group, is $\widetilde{M}$ homeomorphic to $\mathbb{R}^3$? And if the answer is yes, how does $\pi_1(M)$ act? When M is a Seifert fibered space, the answer to the first question is positive. Any Seifert fibered space with infinite fundamental group has a finite cover which is diffeomorphic to the product of a closed surface by the circle [**Sei**]. Therefore, the universal cover of an irreducible

Seifert fibered space M with infinite fundamental group is homeomorphic to $\mathbb{R}^3$. Before Thurston intervened, it seems not to have been realized that the corresponding action of $\pi_1(M)$ is always geometric. Excepting the Seifert fibered spaces which were thoroughly studied in the 1930's [**Sei**], nothing was known about the topological type of a contractible cover of an irreducible manifold. For example, it was plausible that the Whitehead manifold (a contractible open 3-manifold which is not homeomorphic to $\mathbb{R}^3$ [**Wh**]) should cover a compact 3-manifold (as a matter of fact, this was shown only in the late 1980's: the Whitehead manifold covers no manifolds but itself [**My**]). Great progress on this question was made in the late 1960's by F. Waldhausen [**Wa2**]. His methods provided a deep understanding of the vast class of Haken manifolds. *A Haken manifold* is an irreducible manifold M which contains *an incompressible surface*, i.e. a properly embedded connected surface S such that

(i) the fundamental group $\pi_1(S)$ injects into $\pi_1(M)$ and the relative fundamental group $\pi_1(S, \partial S)$ injects into $\pi_1(M, \partial M)$, and

(ii) S cannot be isotoped into a component of ∂M (cf. §7).

Among other results, Waldhausen proved that the universal cover of a Haken manifold M is homeomorphic to $\mathbb{B}^3 - F$, where F is a closed subset of $\mathbb{S}^2 = \partial\mathbb{B}^3$ [**Wa2**, p. 86]. In reality, he offered a picture of $\widetilde{M}$ which is reminiscent of the complement of the limit set of a Kleinian group in the compactified hyperbolic space.

A new light on 3-manifolds came through *the Torus decomposition*. Any compact irreducible orientable 3-manifold M contains a (possibly empty) finite collection $\mathcal{T}$ of disjoint incompressible tori such that any component V of the manifold obtained by splitting M along $\mathcal{T}$ either is a Seifert fibered space or does not contain any incompressible torus. This collection $\mathcal{T}$ is well defined up to isotopy, once we require that it satisfy a minimality condition and the pieces obtained by splitting M along $\mathcal{T}$ form *the Torus decomposition of* M. The existence of the Torus decomposition appears in a cryptical announcement of Waldhausen [**Wa3**] which to some extent guided the development. The complete proof along with important applications was established by W. Jaco and P. Shalen ([**JS1**], [**JS2**]), and independently by K. Johannson ([**Joh1**], [**Joh2**]). Although the Seifert fibered spaces were well understood topologically [**Sei**], there were no general methods to describe the pieces of other type in the Torus decomposition.

Towards the mid 1970's, a number of algebraic properties of Haken manifold fundamental groups were established which may well have helped to convince Thurston that the non-Seifert pieces in the Torus decomposition of a Haken manifold are in fact hyperbolic. The most important is probably *the Torus theorem* ([**Wa3**], [**Fe**]). This theorem asserts that, if M is Haken, any non-Seifert piece V in the Torus decomposition of M is *atoroidal*, i.e. any $\mathbb{Z}+\mathbb{Z}$-subgroup of $\pi_1(V)$ can be conjugated into the fundamental group of a component of ∂V. Any $\mathbb{Z}+\mathbb{Z}$-subgroup of a Kleinian group is *parabolic* (cf. §1); it follows that any hyperbolic manifold is atoroidal. The fundamental group of a Haken manifold M shares other common properties with Kleinian groups.

(i) Each non-zero element of $\pi_1(M)$ is *uniquely divisible*, i.e. it has a unique positive root of maximal order **[Sh]**, and

(ii) if M is atoroidal and *acylindrical* (see below), then the group of outer automorphisms of $\pi_1(M)$ is finite **[Joh2]**.

Property (i) is easily seen to be satisfied when M is a hyperbolic manifold. When M is a closed hyperbolic manifold, (ii) follows from the Mostow rigidity theorem **[Mos]**.

In the spring of 1977, in his lectures on hyperbolic 3-manifolds, Thurston announced his *Hyperbolization theorem for Haken manifolds.* The printed announcement came several years later, along with generalizations that won't be treated in the present article **[Thu3]**. This theorem gives a necessary and sufficient condition for a Haken manifold to be hyperbolic.

Thurston's hyperbolization theorem [Thu3].— *Let M be an irreducible and atoroidal manifold. If M is Haken, then M is hyperbolic.*

Furthermore, at about the same time, Thurston formulated his Geometrization conjecture. This conjecture says that each piece of the Torus decomposition of an irreducible manifold is modeled locally on one of the following *eight geometries*: the three constant curvature ones: $\mathbb{H}^3$, $\mathbb{R}^3$, $\mathbb{S}^3$, and the five fibered ones: $\mathbb{H}^2 \times \mathbb{R}$, $\mathbb{S}^2 \times \mathbb{R}$, $\widetilde{\mathrm{PSL}_2(\mathbb{R})}$, Nil and Sol (cf. **[Sco]** for a detailed survey on these geometries). This conjecture is satisfied by Haken manifolds: in view of the Hyperbolization theorem above, the proof amounts to merely observing geometric structures on Seifert fibered spaces. Thus, Thurston's hyperbolization theorem, as a particular case of the Geometrization conjecture was in harmony with the recent Torus decomposition theorem.

Thus, for Haken manifolds, Thurston completely settled the Geometrization conjecture. Recall that every compact irreducible manifold whose boundary is non-empty is necessarily Haken; again, every compact irreducible manifold with infinite first homology group is Haken (cf. **[He1]**, **[Ja]**). However, some of the closed hyperbolic manifolds that had been considered before Thurston were already known to be non-Haken: for instance, Haken himself **[He2]**, and also Waldhausen knew this for the hyperbolic dodecahedral space. Thurston went on to prove that in some sense, most 3-manifolds with finite first homology group are non-Haken. His *Hyperbolic Dehn's surgery theorem* implies that, with only a finite number of exceptions, the manifolds obtained by Dehn surgery on a knot in $\mathbb{S}^3$ whose complement is hyperbolic and does not contain any closed incompressible surface are hyperbolic. At the same time, only finitely many of them are Haken (**[Thu1]**, **[Hat]**). (For the test case of 2-bridge knots which are not torus knots, see **[HT]**.)

According to the Geometrization conjecture, one should be able to replace, in the above Hyperbolization theorem, the phrase "is Haken" by "has torsion-free fundamental group". The later is clearly a necessary hypothesis.

The proof of Thurston's hyperbolization theorem distinguishes two cases according as M is fibered over the circle or not. The difference arises from the fact that, in the fibered case, the quasi-Fuchsian groups appearing in the construction degenerate, while in the non-fibered case, they remain quasi-Fuchsian. The fibered case can be formulated as the beautiful theorem below. Recall that a 3-manifold which fibers over the circle with fiber a surface S, is determined up to diffeomorphism by the isotopy class of its *monodromy*, which is an element of *the mapping class group* $\mathrm{Mod}(S)$ (cf. §1). To any diffeomorphism $\phi \in \mathrm{Mod}(S)$, one can associate a 3-manifold M_ϕ, called *its mapping torus* which is defined as the quotient space of $S \times [0,1]$ by the relation which identifies $(x,1)$ with $(\phi(x),0)$. We say that $\phi \in \mathrm{Mod}(S)$ is *pseudo-Anosov* when its action on the set of conjugacy classes of $\pi_1(S)$ does not act periodically on any non-trivial element [**Thu2**].

Hyperbolization theorem for manifolds which fiber over the circle ([Thu5], [Su]). — *Let S be a closed surface of genus greater than 2 and let $\phi \in Mod(S)$. Then M_ϕ is hyperbolic if and only if ϕ is pseudo-Anosov.*

The proof of this particular case of Thurston's hyperbolization theorem is completely different from the proof in the non-fibered case and it has been already quite well explored (cf. [**McM3**], [**O**]). For this reason, we will restrict our attention in this article to the manifolds which are not fibered.

The proofs of the two halves of the Hyperbolization theorem may nevertheless overlap, as Thurston himself observed. For example, there is the following still unsettled question: does every compact 3-manifold fibered over the circle have a finite cover that contains an incompressible surface which is not a fiber of a fibration over the circle? If this were true, the results of the present article alone would suffice to completely prove the Hyperbolization theorem.

In the proceedings of the Smith Conjecture Symposium [**BM**], J. Morgan gave a survey of a part of Thurston's original proof and the book of M. Kapovitch [**Ka**] is also devoted to the original approach. However, in this article, we will detail the proof from another viewpoint. The crucial part relies more on Teichmüller theory than Thurston's proof. It is due to C. McMullen [**McM2**].

We will here prove the Hyperbolization theorem only under an assumption which is stronger than the assumption "atoroidal". Namely, we will assume that $\pi_1(M)$ *does not contain any* $\mathbb{Z}+\mathbb{Z}$*-subgroups.* The main consequence is that we can completely avoid the study of Kleinian groups containing parabolic elements. (The general case is more complex, partly due to heavier notations, but it is not much more difficult.) It is in a similar spirit that this article excludes all mentions of non-orientable manifolds (see [**To**]).

A crucial property of a Haken manifold is the existence of *hierarchies*, discovered by Haken. By definition, a Haken manifold M contains an incompressible surface S, and by splitting M along S, we obtain a new manifold M_S. The incompressibility of S implies that M_S is irreducible. Moreover, if M_S is not a disjoint union of 3-balls, it is Haken (cf. §7). Therefore, the splitting process can be iterated. It is a fundamental observation of W. Haken that this process ends up, after a finite number of steps, with a disjoint union of 3-balls (cf. [**Ha**]). This sequence of manifolds is called *a hierarchy for M* and the number of terms in this sequence is called *the length* *of the hierarchy.* For technical reasons, we prefer to use hierarchies of another type, called *special hierarchies* (cf. [**Ja**]): this will allow us to prove the Hyperbolization theorem by induction on an integer, which is defined as the greatest length of a special hierarchy, and denoted by $\ell(M)$ (cf. §7).

Like other theorems on Haken manifolds, such as Waldhausen's well-known theorem that homotopy equivalences between Haken manifolds respecting boundary can be deformed respecting boundary to diffeomorphisms, the proof of Thurston's hyperbolization theorem in the non-fibered case involves a finite induction using a hierarchy for M. There is a further parallel between the hyperbolization procedure of Thurston and the proof of this theorem of Waldhausen. Both proofs consist of two distincts parts, one combinatorial and 3-dimensional, and the other 2-dimensional and more geometrical. In each case, the 3-dimensional part is a hierarchical induction. In Waldhausen's theorem, the 2-dimensional part is a known theorem of Nielsen [**Nie**]. The 2-dimensional part of Thurston's theorem however, was an entirely new Fixed point theorem, involving the Teichmüller spaces of the gluing surfaces.

To carry out the proof of Thurston's theorem by induction on the length of M, we will enunciate before long a more general theorem, which applies to *manifolds-with-corners.* First, we explain the basic gluing procedure to which the inductive step will reduce.

Final gluing theorem.— *Let N be a hyperbolic manifold with incompressible boundary. Let τ be an orientation reversing involution of ∂N which exchanges the boundary components by pairs. Suppose that N is not an interval bundle. Then if N/τ is atoroidal, it is hyperbolic.*

This theorem still holds if N is an interval bundle but the proof is entirely different. It corresponds to the case when N/τ is fibered and the hyperbolic structure is obtained as a degeneration of a certain sequence of quasi-Fuchsian structures on N.

Note that the Final gluing theorem never directly provides a hyperbolic structure on a compact manifold with non-empty boundary. However it does so indirectly by a trick of Thurston. This trick makes the boundary of a 3-manifold invisible by covering it with mirrors: in some sense, it converts boundary points to interior points. We present an explication of this trick which is due to F. Bonahon who was first to observe that right-angled corners are sufficient. The following notion of manifold-with-corners is essentially equivalent to the notion of "manifold with (useful) boundary pattern", introduced by Johannson in his work on homotopy equivalences ([**Jo1**], [**Jo2**]). *A manifold-with-corners* is a triple $(M, \mathcal{G}, \partial^0 M)$, where M is a 3-manifold, and $\mathcal{G} \subset \partial M$ is a smooth trivalent graph such that

(i) each component of $\partial M - \mathcal{G}$ equals the interior of its closure, and

(ii) each component of $\partial^0 M$ is the closure of a component of $\partial M - \mathcal{G}$.

The closure of a component of $\partial M - \mathcal{G}$ which is not in $\partial^0 M$ is called *a mirror of* $(M, \mathcal{G}, \partial^0 M)$, and $\partial^0 M$ is called *the boundary of* $(M, \mathcal{G}, \partial^0 M)$. One should think of $(M, \mathcal{G}, \partial^0 M)$ as *a differentiable structure with corners on* M, i.e. an atlas of class C^1 on M with charts modeled on open subsets of $(\mathbb{R}^+)^3$. Then, the graph $\mathcal{G}$ corresponds to points which have a neighborhood diffeomorphic to the neighborhood of a point of $(\mathbb{R}^+)^3$ with 2 or 3 coordinates equal to 0, and the boundary of $(M, \mathcal{G}, \partial^0 M)$ corresponds to distinguishing a set of disjoint mirrors. Such a differentiable structure depends only on the pair $(M, \mathcal{G})$: this follows from **[Ce]** and **[Do]**.

The notions of irreducibility and atoroidality can be extended to manifolds-with-corners; one can also define the notion of a manifold-with-corners with incompressible boundary (cf §7). Rather than directly prove the Hyperbolization theorem for manifolds possibly with boundary, we will proceed by proving a Hyperbolization theorem for manifolds-with-corners that have empty boundary —which will turn out to be just as strong.

A manifold-with-corners $(M, \mathcal{G}, \partial^0 M)$ is *hyperbolic* when there is a hyperbolic metric on M such that

(i) the mirrors are totally geodesic,

(ii) M is locally outwardly convex along $\partial^0 M$, and

(iii) the components of $\partial M - \mathcal{G}$ meet at right-angles along the edges of $\mathcal{G}$.

Let $(M, \mathcal{G})$ be a hyperbolic manifold-with-corners having empty boundary and let $S' \subset \partial M$ be a surface which is a disjoint union of mirrors. Let $\mathcal{G}'$ be the graph obtained from $\mathcal{G}$ by erasing the edges whose interior is contained in the interior of S'. Then, after rounding the corners along the erased edges, $(M, \mathcal{G}', S')$ becomes a manifold-with-corners and it follows almost directly by taking δ-neighborhoods in the ambient complete hyperbolic manifold that $(M, \mathcal{G}', S')$ is hyperbolic. In particular, if $(M, \mathcal{G})$ is a hyperbolic manifold-with-corners having empty boundary, then M is hyperbolic.

Hyperbolization theorem for manifolds-with-corners.— *Let $(M, \mathcal{G})$ be a compact irreducible oriented and atoroidal manifold-with-corners having empty boundary. If M is Haken, then $(M, \mathcal{G})$ is hyperbolic.*

Any irreducible and atoroidal manifold M having non-empty boundary can be easily promoted to a manifold-with-corners $(M, \mathcal{G})$ that has empty boundary and is irreducible and atoroidal: this is the mirror trick (absolute version) see §7. Therefore, Thurston's hyperbolization theorem is a consequence of the theorem above. This theorem is proven by induction on the special length of M, viewed as a manifold without corners. The most important advantage of the introduction of manifolds-with-corners is above all that the proof of the inductive step will be in strict parallel with that of the Final gluing theorem and is in the final analysis a consequence of it.

Suppose that $(M, \mathcal{G})$ is an irreducible and atoroidal manifold-with-corners with empty boundary. When M is Haken, we show in §7 the existence of an incompressible surface S which is *a good splitting surface*. This means that the manifold-with-corners $(M_S, \mathcal{G}_S, S')$ obtained by splitting $(M, \mathcal{G})$ along S has incompressible boundary. We show also that $\mathcal{G}_S$ can be extended to a graph $\mathcal{G}'_S$ by adding edges contained in S' so that $(M_S, \mathcal{G}'_S)$ is an irreducible and atoroidal manifold-with-corners with empty boundary. Suppose that $(M_S, \mathcal{G}'_S)$ is hyperbolic. Then, as observed above, $(M_S, \mathcal{G}'_S, S')$ is also hyperbolic, and further the hypotheses of the next theorem are satisfied.

Gluing theorem for manifolds-with-corners.— *Let $(M, \mathcal{G})$ be an irreducible and atoroidal manifold-with-corners having empty boundary. Let S be a good splitting surface for $(M, \mathcal{G})$ and let $(M_S, \mathcal{G}_S, S')$ be the manifold-with-corners obtained from $(M, \mathcal{G})$ by splitting along S. Suppose that $(M_S, \mathcal{G}_S, S')$ is hyperbolic. Then $(M, \mathcal{G})$ is hyperbolic.*

This gluing theorem can be deduced from the statement of the Final gluing theorem (cf. §8). We now apply the Gluing theorem for manifolds-with-corners to prove the Hyperbolization theorem for manifolds-with-corners. The proof is by induction on the special length $\ell(M)$ of M.

The induction starts at $\ell(M) = 0$. Then M is an handlebody and M_S is diffeomorphic to $\mathbb{B}^3$ (cf. §7). Thus $(M_S, \mathcal{G}'_S)$ can be interpreted as a polyhedron. Saying that $(M_S, \mathcal{G}'_S)$ is hyperbolic means that this polyhedron can be realized in $\mathbb{H}^3$ with all dihedral angles equal to $\pi/2$. A characterization of the compact polyhedra which can be embedded in $\mathbb{H}^3$ with prescribed acute dihedral angles is provided by the theorem of Andreev already mentioned **[An]**. In the case of a right-angled polyhedron, the hypothesis of this theorem turn out to be equivalent to the irreducibility and atoroidality of $(M_S, \mathcal{G}'_S)$. Therefore the Andreev theorem asserts that $(M_S, \mathcal{G}'_S)$ is hyperbolic. (Incidently, note that the problem of realizability of a polyhedron in $\mathbb{H}^3$ with various sorts of prescribed data is currently a field of intensive study **[HR]**.) Thus, by the Gluing theorem for manifolds-with-corners, $(M, \mathcal{G})$ is hyperbolic when $\ell(M) = 0$. The inductive step reduces similarly to the Gluing theorem. This proves the Hyperbolization theorem for manifolds-with-corners.

Now we sketch the logic of the proof of the Final gluing theorem. Let N be a hyperbolic manifold. Let G be a geometrically finite group such that N is diffeomorphic to the Kleinian manifold $\overline{M}(G)$ associated to G (cf. §1). The space $\mathcal{GF}(G)$ of the geometrically finite groups isomorphic to G can be parametrized by the Teichmüller space $\mathcal{T}(\partial N)$: this is one important application of Ahlfors-Bers theorem on the existence and uniqueness of solutions to the Beltrami equation. Using this parametrization and the Maskit combination theorem (cf. §2), Thurston showed how to reduce the Final gluing theorem to the problem of finding a fixed point for a certain map on Teichmüller space. This map is the composition $\tau^* \circ \sigma$ of two maps: $\tau^* : \mathcal{T}(\overline{\partial N}) \to \mathcal{T}(\partial N)$ is the action induced by the (orientation reversing) diffeomorphism τ and $\sigma : \mathcal{T}(\partial N) \to \mathcal{T}(\overline{\partial N})$ is *the skinning map* (see below). This translates the final gluing theorem into the following:

Thurston's fixed point theorem.— *Let N be a hyperbolic manifold with incompressible boundary which is not an interval bundle. Let τ be an orientation reversing involution of ∂N which permutes the components by pairs. Then, if N/τ is atoroidal, $\tau^* \circ \sigma$ has a fixed point.*

To prove this theorem, we will adopt the approach which was given by McMullen [**McM2**]. This approach originated in an observation that J. Hubbard made shortly after Thurston enunciated the Fixed point theorem. Hubbard noticed that the formula for the coderivative of the skinning map involved a well-known operator in Teichmüller theory, *the Poincaré series operator*. As the proof of Thurston's fixed point theorem presented here relies mostly on complex analysis, we must recall briefly the definition of the Poincaré series.

Let $Y \to X$ be a cover of Riemann surfaces. A holomorphic L^1-integrable quadratic differential ϕ on Y can be summed over the sheets of the cover to define a holomorphic integrable quadratic differential $\Theta_{Y/X}\phi$ on X. If we denote by $\mathcal{Q}(X)$, $\mathcal{Q}(Y)$ the space of integrable holomorphic quadratic differentials on X and Y respectively, this defines a map $\Theta_{Y/X} : \mathcal{Q}(Y) \to \mathcal{Q}(X)$, called *the Poincaré series operator* or *Theta operator*. When $\mathcal{Q}(X)$ and $\mathcal{Q}(Y)$ are endowed with their respective L^1-norms, the norm of $\Theta_{Y/X}$ is less than or equal to 1.

In view of the formula for the coderivative of σ, Hubbard suggested that the existence of a fixed point for $\tau^* \circ \sigma$ would be easier to establish if one could prove a conjecture of Kra [**Kr**]. This conjecture asserts that the norm of the Theta operator associated to the universal cover of a finite volume hyperbolic Riemann surface X is *strictly* less than 1. It was McMullen who, in 1989, succeeded in carrying out the program of Hubbard. In [**McM1**] he proves a generalized version of Kra's conjecture, giving a necessary and sufficient condition on a cover $Y \to X$ for the norm of $\Theta_{Y/X}$ to be strictly less than 1. In [**McM2**] he shows how this result applies to give a new proof of Thurston's fixed point theorem.

§2 begins with a proof of the particular case of the Maskit combination theorem we need in order to show the equivalence between the Final gluing theorem and Thurston's fixed point theorem. Next, we study the skinning map which is defined as follows. By the Ahlfors-Bers theorem, there corresponds to any point $s = (s_1, \cdots, s_k) \in \mathcal{T}(\partial N)$ a Kleinian manifold N^s diffeomorphic to N, such that $\partial N^s = s$ (cf. §1). By taking the cover of N^s associated to the component S_i of ∂N, we obtain a quasi-Fuchsian structure on $S_i \times [0,1]$: the Ahlfors-Bers parameters of this structure are (s_i, s_i'), where s_i' is a complex structure on $\overline{S_i}$, the surface S_i with the reversed orientation. Then, the skinning map assigns to ∂N^s the point $\sigma(\partial N^s) \in \mathcal{T}(\overline{\partial N})$, whose i-th coordinate is s_i'. We see *spots* on the Riemann surface $\sigma(\partial N^s)$: in the cover of N^s associated to S_i they are the components of the preimage of ∂N^s others than the canonical lift of S_i. In particular, to each spot U is associated a cover $U \to X_U$ of a component X_U of ∂N^s. The topological configuration of the spots on ∂N^s and the topological type of the covers associated to them are independent of s. The cover $U \to X_U$ associated to a spot is *geometric*: it arises from a compact incompressible surface contained in X_U. Also the shape of the spots reflects some important topological properties of N. The basic one is that each curve on U which is not homotopically trivial projects under the covering $U \to X_U$ to a curve on ∂N which is one boundary component of an essential annulus in N. In particular, the spots are all simply connected if (and only if) N is *acylindrical*, i.e. if N does not contain essential annuli. In §2, we compute also the coderivative of σ: it is a convex linear combination of the Theta operators associated to the spots.

§4 and 5 are devoted to prove the McMullen theorem that the Theta operator $\Theta_{Y/X}$ associated to a geometric cover $Y \to X$ has norm bounded away from 1 by a constant depending only on $\chi(X)$ and on *the systole of* X, i.e. the length of the shortest closed geodesic of X. We will follow the approach given by Barrett and Diller **[BD]**.

§3 contains auxiliary results on Riemann surfaces which are used during the proof. The most important is Theorem 3.1 which is due to McMullen **[McM1]**. It establishes a property of convergence for a sequence of triples (X_i, x_i, ϕ_i) where ϕ_i is an integrable holomorphic quadratic differentials on the Riemann surface X_i and $x_i \in X_i$. Recall first what it means for a sequence of pointed hyperbolic Riemann surfaces (X_i, x_i) having a fixed topological type to converge to a Riemann surface (X, x). There are two cases to consider, according the behaviour of the injectivity radius $\text{inj}(x_i)$ at x_i. If $\text{inj}(x_i)$ remains bounded away from 0, it means that (X_i, x_i) converges to (X, x) for the Hausdorff-Gromov topology on pointed metric spaces: in this case, X is a hyperbolic Riemann surface with finite volume. If $\text{inj}(x_i)$ tends to 0, it means that (X_i, x_i) with the hyperbolic metric rescaled by $1/\text{inj}(x_i)$ converges to (X, x): in that case, $X = \mathbb{C}^*$ with a complete flat metric. Then, if $\phi_i \in \mathcal{Q}(X_i)$ and if ϕ is a holomorphic quadratic differential on X, we say that (ϕ_i) *converges uniformly to* ϕ, when the local expression of ϕ_i (in a chart for X_i which converges to a chart for X) converges uniformly to the local expression of ϕ. Theorem 3.1 asserts that, if (X_i, x_i) converges to (X, x) and if (ϕ_i) is a sequence in $\mathcal{Q}(X_i)$ with $||\phi_i|| = 1$, (ϕ_i) *converges uniformly* to a non-zero holomorphic quadratic differential ϕ on X, up to multiplying ϕ_i by a constant and up to extracting a subsequence. This theorem is proved in two steps. First, we produce a sequence of non-zero $\theta_i \in \mathcal{Q}(X_i)$ which converges uniformly to a holomorphic quadratic differential ϕ in the above sense. This reduces the uniform convergence of $\phi_i = (\phi_i/\theta_i)\theta_i$ to the uniform convergence of (the functions) (ϕ_i/θ_i). This follows from the classical theorems of Montel and Picard. The proof we give here of the first step is slightly different from the original one and relies on less advanced machinery than **[McM1]**.

§4 concerns the solution of a certain $\bar{\partial}$-problem on *an open Riemann surface* Y; an open Riemann surface is a Riemann surface with finite topological type but with infinite volume. On any open Riemann surface Y, the hyperbolic volume form dv is exact. Moreover, since any open Riemann surface is a Stein manifold, dv is $\bar{\partial}$-exact. Theorem 4.1, which is due to Diller **[Di]**, provides a *well-behaved* solution to the equation $\bar{\partial}\eta = dv$. This solution η is well-behaved in the sense that it is a 1-form of type $(1,0)$ whose hyperbolic norm is finite, bounded by a function of $\chi(Y)$ and of the systole of Y. We consider only the case when Y has no cusps. Then Y is the union along the boundary of a compact surface Y_0 with geodesic boundary and a finite collection of half-infinite annuli. Using the formula for the hyperbolic metric on an annulus, one can define an explicit 1-form η_0 of type $(1,0)$ which is supported on $Y - Y_0$ and which solves $\bar{\partial}\eta_0 = dv$ on a neighborhood of the ends of Y. By construction, the hyperbolic norm of η_0 is independent of Y, and therefore, we are led to find a well-behaved solution η' of $\bar{\partial}\eta' = dv - \bar{\partial}\eta_0$, for then $\eta = \eta_0 + \eta'$ will be the required solution. Since the 2-form $dv - \bar{\partial}\eta_0$ has compact support, this equation can be solved using the Green's function on Y. Estimates on the circular averages of this Green's function provide then the required bound on the hyperbolic norm of η'.

Also in §4, we prove a refinement of Theorem 4.1, in which one we no longer have control on the systole of Y. Let ε a constant smaller than the Margulis constant. Rather than finding a well-behaved solution to the equation $\bar{\partial}\eta = dv$ which is defined on the entire surface Y, we find such a solution on the *unbounded* (i.e. non-compact) components of the ε-thick part $Y^{[\varepsilon,\infty[}$ of Y, and whose norm is bounded by a function of ε and $\chi(Y)$.

In **[BD]**, D. Barrett and J. Diller show how to deduce from Theorem 4.1 the McMullen theorem about the norm of the Theta operator $\Theta_{Y/X}$ associated to a geometric cover $Y \to X$. This short proof is explained in §5. For the applications, one further needs to control how the norm of $\Theta_{Y/X}$ can approach 1, when the topological type of the cover $Y \to X$ is fixed but when there is no information on the systole of X. This control can be formulated in terms of *the ε-amenable part of the cover $Y \to X$*. Recall that the cover $Y \to X$ arises from a proper incompressible surface $S \subset X$. Thus, any component of $X^{]0,\varepsilon]}$ or of $X^{[\varepsilon,\infty[}$ which can be isotoped into S can be lifted homeomorphically to a surface contained in Y. The ε-amenable part of the cover $Y \to X$ is then defined as the union of the preimage of $X^{]0,\varepsilon]}$ and the lifts of the components of $X^{[\varepsilon,\infty[}$ which can be isotoped into S. Theorem 5.1 is made more precise by the next statement (Theorem 5.3): if $||\Theta_{Y/X}\phi||$ is more than $1-\delta$ for a unit norm $\phi \in \mathcal{Q}(Y)$, then the ϕ-mass of *the ε-amenable part* of the cover $Y \to X$ is more than $1 - c(\delta)$, where $c(\delta)$ depends on ε and tends to 0 with δ. This result is also due to McMullen, who deals not only with geometric covers, but also with *non-amenable covers* **[McM1]**.

In §6, we prove, following McMullen **[McM2]**, Thurston's fixed point theorem. The existence of a fixed point for $\tau^* \circ \sigma$ is related to a contraction property for $\tau^* \circ \sigma$ with respect to the Teichmüller distance. Since τ^* is an isometry, the contraction properties of $\tau^* \circ \sigma$ follows from the contraction properties of σ. The results of §5 have direct consequences for the norm of $d^*\sigma$ at $s \in \mathcal{T}(\partial N)$. From Theorem 5.1, it follows that $||d_s^*\sigma||$ is bounded away from 1 by a constant which depends only on $\chi(\partial N)$ and on the systole of s (Proposition 6.1). Proposition 6.2 is also a consequence of Theorem 5.3: it says that if $||d_s^*\phi|| \geq 1-\delta$ for a unit norm $\phi \in \mathcal{Q}(\sigma(\partial N^s))$, then the ϕ-mass of the ε-amenable part of $\sigma(\partial N^s)$ is more than $1 - c'(\delta)$ where $c'(\delta)$ tends to 0 with δ (the ε-amenable part of $\sigma(\partial N^s)$ is the union of the ε-amenable parts of the covers $U \to X_U$ associated to the spots $U \subset \sigma(\partial N^s)$). However, these two results of 2-dimensional nature, don't suffice to solve the Fixed point problem and another argument is needed. For this, we observe that the ε-amenable part of $\sigma(\partial N^s)$ can be decomposed into the union of the simply connected components of the preimage of $X_U^{]0,\varepsilon]}$ and the lifts of the components of $X_U^{]0,\varepsilon]}$ or $X_U^{[\varepsilon,\infty[}$: the (possibly empty) union of these lifts forms a compact surface, called *the ε-liftable part of $\sigma(\partial N^s)$*. With this terminology, Proposition 6.3 asserts that, for sufficiently small ε, if $\|d_s^*\sigma\phi\| \geq 1-\delta$ for a unit norm $\phi \in \mathcal{Q}(\sigma(\partial N^s))$, then the ϕ-mass of the ε-liftable part of $\sigma(\partial N^s)$ is more than $1 - c''(\delta)$ where $c''(\delta)$ only depends on ε and tends to 0 with δ. One deduces Proposition 6.3 from Proposition 6.2 and from Proposition 6.4, whose proof rests on a more global argument: after a normalization of the limit set of G^s in $\overline{\mathbb{C}}$ (which uses the geometry of the 3-dimensional Margulis tubes) it is a consequence of the compacity theorem for holomorphic quadratic differentials (Theorem 3.1).

To prove the Fixed point theorem, we exploit the geometric consequences of Proposition 6.3. There are two cases to consider according as N is acylindrical or not.

When N is acylindrical, all the spots are simply connected, and in particular, the ε-liftable part is empty. This implies that σ contracts uniformly the Teichmüller distance and therefore $\tau^* \circ \sigma$ also (for any gluing data τ). Since Teichmüller space is complete, $\tau^* \circ \sigma$ has a fixed point.

When N is cylindrical, some gluing data τ may produce non-atoroidal manifolds, thereby forbidding the existence of a fixed point for $\tau^* \circ \sigma$. This occurs for instance when τ maps one boundary component of an essential annulus to the other. Therefore, we must take into account τ in proving the Fixed point theorem. In the cylindrical case, we won't prove that $\tau^* \circ \sigma$ is uniformly contracting, but only that some iterate $(\tau^* \circ \sigma)^K$ is uniformly contracting on a certain $\tau^* \circ \sigma$-invariant closed subset of $\mathcal{T}(\partial N)$ (since $(\tau^* \circ \sigma)^K$ and $\tau^* \circ \sigma$ commute, this suffices to prove the Fixed point theorem). The principal geometric consequence of Proposition 6.3 and indeed, the only one necessary to prove the Fixed point theorem is: if $\|d_s^*\sigma\|$ is sufficiently near to 1, there is an essential annulus in N which joins two curves α and γ such that α is shorter than ε for the metric s and such that γ is shorter than ε for the metric $\sigma(s)$. If we suppose that $\|d_s^*(\tau^* \circ \sigma)^K\|$ is near 1, then since σ is contracting and since τ^* is an isometry, the norm of $d^*\sigma$ at $\tau^*(\tau^* \circ \sigma)^k(s)$ is also near one, for all $0 \leq k \leq K-1$. Therefore, by successive applications of Fact 6.14, we produce a sequence of K essential annuli A_i in N with boundary the union of two simple closed curves α_i and γ_i, such that $\alpha_{i+1} = \tau(\gamma_i)$ and such that the curves α_i are shorter for the hyperbolic metric s than the Margulis constant. Then, by the well known Margulis lemma, each of the curves α_i is homotopic to one of finitely many disjoint simple closed curves on ∂N of length less than the Margulis constant. Therefore, if we choose K bigger than the maximal number of disjoint pairwise non-homotopic simple closed curves on ∂N, two of the curves α_i are homotopic on ∂N. It is then easy to produce an essential singular torus in N/τ, contradicting the hypothesis of atoroidality on N/τ.

§3, 4, 5, 6 are completely self-contained. They form the main part of this paper and give a complete and detailed proof of Thurston's fixed point theorem.

§7 develops the theory of manifolds-with-corners which was sketched above. It is the "3-dimensional core" of the proof. Proofs are given with details, but to shorten the exposition, we use the equivariant versions of the Dehn lemma, the Sphere theorem and the Torus theorem.

In §8, we explain the equivariant machinery which allows one to deduce the Final gluing theorem for manifolds-with-corners from the Final gluing theorem. Then, we deduce the Hyperbolization theorem.

Thus, granting the by now standard material on Kleinian groups laid out with appropriate references in §1, the above three equivariant theorems, and the case of Andreev's theorem for right-angled compact polyhedra in $\mathbb{H}^3$, this survey comprises a complete proof of Thurston's hyperbolization theorem for non-fibered Haken manifolds whose fundamental group does not contain $\mathbb{Z}+\mathbb{Z}$. The extension to the general case, i.e. to atoroidal manifolds, or to "pared manifolds" (cf. [**Mor**]) does not encounter any difficulties that are unfamiliar or deep. The reader is invited to find the modifications necessary to establish this general case. For this, he or she needs to extend the results concerning the norm of Theta operators to geometric covers of finite volume Riemann surfaces. The topological results of §7 also need to be extended to deal with the case of "pared manifolds-with-corners". The Hyperbolization theorem for pared manifolds is covered by the (still informal) notes [**OP**].

I wish to thank Peter Shalen and John Stallings who have told me their memories of the Hyperbolization theorem, some of which I have tried to evoke in the first part of this introduction. Larry Siebenmann clarified their points of view and induced me to investigate the earlier literature. I thank him for being so exacting. I felt obliged to reorganize the combinatorial part of this article after he convinced me that the most direct way from Thurston's fixed point theorem to his Hyperbolization theorem goes via the natural generalization to compact manifolds-with-corners of the case for right-angled polyhedra of Andreev's hyperbolization theorem. This approach nicely complements Francis Bonahon's observation [**OP**] that right-angles are sufficient. The pictures which illustrate this article were drawn by Greg McShane who also patiently commented on early drafts. With a lot of criticism, Saar Hersonsky helped to bring the all analytical part of this article to its present form. During the writing, I benefitted also from several fruitful discussions with Frederic Paulin. In particular, §7 and §8 emerged from chapters he contributed to the notes [**OP**].

CHAPTER 1

Kleinian groups and Teichmüller theory

We refer the reader to the books [**Bea**], [**BP**] and [**Ra**] for more details on the first section.

1.1 Kleinian groups

The *hyperbolic space of dimension* n is the complete and simply connected n-dimensional Riemannian manifold of constant curvature -1 (we think of n as being equal to 2 or 3). This manifold has two well-known isometric models, *the Riemann model*

$$\mathbb{D}^n = \{(\xi_1, \cdots, \xi_n) \in \mathbb{R}^n, \quad \sum \xi_i^2 < 1\},$$

endowed with the Riemannian metric

$$ds^2 = \frac{4(d\xi_1^2 + \cdots + d\xi_n^2)}{(1 - (\sum \xi_i^2))^2},$$

and *the upper-half space model*

$$\mathbb{H}^n = \{(x_1, \cdots, x_n) \in \mathbb{R}^n, \quad x_n > 0\},$$

endowed with the Riemannian metric

$$ds^2 = \frac{dx_1^2 + \cdots + dx_n^2}{x_n^2}.$$

Any isometry of $\mathbb{H}^n$ extends continuously to the boundary $\overline{\mathbb{R}^{n-1}}$ where it induces a conformal or anticonformal map according to whether it preserves the orientation or not. Let $\mathrm{Isom}(\mathbb{H}^n)$ be the group of orientation preserving isometries of $\mathbb{H}^n$. Therefore, $\mathrm{Isom}(\mathbb{H}^2)$ and $\mathrm{Isom}(\mathbb{H}^3)$ can be identified with $\mathrm{PSL}_2(\mathbb{R})$ and $\mathrm{PSL}_2(\mathbb{C})$ respectively.

Definition.— *A Kleinian group* is a discrete, finitely generated subgroup of $\mathrm{Isom}(\mathbb{H}^n)$.

Let γ be an isometry of $\mathbb{H}^n$ which is different from the identity. It is well known that γ is either *hyperbolic* (it has exactly two fixed points in $\partial\mathbb{H}^n$) or *parabolic* (it has a unique fixed point in $\partial\mathbb{H}^n$), or *elliptic* (it has a fixed point in $\mathbb{H}^n$). When γ is hyperbolic, it leaves invariant the geodesic of $\mathbb{H}^n$ joining its two fixed points. This geodesic $A(\gamma)$ is called *the axis of* γ. The isometry γ acts on $A(\gamma)$ as a translation of a certain distance $\ell(\gamma)$ called the *translation distance of* γ.

From now on, all the Kleinian groups will be supposed to be torsion-free.

Let G be a Kleinian group. Then G does not contain elliptic elements and its action on $\mathbb{H}^n$ is properly discontinuous. The quotient space $M(G) = \mathbb{H}^n/G$ is a complete Riemannian manifold of constant curvature -1.

The limit set and the domain of discontinuity of a Kleinian group.

Definition.— A group is *elementary* if it contains an Abelian subgroup of finite index.

One can show that elementary Kleinian groups are characterized among all Kleinian groups as those which act on $\partial\mathbb{H}^n$ by fixing one or two points.

Let G be a non-elementary Kleinian group.

Definition.— *The limit set of* G is the smallest non-empty closed subset of $\partial\mathbb{H}^n$ which is invariant under G. It is denoted by $L(G)$.

One can also define $L(G)$ as the closure in $\partial\mathbb{H}^n$ of the set of fixed points of the non-zero elements of G. It is a perfect subset of $\partial\mathbb{H}^n$.

Let $C(G)$ be the smallest closed convex subset of $\mathbb{H}^n$ whose closure in $\overline{\mathbb{H}^n}$ contains $L(G)$. It is a convex subset invariant by G. The quotient space $N(G) = C(G)/G$ is contained in $M(G)$ and is called *the Nielsen core of* $M(G)$. It is the smallest closed convex subset of $M(G)$ such that the inclusion into $M(G)$ is a homotopy equivalence [**Th1**].

For $n = 2$, $\partial N(G)$ is totally geodesic, but for $n = 3$, $N(G)$ is not a differentiable submanifold of $M(G)$ in general: its boundary is "bent" along certain geodesics. One way to avoid this problem is to replace $N(G)$ by its neighborhood of radius δ in $M(G)$. Denote this neighborhood by $N_\delta(G)$. It is not difficult to see that, for any $\delta > 0$, $N_\delta(G)$ is a submanifold of $M(G)$ of class C^1 which is strictly convex (i.e. any geodesic arc of $M(G)$ joining two points of $N_\delta(G)$ is contained in the interior of $N_\delta(G)$, except possibly its endpoints). Furthermore, $N_\delta(G)$ does not depend on $\delta > 0$ up to diffeomorphism. It is called *the thickened Nielsen core.*

Although $\partial N(G)$ is not a differentiable submanifold of $M(G)$, the convexity of $N(G)$ allows to consider the path metric that the metric of $M(G)$ induces. When $n = 3$, a basic property of this distance is that it is "hyperbolic": with this induced metric, $\partial N(G)$ is locally isometric to $\mathbb{H}^2$ (cf. [**Th1**], [**Ro**]). This important property will be used at the end of §6.

Definition.— *The domain of discontinuity of* G is the complement of $L(G)$ in $\partial\mathbb{H}^n$. It is denoted by $\Omega(G)$.

When $\Omega(G) \neq \emptyset$, the action of G on $\mathbb{H}^n \cup \Omega(G)$ is properly discontinuous. This can be seen using *the nearest point retraction.* The map which assigns to $x \in \mathbb{H}^n$ the point of (the closed subset) $C(G)$ which is nearest to x extends continuously to a map $\tilde{r} : \mathbb{H}^n \cup \Omega(G) \to C(G)$ called the nearest point retraction. The map $\tilde{r}$ commutes with the isometries of $\mathbb{H}^n$ which leave $C(G)$ invariant. In particular $\tilde{r}$ commutes with the elements of G. Therefore, since G acts properly and discontinuously on $C(G) \subset \mathbb{H}^n$, G acts properly and discontinuously on $\mathbb{H}^n \cup \Omega(G)$ also and one can form the quotient space of $\mathbb{H}^n \cup \Omega(G)$ by G. It is a smooth (analytical) manifold with boundary denoted by $\overline{M}(G)$ whose interior equals $M(G)$. For $n = 3$, its boundary is a Riemann surface.

Since $\tilde{r}$ commutes with the elements of G, it induces a retraction $r : \overline{M}(G) \to N(G)$. We can define in a similar way a retraction $r_\delta : \overline{M}(G) \to N_\delta(G)$. For $\delta > 0$, it follows from the strict convexity of $N_\delta(G)$ that $r_\delta^{-1}(\partial N_\delta(G))$ is diffeomorphic to $\partial N_\delta(G) \times [0,1]$. Thus, $\overline{M}(G)$ is diffeomorphic to $N_\delta(G)$.

Thus one can associate to G three manifolds:

(i) the manifold with boundary $\overline{M}(G)$, whose interior is

(ii) the complete Riemannian manifold $M(G)$ with constant sectional curvature -1, and

(iii) the Nielsen core $N(G)$ or its δ-neighborhood $N_\delta(G)$.

The Margulis decomposition.

Let G be a non-elementary Kleinian group.

Definition.— Let $\varepsilon > 0$. *The ε-thin part of* $M(G)$ is the set of points $x \in M(G)$ through which goes a geodesic loop of length less than or equal to ε. We denote the ε-thin part by $M(G)^{]0,\varepsilon]}$. Equivalently $M(G)^{]0,\varepsilon]}$ is the set of points where the injectivity radius is less than $\varepsilon/2$. The closure of the complement of $M(G)^{]0,\varepsilon]}$ in $M(G)$, denoted by $M(G)^{[\varepsilon,\infty[}$, is called *the ε-thick part of* $M(G)$.

Margulis lemma [Marg].— *There exists a constant $\varepsilon(n) > 0$ such that, if $G \subset$* $\mathrm{Isom}(\mathbb{H}^n)$ *is a Kleinian group and if $x \in \mathbb{H}^n$, then the subgroup of G generated by the elements which move x a distance smaller than $\varepsilon(n)$ is elementary.*

The constant $\varepsilon(n)$ depends on n but not on G. It is called *the Margulis constant.*

As a consequence of Margulis lemma, we describe now the geometry of $M(G)^{]0,\varepsilon]}$ for $n = 3$ when G has no parabolic elements and for $n = 2$ (cf. **[Th1]**).

Let $G \subset \mathrm{PSL}_2(\mathbb{C})$ be a Kleinian group without parabolic elements. Any elementary subgroup of G is then a cyclic group generated by a hyperbolic isometry. Let $x \in M(G)^{]0,\varepsilon]}$. For any $\tilde{x}$ in the preimage of x in $\mathbb{H}^3$, there is an isometry in G other than the identity, which moves $\tilde{x}$ a distance less than or equal to ε. Let $g \in G$ be a hyperbolic isometry. The set of points in $\mathbb{H}^3$ which are moved by g a distance less than or equal to ε is non-empty only when the translation distance of g is less than or equal to ε. Then, by reasons of symmetry, it is a neighborhood of constant radius of the invariant axis $A(g)$ of g. We denote this neighborhood by $n^\varepsilon(g)$. Let $\langle\langle g \rangle\rangle$ be the maximal Abelian subgroup of G containing g. Since

g is an hyperbolic isometry $\langle\langle g\rangle\rangle$ is a cyclic group generated by a root of g. Let $\mathcal{N}^\varepsilon(g)$ be the union of the neighborhoods $n^\varepsilon(h)$ over all non-zero elements h in $\langle\langle g\rangle\rangle$. Suppose $\varepsilon \leq \varepsilon(3)$. By the Margulis lemma, the restriction of the covering map $p : \mathbb{H}^3 \to M(G)$ to $\mathcal{N}^\varepsilon(g)$ identifies points only when they are in the same orbit of $\langle\langle g\rangle\rangle$. Hence $p(\mathcal{N}^\varepsilon(g)) \subset M(G)$ is a solid tube diffeomorphic to $\mathcal{N}^\varepsilon(G)/\langle\langle g\rangle\rangle$. This image is a neighborhood of constant radius of the (embedded) closed geodesic $p(A(g)) = A(g)/\langle\langle g\rangle\rangle$. It is called *the Margulis tube around* $p(A(g))$. By Margulis lemma again, $M(G)^{]0,\varepsilon]}$ is a disjoint union of Margulis tubes. It is a nice exercice to show that the radius of the Margulis tube around a (very short) geodesic of length $\ell(g)$ is equivalent to $\log(\varepsilon/\ell(g))$, independently of G (**[Th5]**, **[OP]**). Qualitatively, the shorter the geodesic, the larger the Margulis tube around it.

When G is a Kleinian group contained in $\mathrm{PSL}_2(\mathbb{R})$, $M(G)^{]0,\varepsilon]}$ can be similarly described. The set of points which are moved a distance less than or equal to ε by a hyperbolic isometry $g \in G$ is a regular neighborhood of the axis of g. For $\varepsilon \leq \varepsilon(2)$ the quotient of this neighborhood by $\langle\langle g\rangle\rangle$ embeds in $M(G)$. Its image, called a Margulis tube, is diffeomorphic to an annulus. The set of points which are moved a distance less than or equal to ε by a parabolic isometry $g \in \mathrm{PSL}_2(\mathbb{R})$ is an horoball. For $\varepsilon \leq \varepsilon(2)$, the quotient of this horoball by $\langle\langle g\rangle\rangle$ embeds in $M(G)$. Its image is called *a cusp*. A cusp is conformally equivalent to the punctured closed unit disc.

When $n = 2$, $M(G)^{]0,\varepsilon]}$ is a disjoint union of Margulis tubes and cusps and by Margulis lemma, no two of them are homotopic. Therefore $M(G)^{]0,\varepsilon]}$ has finitely many components. When $n = 3$, this is not true in general, but will be satisfied by the groups we will consider.

1.2 Quasi-conformal homeomorphisms

We refer to the books **[Ahl]**, **[Ga]** and **[LV]** for more details on this section.

Definition. — Let $\varphi : U \to V$ be an homeomorphism between two open sets in the complex plane. The map φ is called *quasi-conformal* when the following three conditions hold:

(i) φ is orientation preserving,

(ii) the derivatives $\partial\varphi/\partial x$ and $\partial\varphi/\partial y$ in the sense of distributions exist and are locally square-integrable, and

(iii) there exists $\mu \in L^\infty(U, \mathbb{C})$ with $||\mu||_\infty < 1$, such that, for almost all $z \in U$, $\bar\partial\varphi(z) = \mu(z)\partial\varphi(z)$, where

$$\partial\varphi = \frac{1}{2}\left(\frac{\partial\varphi}{\partial x} - i\frac{\partial\varphi}{\partial y}\right) \quad \text{and} \quad \bar\partial\varphi = \frac{1}{2}\left(\frac{\partial\varphi}{\partial x} + i\frac{\partial\varphi}{\partial y}\right).$$

The quasi-conformal homeomorphism φ is said to be K*-quasi-conformal* for

$$K = K(\varphi) = \frac{1 + ||\mu||_\infty}{1 - ||\mu||_\infty}.$$

The function μ is called *the Beltrami coefficient of* φ and $K(\varphi)$ *the eccentricity of* φ.

A 1-quasi-conformal homeomorphism is conformal. Therefore $\log K$ measures the deviation of φ from being conformal.

When the frontier of U in $\mathbb{C}$ is locally connected, any quasi-conformal homeomorphism φ of U extends continuously to ∂U. In particular, any quasi-conformal homeomorphism of the upper half-space extends continuously to the boundary.

Note that the right or left composition of a K-quasi-conformal homeomorphism with a conformal homeomorphism is K-quasi-conformal. Therefore one can define the notion of being K-quasi-conformal for a homeomorphism between two Riemann surfaces.

The fundamental result about quasi-conformal homeomorphisms is due to L. Ahlfors and L. Bers.

Ahlfors-Bers theorem [AB].— *Let* $\mu \in L^\infty(\overline{\mathbb{C}})$ *with* $||\mu||_\infty < 1$. *Then, there exists a unique quasi-conformal homeomorphism* φ_μ *of* $\overline{\mathbb{C}}$ *such that*

(i) $\dfrac{\bar\partial\varphi_\mu}{\partial\varphi_\mu}(z) = \mu(z)$ *for almost all points* z,

(ii) φ_μ *fixes* 0, 1 *and* ∞.

The function $\mu \to \varphi_\mu$ *is continuous for the topology of the uniform convergence over compact sets. Furthermore, for* $t \leq 1$ *we have*

$$\varphi_{t\mu}(z) = z + tF_\mu(z) + O(t^2),$$

where

$$F_\mu(z) = -\frac{z(z-1)}{\pi}\iint_{\mathbb{H}^2}\frac{\mu(\zeta)}{\zeta(\zeta-1)(\zeta-z)}d\xi d\eta.$$

Equation (i) is called *the Beltrami equation.* A quasi-conformal homeomorphism of $\mathbb{C}$ (resp. of $\overline{\mathbb{H}^2}$) which satisfies (ii) is said to be *normalized.*

The following is a corollary of Ahlfors-Bers theorem (existence and unicity).

Theorem [AB].— *Let* $\mu \in L^\infty(\mathbb{H}^2)$ *with* $||\mu||_\infty < 1$. *Then there exists a unique normalized quasi-conformal homeomorphism* φ_μ *of* $\mathbb{H}^2$ *which satisfies*

$$\frac{\bar\partial\varphi_\mu}{\partial\varphi_\mu}(z) = \mu(z).$$

1.3 Teichmüller space

Definition.— *A Fuchsian group* is a Kleinian group $\Gamma \subset \mathrm{PSL}_2(\mathbb{R})$. We say that Γ is *cocompact* (resp. *has finite covolume*) if $\mathbb{H}^2/\Gamma$ is compact (has finite volume). If $\mathbb{H}^2/\Gamma$ has finite volume, it is conformally equivalent to the complement of a finite number of points in a compact Riemann surface. A point in this finite set is called a *puncture*. A connected Riemann surface X is *hyperbolic* when its universal cover is conformally equivalent to $\mathbb{H}^2$. A Riemann surface is hyperbolic if each of its components is hyperbolic.

Let X be a connected Riemann surface of negative Euler characteristic. By the Poincaré-Koebe uniformization theorem, the universal cover of X is conformally equivalent to $\mathbb{H}^2$. Hence X is conformally equivalent to the quotient of $\mathbb{H}^2$ by a group of conformal automorphisms. The group of conformal automorphisms of $\mathbb{H}^2$ equals $\mathrm{PSL}_2(\mathbb{R})$ acting by homographies. Therefore X is hyperbolic. Since $\mathrm{PSL}_2(\mathbb{R})$ is the group of isometries of $\mathbb{H}^2$, the hyperbolic metric on $\mathbb{H}^2$ projects to a complete Riemannian metric of constant curvature -1 on X which is in the given conformal class. This Riemannian metric is called *the hyperbolic metric on* X. This construction gives us our first examples of Kleinian groups.

The goal of Teichmüller theory is to describe all hyperbolic metrics on a fixed surface or equivalently, all the "deformations" of a given Fuchsian group Γ. We will restrict our attention to cocompact Fuchsian groups.

Let $\Gamma \subset \mathrm{PSL}_2(\mathbb{R})$ be a cocompact Fuchsian group. Let $X = \mathbb{H}^2/\Gamma$.

Definition.— *A Fuchsian deformation of* Γ is a couple $(\rho, \widetilde{\varphi})$, where ρ is a representation of Γ in $\mathrm{PSL}_2(\mathbb{R})$ and where $\widetilde{\varphi}$ is a normalized quasi-conformal homeomorphism of $\mathbb{H}^2$ which conjugates Γ and $\rho(\Gamma)$, i.e. such that

(i) for all $\gamma \in \Gamma$,

$$\rho(\gamma) = \widetilde{\varphi} \circ \gamma \circ \widetilde{\varphi}^{-1}, \quad \text{and}$$

(ii) the continuous extension of $\widetilde{\varphi}$ to $\overline{\mathbb{R}}$ fixes 0, 1 and ∞.

Define an equivalence relation on the set of Fuchsian deformations of Γ by $(\rho, \widetilde{\varphi}) \simeq (\rho', \widetilde{\varphi}')$ if and only if $\rho = \rho'$.

Observe that the extension of $\widetilde{\varphi}$ (resp. $\widetilde{\varphi}'$) to the real axis conjugates the action of Γ to the one of $\rho(\Gamma)$ (resp. $\rho'(\Gamma)$). Thus, the density of the set of fixed points of elements of Γ in $\mathbb{R}$ implies that: $\rho = \rho'$ if and only if $\widetilde{\varphi}|\mathbb{R} = \widetilde{\varphi}'|\mathbb{R}$. The quotient of the set of Fuchsian deformations of Γ by this equivalence relation is called *the Teichmüller space of* Γ. We denote it by $\mathcal{T}(\Gamma)$ or by $\mathcal{T}(X)$.

If $(\rho, \widetilde{\varphi})$ is a Fuchsian deformation of Γ, $\rho(\Gamma)$ is discrete because its action on $\mathbb{H}^2$ is conjugate to that of Γ. So $\widetilde{\varphi}$ projects to a homeomorphism φ between $\mathbb{H}^2/\Gamma$ and $\mathbb{H}^2/\rho(\Gamma)$. One checks that the equivalence relation which defines Teichmüller space identifies two Fuchsian deformations $(\rho, \widetilde{\varphi})$ and $(\rho', \widetilde{\varphi}')$ when the composed homeomorphism $\varphi^{-1} \circ \varphi'$ of $\mathbb{H}^2/\Gamma$ is homotopic to the identity. Therefore a point in $\mathcal{T}(X)$ can be interpreted as a hyperbolic surface with a homeomorphism from X, which is well defined up to homotopy.

The Teichmüller distance.

For two points σ_1, σ_2 in $\mathcal{T}(\Gamma)$ we define

$$d(\sigma_1, \sigma_2) = \frac{1}{2} \inf \log K(\widetilde{\varphi}_2 \circ \widetilde{\varphi}_1^{-1}),$$

where the infimum is taken over all representatives $(\rho_1, \widetilde{\varphi}_1)$ (resp. $(\rho_2, \widetilde{\varphi}_2)$) of σ_1 (resp. σ_2).

Then d is a distance, called *the Teichmüller distance*, which turns $\mathcal{T}(\Gamma)$ into a complete metric space.

In §6, we will need the following result (cf. **[Ga]**).

Distortion lemma.— *Let* $\sigma_1 = (\rho_1, \widetilde{\varphi}_1)$, $\sigma_2 = (\rho_2, \widetilde{\varphi}_2)$ *be two points in* $\mathcal{T}(\Gamma)$. *Let* $\gamma \in \Gamma$ *be a hyperbolic element. Then, we have*

$$e^{-2d(\sigma_1,\sigma_2)}\ell(\rho_1(\gamma)) \leq \ell(\rho_2(\gamma)) \leq e^{2d(\sigma_1,\sigma_2)}\ell(\rho_1(\gamma)).$$

The modular group.

Definition.— *The modular group of* X is the group of homotopy equivalences of orientation preserving diffeomorphisms of X. It is denoted by $\mathrm{Mod}(X)$.

We will denote by $\overline{X}$ the Riemann surface X with the opposite orientation, i.e. the quotient of the lower half-plane $\overline{\mathbb{H}^2}$ by Γ. The space of equivalences classes of Fuchsian deformations of the action of Γ on $\overline{\mathbb{H}^2}$ is denoted by $\mathcal{T}(\overline{\Gamma})$ (or by $\mathcal{T}(\overline{X})$). There is a natural map from from $\mathcal{T}(\Gamma)$ to $\mathcal{T}(\overline{\Gamma})$ induced by the map which assigns to the Fuchsian deformation $(\rho, \widetilde{\varphi})$ of Γ the deformation $(\rho, \bar{\varphi})$ of Γ acting on $\overline{\mathbb{H}^2}$, where $\bar{\varphi}$ is the conjugate of $\widetilde{\varphi}$ by the complex conjugation. It is denoted by $s \to \bar{s}$ and called *the complex conjugation*. Its inverse is a map from $\mathcal{T}(\overline{\Gamma})$ to $\mathcal{T}(\Gamma)$ which is defined similarly.

Let f be an orientation preserving diffeomorphism of $X = \mathbb{H}^2/\Gamma$. Choose a lift $\widetilde{f}$ of f to the universal cover $\mathbb{H}^2$. Since X is compact, f is K-quasi-conformal for some constant K and $\widetilde{f}$ is K-quasi-conformal also. Let $(\rho, \widetilde{\varphi})$ be a Fuchsian deformation of Γ. Then $\widetilde{\varphi} \circ \widetilde{f}^{-1}$ is a quasi-conformal homeomorphism which conjugates Γ to a certain Fuchsian group. Let $a^{-1} \in \mathrm{PSL}_2(\mathbf{R})$ be an element which takes the same values as $\widetilde{\varphi} \circ \widetilde{f}^{-1}$ on the points 0, 1 and ∞. Then $a \circ \widetilde{\varphi} \circ \widetilde{f}^{-1}$ is a normalized quasi-conformal homeomorphism which conjugates the trivial representation of Γ to a representation ρ_f. Therefore, the couple $(\rho_f, a \circ \widetilde{\varphi} \circ \widetilde{f}^{-1})$ is a Fuchsian deformation of Γ. One checks easily that the equivalence class of this deformation depends only of the equivalence class of $(\rho, \widetilde{\varphi})$, and defines therefore a map of $\mathcal{T}(\Gamma)$ to itself, which depends only on the homotopy class of f. We denote this map by f^*: it is an isometry for the Teichmüller distance. Thus $f \to f^*$ defines an isometric action of $\mathrm{Mod}\,(X)$ on $\mathcal{T}(\Gamma)$.

An orientation reversing diffeomorphism f of X induces also a map f^* from $\mathcal{T}(\Gamma)$ to $\mathcal{T}(\overline{\Gamma})$ (and also from $\mathcal{T}(\overline{\Gamma})$ to $\mathcal{T}(\Gamma)$). This map is an isometry which commutes with the complex conjugation.

The differentiable structure of Teichmüller space.

Teichmüller space has been defined so far as a metric space. In fact it is a smooth manifold, a property which is crucial in McMullen's proof of Thurston's fixed point theorem. We sketch below how Teichmüller space can be viewed as a complex manifold (cf. **[Ga]**). In order to do this, we first introduce two objects which are fundamental for understanding Teichmüller space from an infinitesimal viewpoint, *the Beltrami forms* and *the holomorphic quadratic differentials.*

Beltrami forms and holomorphic quadratic differentials.

Let Γ be a (not necessarily cocompact) Fuchsian group. Let $X = \mathbb{H}^2/\Gamma$.

Definition.— *A Beltrami form* μ on $\mathbb{H}^2/\Gamma$ is a measurable function $\widetilde{\mu} : \mathbb{H}^2 \to \mathbb{C}$, with finite L^∞-norm and such that

$$\forall \gamma \in \Gamma, \qquad \widetilde{\mu}(\gamma(z))\frac{\overline{\gamma'(z)}}{\gamma'(z)} = \widetilde{\mu}(z)$$

for almost all $z \in \mathbb{H}^2$. We denote by $\mathcal{B}(\Gamma)$ (or $\mathcal{B}(X)$) the Banach space of Beltrami forms on $\mathbb{H}^2/\Gamma$ endowed with the L^∞-norm.

Let $\mathcal{B}^1(\Gamma)$ be the open unit ball in $\mathcal{B}(\Gamma)$. Let $\mu \in \mathcal{B}^1(\Gamma)$. By Ahlfors-Bers theorem, there is a quasi-conformal homeomorphism φ_μ of $\mathbb{H}^2$ with Beltrami coefficient $\widetilde{\mu}$. A short computation shows that the Beltrami coefficient of the quasi-conformal homeomorphism $\rho_\mu(\gamma) = \varphi_\mu^{-1} \circ \gamma \circ \varphi_\mu$ vanishes so that $\rho_\mu(\gamma)$ belongs to $\mathrm{PSL}_2(\mathbb{R})$. Clearly, $\gamma \to \rho_\mu(\gamma)$ is a homomorphism of Γ into $\mathrm{PSL}_2(\mathbb{R})$. On this way, we define a map $\Pi : \mathcal{B}^1(\Gamma) \to \mathcal{T}(\Gamma)$ by assigning to μ the equivalence class of the couple (ρ_μ, φ_μ). By definition of $\mathcal{T}(\Gamma)$, Π is onto.

Definition.— *A holomorphic quadratic differential* ϕ on $\mathbb{H}^2/\Gamma$ is a holomorphic function $\widetilde{\phi}$ on $\mathbb{H}^2$ which satisfies

$$\forall \gamma \in \Gamma, \quad \forall z \in \mathbb{H}^2, \quad \widetilde{\phi}(\gamma(z))(\gamma'(z))^2 = \widetilde{\phi}(z).$$

This transformation rule means that the tensor $\widetilde{\phi}(z)dz^2$ is invariant under the action of Γ and projects therefore to a tensor on X. In particular the expression $|\widetilde{\phi}(z)||dz^2|$ is invariant under Γ. It defines a measure on X. For $\phi \in \mathcal{Q}(X)$ and for $E \subset X$ a measurable set, the measure of E, $\int_E |\phi|$ is called *the* ϕ*-mass of* E. When the ϕ-mass of X is finite, ϕ is said to be *integrable*. We denote by $\mathcal{Q}(\Gamma)$ (or $\mathcal{Q}(X)$) the Banach space of integrable holomorphic quadratic differentials endowed with the L^1-norm.

When Γ is cocompact, any holomorphic quadratic differential is of course integrable. When Γ is not cocompact but has finite covolume, the integrability condition is not necessarily satisfied. It means precisely that when ϕ is expressed in a conformal chart around each puncture, it has at worst a simple pole. When Γ has finite covolume, it follows then (for instance from Riemann-Roch theorem) that $\mathcal{Q}(\Gamma)$ is a finite dimensional complex vector space. Its dimension is $3g - 3 + p$, where g is the genus of X and p is the number of punctures of X. When Γ has infinite covolume, it is not hard to see that $\mathcal{Q}(\Gamma)$ is infinite dimensional.

There is a natural pairing between $\mathcal{B}(\Gamma)$ and $\mathcal{Q}(\Gamma)$. If $\phi \in \mathcal{Q}(\Gamma)$ and $\mu \in \mathcal{B}(\Gamma)$, the local expression $\phi(z)\widetilde{\mu}(z)|dz^2|$ is invariant under Γ. It projects to a complex measure on X which has finite total mass. One defines:

$$\langle \phi, \mu \rangle = \Re\Big(\int_X \phi(z)\mu(z)|dz^2|\Big).$$

We denote by $\mathcal{N}(\Gamma)$ the kernel of this pairing:

$$\mathcal{N}(\Gamma) = \{\mu \in \mathcal{B}(\Gamma) \mid \forall \phi \in \mathcal{Q}(\Gamma), \quad \langle \phi, \mu \rangle = 0\}.$$

It follows from the Hahn-Banach theorem and from the Riesz representation theorem that $\langle .,. \rangle$ induces a duality between $\mathcal{Q}(\Gamma)$ and the quotient space $\mathcal{B}(\Gamma)/\mathcal{N}(\Gamma)$.

Definition.— Let Y be a hyperbolic Riemann surface with finitely many components. *The space of Beltrami forms* $\mathcal{B}(Y)$ is defined as the product of the spaces $\mathcal{B}(X)$ where X varies over the components of Y. *The space of integrable holomorphic quadratic differentials* $\mathcal{Q}(Y)$ is defined as the product of the spaces $\mathcal{Q}(X)$, where X varies over the components of Y. The norm on $\mathcal{B}(Y)$ (resp. on $\mathcal{Q}(Y)$) is the supremum norm of the norms on the spaces $\mathcal{B}(X)$ (resp. the sum of the norms on the spaces $\mathcal{Q}(X)$). The pairing between $\mathcal{Q}(Y)$ and $\mathcal{B}(Y)$ is defined as the sum of the pairings on the components of Y.

Poincaré series.

Consider a covering of Riemann surfaces $\pi : Y \to X$. Let $\phi \in \mathcal{Q}(Y)$. If $U \subset X$ is an embedded disc, the cover $\pi^{-1}(U) \to U$ is trivial so that for each component V_i of $\pi^{-1}(U)$, the map π admits a section $s_i : U \to V_i$. In any holomorphic chart, the restriction of ϕ to V_i can be expressed in the form $\phi_i(z)dz^2$. By Cauchy's formula and since ϕ is integrable, the series

$$\sum_i \phi_i \circ s_i(u)(s_i'(u))^2$$

is absolutely convergent on compact subsets of U. It defines therefore a holomorphic function on X. This function transforms under changes of charts like a quadratic differential and defines an element in $\mathcal{Q}(X)$ denoted by $\Theta_{Y/X}\phi$. One has: $||\Theta_{Y/X}\phi|| \leq ||\phi||$. The operator $\Theta_{Y/X} : \mathcal{Q}(Y) \to \mathcal{Q}(X)$ is called *the Theta operator associated to the cover* $Y \to X$. Its norm is less than or equal to 1. The study of the contraction properties of the operator $\Theta_{Y/X}$ will be the main theme of §5. For the moment, we note that $\Theta_{Y/X}$ is dual to *the pull-back operator on Beltrami forms* for the pairing between Beltrami forms and quadratic differentials. For $\mu \in \mathcal{B}(X)$, one can define a Beltrami form $\pi^*(\mu) \in \mathcal{B}(Y)$ by setting $\widetilde{\pi^*(\mu)} = \widetilde{\mu}$. This form is called *the pull-back of* μ. For all $\phi \in \mathcal{Q}(Y)$ and for all $\mu \in \mathcal{B}(X)$, we have

$$\langle \Theta_{Y/X}\phi, \mu \rangle = \langle \phi, \pi^*(\mu) \rangle.$$

The Bers embedding.

The key ingredient in the proof that the Teichmüller space carries a complex structure is a result of L. Bers which allows us to embed $\mathcal{T}(\Gamma)$ in the complex vector space $\mathcal{Q}(\overline{\Gamma})$. We outline below this construction.

Let Γ be a cocompact Fuchsian group. Let $\mu \in \mathcal{B}^1(\Gamma)$. Consider the measurable function on $\overline{\mathbb{C}}$ which agrees with $\widetilde{\mu}$ on $\mathbb{H}^2$ and which vanishes identically on the lower half-space $\overline{\mathbb{H}^2}$. This function has a L^∞-norm strictly less than 1, and by the Ahlfors-Bers theorem it is the Beltrami coefficient of a unique normalized quasiconformal homeomorphism of $\overline{\mathbb{C}}$. We denote this homeomorphism by φ^μ.

Lemma. — *For any two elements μ, μ' in $\mathcal{B}^1(\Gamma)$, we have:*

$$\varphi_\mu = \varphi_{\mu'} \Leftrightarrow \varphi^\mu|\overline{\mathbb{H}^2} = \varphi^{\mu'}|\overline{\mathbb{H}^2}.$$

By construction, the restriction $\varphi^\mu|\overline{\mathbb{H}^2}$ is a univalent map. Recall that *the Schwarzian derivative* of a univalent map f is the holomorphic function defined by

$$S(f) = (\frac{f''}{f'})' - \frac{1}{2}(\frac{f''}{f'})^2.$$

The transformation rule of μ under Γ implies that $S(\varphi^\mu)$ transforms under the action of Γ on $\overline{\mathbb{H}^2}$ like a holomorphic quadratic differential. Hence $S(\varphi^\mu) \in \mathcal{Q}(\overline{\Gamma})$, the space of integrable holomorphic quadratic differentials on $\overline{\mathbb{H}^2}/\Gamma$. Consider the map $\Phi : \mathcal{B}^1(\Gamma) \to \mathcal{Q}(\overline{\Gamma})$ which assigns to $\mu \in \mathcal{B}^1(\Gamma)$ the holomorphic quadratic differential $S(\varphi^\mu) \in \mathcal{Q}(\overline{\Gamma})$. By the lemma above, Φ can be factorized through a map $\mathcal{B} : \mathcal{T}(\Gamma) \to \mathcal{Q}(\overline{\Gamma})$: $\Phi = \mathcal{B} \circ \Pi$. By this lemma also, $\mathcal{B}$ is injective since any univalent map is characterized by its Schwarzian derivative, up to post-composition with a Möbius map. The map $\mathcal{B}$ is called *the Bers embedding.* We identify $\mathcal{T}(\Gamma)$ with its image under $\mathcal{B}$ and we continue to denote this image by $\mathcal{T}(\Gamma)$. By the Ahlfors-Bers theorem, Φ is holomorphic and its derivative $D\Phi_0$ at the origin is the map

$$D\Phi_0(\nu)(z) = \frac{-6}{\pi} \iint_{\mathbb{H}^2} \frac{\nu(\zeta)}{(\zeta - z)^4} d\xi d\eta, \quad \text{for} \quad z \in \overline{\mathbb{H}^2}.$$

Since Γ is cocompact, $\mathcal{Q}(\overline{\Gamma})$ is a finite dimensional vector space, and the following result is easy to prove (when Γ is an arbitrary Fuchsian group, it is still true but is a deep theorem due to Bers [**Ber**]).

Fact. — *The map $D\Phi_0$ is surjective and induces an isomorphism from $\mathcal{B}(\Gamma)/\mathcal{N}(\Gamma)$ to $\mathcal{Q}(\overline{\Gamma})$.*

Thus, by the implicit function theorem, $\mathcal{T}(\Gamma)$ contains a neighborhood of 0 in $\mathcal{Q}(\overline{\Gamma})$. It says also that in any sufficiently small neighborhood of $0 \in \mathcal{T}(\Gamma)$, Φ admits local sections.

Let $\mu \in \mathcal{B}^1(\Gamma)$ and denote by Γ_μ the group conjugated to Γ by φ_μ. Define a holomorphic map $\alpha_\mu : \mathcal{B}^1(\Gamma) \to \mathcal{B}^1(\Gamma_\mu)$ by

$$\alpha_\mu(\nu) = (\frac{\nu - \mu}{1 - \nu\bar{\mu}})(\frac{\partial\varphi_\mu}{|\partial\varphi_\mu|}) \circ \varphi_\mu^{-1}.$$

Then α_μ is biholomorphic. Furthermore it conjugates Φ and $\Phi_\mu : \mathcal{B}^1(\Gamma_\mu) \to \mathcal{T}(\Gamma_\mu)$. In this way, a neighborhood of Γ_μ in $\mathcal{T}(\Gamma)$ becomes biholomorphically equivalent to a neighborhood of $0 \in \mathcal{Q}(\overline{\Gamma_\mu})$. In particular, $\mathcal{T}(\Gamma)$ is an open subset of $\mathcal{Q}(\overline{\Gamma})$ and so it inherits the complex structure from $\mathcal{Q}(\overline{\Gamma})$. The derivative $D\Phi_0$ induces an isomorphism from $\mathcal{B}(\Gamma)/\mathcal{N}(\Gamma)$ to the tangent space of $\mathcal{T}(\Gamma)$ at point Γ (identified with $0 \in \mathcal{Q}(\overline{\Gamma})$). The map α_μ induces an identification between the tangent space of $\mathcal{T}(\Gamma)$ at Γ_μ with the tangent space of $\mathcal{T}(\Gamma_\mu)$ at the origin. Hence the tangent space to $\mathcal{T}(\Gamma)$ at Γ_μ is isomorphic to $\mathcal{B}(\Gamma_\mu)/\mathcal{N}(\Gamma_\mu)$.

The infinitesimal form of Teichmüller distance.

For the smooth structure constructed above, the Teichmüller distance has an infinitesimal form, that of a Finsler metric.

On $\mathcal{B}^1(\Gamma)$, consider the distance

$$\bar{d}(\mu_1, \mu_2) = -\frac{1}{2} \log K(\phi_{\mu_2} \circ \phi_{\mu_1}^{-1}).$$

A short computation gives

$$K(\phi_{\mu_2} \circ \phi_{\mu_1}^{-1}) = (\frac{\mu_2 - \mu_1}{1 - \mu_2 \overline{\mu_1}})(\frac{\partial \phi_{\mu_1}}{|\partial \phi_{\mu_1}|}) \circ \phi_{\mu_1},$$

from which we deduce the nice formula $\bar{d}(\mu_1, \mu_2) = \sup_{z \in \mathbf{H}^2} d_{\mathbb{D}^2}(\mu_1(z), \mu_2(z))$, where $d_{\mathbb{D}^2}$ is the hyperbolic distance on $\mathbb{D}^2$.

By definition, the Teichmüller distance d satisfies: $d(\sigma_1, \sigma_2) = \inf \bar{d}(\mu_1, \mu_2)$, where the infimum is taken over all pairs (μ_1, μ_2) such that $\Pi(\mu_i) = \sigma_i$. This means that d is the quotient distance of $\bar{d}$ by Π. Now we construct *a Finsler metric* on $\mathcal{T}(\Gamma)$ with associated distance d. It will appear as the quotient of a Finsler metric on $\mathcal{B}^1(\Gamma)$ with associated distance $\bar{d}$.

Definition.— *A Finsler metric* on a Banach manifold $\mathcal{B}$ (like for instance an open set in a Banach space) is a continuous function on the tangent space $T\mathcal{B}$ which induces a Banach norm on the tangent space at each point and which is locally equivalent to the (ambient) Banach norm.

When $\mathcal{B}$ is connected, one can associate a distance to a Finsler metric on $\mathcal{B}$: the distance between two given points is equal to the infimum of the length of a smooth arc joining these two points.

On the tangent space of $\mathcal{B}^1(\Gamma)$ ($\subset \mathcal{B}(\Gamma)$) consider the function

$$\bar{\beta}_\mu(\nu) = 2 \|\frac{\nu}{1 - |\mu|^2}\|,$$

where $\nu \in \mathcal{B}(\Gamma)$ is a tangent vector at $\mu \in \mathcal{B}^1(\Gamma)$. It is a Finsler metric on $\mathcal{B}^1(\Gamma)$ (which is formally reminiscent of the hyperbolic metric on $\mathbb{D}^2$). From the formula

$$\bar{d}(\mu_1, \mu_2) = \sup_{z \in \mathbf{H}^2} d_{\mathbb{D}^2}(\mu_1(z), \mu_2(z)),$$

one deduces that the distance on $\mathcal{B}^1(\Gamma)$ associated to $\bar{\beta}$ is $\bar{d}$.

The Finsler metric $\bar{\beta}$ projects to a Finsler metric on $\mathcal{T}(\Gamma)$. Let $s \in \mathcal{T}(\Gamma)$. Let $\mu \in \mathcal{B}^1(\Gamma)$ be such that $s = \Pi(\mu)$. Let us define

$$\beta_s(v) = \inf \bar{\beta}_\mu(\nu),$$

where the infimum is taken over all vectors ν such that $D\Pi_\mu(\nu) = v$. One shows that $\beta_s(v)$ does not depend on μ and that β is a Finsler metric on $\mathcal{T}(\Gamma)$. Since d is the quotient distance of $\bar{d}$ by Π, it follows that d is the distance associated to β. This is a general result on Finsler metrics due to O'Byrne **[O'B]**.

On the tangent space at the origin of Teichmüller space $\mathcal{T}(\Gamma)$ (i.e. at point Γ), β equals the quotient norm of the L^∞-norm on $\mathcal{B}(\Gamma)/\mathcal{N}(\Gamma)$. The identification of the tangent space to $\mathcal{T}(\Gamma)$ at an arbitrary point Γ_μ with $\mathcal{B}(\Gamma_\mu)/\mathcal{N}(\Gamma_\mu)$ implies that β equals on this tangent space the quotient of the L^∞-norm on $\mathcal{B}(\Gamma_\mu)/\mathcal{N}(\Gamma_\mu)$.

Since the cotangent space of $\mathcal{T}(\Gamma)$ at Γ_μ is identified with $\mathcal{Q}(\Gamma_\mu)$ via the pairing $\langle .,. \rangle$, the norm dual to β equal the L^1-norm on $\mathcal{Q}(\Gamma_\mu)$.

1.4 Geometrically finite groups

An important feature of the proof of the Hyperbolization theorem for non-fibered manifolds is that we need only to consider Kleinian groups which are *geometrically finite.* Since we will restrict to Kleinian groups without parabolic elements, we adopt the following definition (for a detailed discussion of geometrically finite groups in $\mathrm{PSL}_2(\mathbb{C})$, see [**Mor**]).

Definition. — *A geometrically finite group* is a non-elementary Kleinian group $G \subset \mathrm{PSL}_2(\mathbb{C})$ without parabolic elements, such that $N(G)$ is compact.

A cocompact Fuchsian group Γ is geometrically finite and $N_\delta(G)$ is diffeomorphic to the product $\mathbb{H}^2/\Gamma \times [0,1]$. If G is a geometrically finite group which is not Fuchsian, $N(G)$ is a 3-manifold with compact boundary and, for all $\delta > 0$, $N_\delta(G)$ is a compact 3-manifold (of class C^1).

Hyperbolic manifolds.

Definition. — Suppose that M is a compact 3-manifold which admits an atlas $\mathcal{A}$ of class C^1 with charts modelled on convex subsets of $\mathbb{H}^3$ and with coordinate changes in $\mathrm{PSL}_2(\mathbb{C})$. Then the constant curvature -1 Riemannian metric of $\mathbb{H}^3$ can be pulled back to a Riemannian metric $m_{\mathcal{A}}$ on M (of class C^1).

We say that M is *hyperbolic* if M admits an atlas $\mathcal{A}$ such that

(i) the distance associated to $m_{\mathcal{A}}$ is complete, and

(ii) the volume of $m_{\mathcal{A}}$ is finite.

A Riemannian metric $m_{\mathcal{A}}$ with these properties is called *a hyperbolic metric on* M.

To a hyperbolic metric $m_{\mathcal{A}}$ on M, one can associate *its developping map* ([**Th1**], [**CEG**]): it is a local isometry Dev from the universal cover $\widetilde{M}$ of M (endowed with the lift of the metric $m_{\mathcal{A}}$) to $\mathbb{H}^3$ which conjugates the canonical action of $\pi_1(M)$ to the action of a subgroup of $\mathrm{PSL}_2(\mathbb{C})$ denoted by $G_{\mathcal{A}}$, and called *the holonomy group of* $m_{\mathcal{A}}$.

Let $c(p,q)$ be a length-minimizing geodesic connecting the points p and q of $\widetilde{M}$. It follows from the fact that the charts in $\mathcal{A}$ are modelled on convex subsets of $\mathbb{H}^3$ that $\mathrm{Dev}(c(p,q))$ is a geodesic of $\mathbb{H}^3$. The uniqueness of the geodesic connecting two points of $\mathbb{H}^3$ implies then that Dev is a diffeomorphism onto its image. Therefore $G_{\mathcal{A}}$ is a Kleinian group. Furthermore, Dev induces an isometric embedding from

M into $M(G_A)$, which identifies isometrically M with a closed convex subset of $M(G_A)$.

Examples.— 1) The manifold $(\mathbb{S}^1 \times \mathbb{B}^2)$ is hyperbolic: $\mathbb{S}^1 \times \mathbb{B}^2$ is diffeomorphic to the quotient of a constant radius neighborhood of a geodesic in $\mathbb{H}^3$ by a hyperbolic element fixing this geodesic.

2) Let G be a geometrically finite group. Then, for $\delta > 0$, $N_\delta(G)$ is a hyperbolic manifold.

Fact.— *Every hyperbolic manifold is diffeomorphic to one of the above examples.*

Proof.— Let m_A be a hyperbolic metric on M. Let G_A $(\simeq \pi_1(M))$ be its holonomy group. Since M is compact, each element of $\pi_1(M)$ is represented by a closed geodesic. Therefore, if G_A is elementary, it is a cyclic group generated by a hyperbolic isometry and M is diffeomorphic to the first example.

Suppose that G_A is non-elementary. Since $C(G_A)$ is the smallest closed convex subset of $\mathbb{H}^3$ invariant under G_A, it is contained in the image of Dev. Therefore $N(G_A)$ is naturally contained in M. Since M is compact, G_A has no parabolic elements, and $N(G_A)$ is compact. Therefore, G_A is geometrically finite. Since M is convex in $M(G_A)$, M is diffeomorphic to $N_\delta(G_A)$. □

Thurston's hyperbolization theorem gives a sufficient condition on the topology of M that guarantees that M is hyperbolic.

Thurston's hyperbolization theorem.— *Let M be a Haken manifold whose fundamental group does not contain $\mathbb{Z}+\mathbb{Z}$-subgroups. Then M is hyperbolic.*

The definition of a Haken manifold will be given in §7.

Except the arithmetic constructions which give little control on the topology of the resulting quotient manifold [**Bo**], there are essentially three ways to construct Kleinian groups:

(i) *Andreev's theorem* yields a lot of examples of hyperbolic polyhedra in $\mathbb{H}^3$ [**An**] (cf. introduction);

(ii) *the deformation theory*, using quasi-conformal maps, gives us a way to deform a given geometrically finite group;

(iii) *Maskit's combination theorem* give conditions under which two geometrically finite groups can be amalgamated to form a new geometrically finite group [**Mas2**].

All of these techniques are used in the proof of Thurston's hyperbolization theorem. Andreev's theorem will be discussed in §8. Maskit's combination theorem will be explained in §2. Now, we describe how quasi-conformal homeomorphisms can be used to deform a given geometrically finite group.

The deformation space of a geometrically finite group.

The deformation theory of a geometrically finite group in $\mathrm{PSL}_2(\mathbb{C})$ can be formulated in exactly the same terms as the deformation theory of a Fuchsian group in $\mathrm{PSL}_2(\mathbb{R})$.

Let G be a geometrically finite group such that $\Omega(G) \neq \emptyset$ and such that each component of $\Omega(G)$ is conformally equivalent to $\mathbb{D}^2$. This is equivalent to say that $M = \overline{M}(G)$ is a compact manifold with non-empty and *incompressible boundary* (i.e. the fundamental group of any component of ∂M injects into $\pi_1(M)$, see §7). We will suppose also that $L(G) \subset \overline{\mathbb{C}}$ contains 0, 1 and ∞. This situation can always be achieved by conjugating G in $\mathrm{PSL}_2(\mathbb{C})$.

Definition. — *A quasi-conformal deformation of* G is a couple $(\rho, \widetilde{\varphi})$, where ρ is a representation of G into $\mathrm{PSL}_2(\mathbb{C})$ and where $\widetilde{\varphi}$ is a normalized quasi-conformal homeomorphism of $\overline{\mathbb{C}}$ which conjugates the actions of G and $\rho(G)$ on $\overline{\mathbb{C}}$, i.e. which satisfies, for all $g \in G$

$$\rho(g) = \widetilde{\varphi} \circ g \circ \widetilde{\varphi}^{-1}.$$

Consider the equivalence relation on the set of quasi-conformal deformations of G defined by $(\rho, \varphi) \simeq (\rho', \varphi')$ if and only if $\rho = \rho'$. The set of equivalence classes of quasi-conformal deformations of G is denoted by $\mathcal{GF}(G)$ (or $\mathcal{GF}(M)$).

(i) in the next Proposition is one reason for this notation.

Proposition. — *Let* $(\rho, \widetilde{\varphi})$ *be a quasi-conformal deformation of* G. *Then*

(i) $\rho(G)$ *is geometrically finite, and*

(ii) $\widetilde{\varphi}$ *induces a quasi-conformal homeomorphism* φ *between* $\Omega(G)/G$ *and* $\Omega(\rho(G))/\rho(G)$ *which extends to a homeomorphism* Φ *between* $\overline{M}(G)$ *and* $\overline{M}(\rho(G))$.

One way for proving this Proposition is to use the following result of Thurston which provides a natural extension of any quasi-conformal homeomorphism of $\overline{\mathbb{C}}$ to a homeomorphism of $\mathbb{H}^3 \cup \overline{\mathbb{C}}$ (**[Th1]**, **[Re]**). Let $\widetilde{\varphi}$ be a quasi-conformal homeomorphism of $\overline{\mathbb{C}}$. Then there is a homeomorphism $\widetilde{\Phi} = \widetilde{\Phi}(\widetilde{\varphi})$ such that

(i) $\widetilde{\Phi}(\widetilde{\varphi})$ extends continuously to a homeomorphism of $\mathbb{H}^3 \cup \overline{\mathbb{C}}$ and restricts to $\widetilde{\varphi}$ on $\overline{\mathbb{C}}$, and

(ii) for any isometries g, h of $\mathbb{H}^3$, $\widetilde{\Phi}(g \circ \widetilde{\varphi} \circ h) = g \circ \widetilde{\Phi}(\widetilde{\varphi}) \circ h$.

Definition. — The homeomorphism $\widetilde{\Phi}(\widetilde{\varphi})$ is called *the natural extension of* $\widetilde{\varphi}$.

Let us go back to the Proposition. The natural extension $\widetilde{\Phi}(\widetilde{\varphi})$ induces a homeomorphism $\Phi = \Phi(\varphi)$ from $\overline{M}(G)$ to $\overline{M}(\rho(G))$, called also the natural extension of φ. This proves (ii). Under our hypothesis, $\overline{M}(G)$ is compact. Therefore (ii) implies that $\overline{M}(\rho(G))$ is compact also. Thus, $\rho(G)$ is geometrically finite.

The Ahlfors-Bers map.

The Ahlfors-Bers theorem gives us a way to parametrize $\mathcal{GF}(M)$ by a product of Teichmüller spaces.

Notations. — Let G be a geometrically finite group such that the manifold $M = \overline{M}(G)$ has a non-empty and incompressible boundary. Denote by $(S_i)_{i=1,..,k}$ the components of the Riemann surface $\partial M = \Omega(G)/G$. For $1 \leq i \leq k$, choose a component Ω_i of the preimage of S_i in $\Omega(G)$. Denote by Γ_i' the stabilizer of Ω_i

in G. By Koebe's uniformization theorem, there is a conformal homeomorphism $f_i : \Omega_i \to \mathbf{H}^2$ which conjugates Γ_i' to a Fuchsian group $\Gamma_i \subset \mathrm{PSL}_2(\mathbf{R})$.

Let $(\rho, \widetilde{\varphi})$ be a quasi-conformal deformation of G. For $1 \leq i \leq k$, let $f_i' : \widetilde{\varphi}(\Omega_i) \to \mathbf{H}^2$ be a conformal homeomorphism which conjugates the action of $\rho(\Gamma_i')$ on Ω_i with the action of some Fuchsian group $\rho_i'(\Gamma_i)$ on $\mathbf{H}^2$. Then $f_i' \circ \widetilde{\varphi} \circ f_i^{-1}$ is a quasi-conformal homeomorphism. Up to post-composing f_i' with a Möbius map, we can suppose that $f_i' \circ \widetilde{\varphi} \circ f_i^{-1}$ is normalized. Then $(\rho_i', f_i' \circ \widetilde{\varphi} \circ f_i^{-1})$ is a Fuchsian deformation of Γ_i. One checks easily that the class of $(\rho_i', f_i' \circ \widetilde{\varphi} \circ f_i^{-1})$ in $\mathcal{T}(\Gamma_i)$ depends only on the class of $(\rho, \widetilde{\varphi})$ in $\mathcal{GF}(M)$.

Set $\mathcal{T}(\partial M) = \times_i \mathcal{T}(\Gamma_i)$.

Definition.— *The Ahlfors-Bers map* is the map from $\mathcal{GF}(M)$ to $\mathcal{T}(\partial M)$ which assigns to $(\rho, \widetilde{\varphi})$ the k-tuple whose i-th coordinate is the class of $(\rho_i', f_i' \circ \widetilde{\varphi} \circ f_i^{-1})$. It is denoted by ∂.

Theorem.— *∂ is a bijection.*

Proof.— Let $s = (s_1, \cdots, s_k) \in \mathcal{T}(\partial M)$. Choose $\mu_i \in \mathcal{B}^1(S_i)$ such that $s_i = (\rho_i', \varphi_{\mu_i})$. By taking the pull-back of each Beltrami form μ_i to the preimage of S_i in $\Omega(G)$, we define a function $\widetilde{\mu} \in L^\infty(\Omega(G))$ which satisfies

$$\widetilde{\mu}(\gamma(z)) \frac{\overline{\gamma'(z)}}{\gamma'(z)} = \widetilde{\mu}(z),$$

for all $\gamma \in G$. Extend $\widetilde{\mu}$ to a function $\bar{\mu} \in L^\infty(\mathbb{C})$ with norm strictly smaller than 1 by setting it equal to 0 on $L(G)$. Then the quasi-conformal homeomorphism $\phi_{\bar{\mu}}$ provided by the Ahlfors-Bers theorem conjugates G to a group $\rho(G)$. The uniqueness of the solution of the Beltrami equation implies

$$\partial(\rho, \phi_{\bar{\mu}}) = (s_1, \cdots, s_k).$$

This proves that ∂ is onto. The injectivity is more delicate. The proof uses the Ahlfors measure 0 theorem, which states that the limit set of any geometrically finite group has Lebesgue measure 0 or equals the whole sphere **[Ahl]**. □

Notations.— Let $s \in \mathcal{T}(\partial M)$. Let $(\rho, \widetilde{\varphi})$ such that $\partial(\rho, \widetilde{\varphi}) = s$. We will denote by G^s the group $\rho(G)$ and by M^s the manifold $\overline{M}(G^s)$. The point s will be then identified with ∂M^s, thought of as a hyperbolic metric on ∂M.

Ahlfors' lemma.

The following (particular case of a) lemma of Ahlfors compares the lengths of closed geodesics in ∂M^s and in $M(G^s)$. Its proof is an application of Koebe's $1/4$-theorem. We keep the same hypothesis and notations as for the definition of the Ahlfors-Bers map.

Lemma [Ahl].— *Let g be a closed curve on ∂M^s which is homotopic to a geodesic of length $\ell^s(g)$ for the hyperbolic metric ∂M^s. Then g is homotopic in M^s to a geodesic contained in $M(G^s)$ of length smaller than $2\ell^s(g)$.*

Quasi-Fuchsian groups.

Let Γ be a cocompact Fuchsian group. Then $L(\Gamma) = \overline{\mathbb{R}}$ and $\Omega(\Gamma)$ has two invariant components, the upper-half plane $\mathbb{H}^2$ and the lower half-plane $\overline{\mathbb{H}^2}$. Clearly, $\overline{M}(\Gamma)$ is homeomorphic to $(\mathbb{H}^2/\Gamma) \times [0,1]$; also, one component of $\partial\overline{M}(\Gamma)$ is conformally equivalent to $\mathbb{H}^2/\Gamma$ and the other to $\overline{\mathbb{H}^2}/\Gamma$. Thus the Ahlfors-Bers map leads to a new class of Kleinian groups, the quasi-Fuchsian groups.

Definition. — *A quasi-Fuchsian group is a geometrically finite G such that for some cocompact Fuchsian group Γ, there is a quasi-conformal deformation $(\rho, \widetilde{\varphi})$ of Γ with $G = \rho(\Gamma)$.*

If G is quasi-Fuchsian, $\overline{M}(G)$ is diffeomorphic to the product of a closed surface by an interval.

Maskit's theorem.

Same hypothesis and notations as for the definition of the Ahlfors-Bers map.

Let S be a component of ∂M. Then the inclusion $\pi_1(S) \subset \pi_1(M) \simeq G$ gives a representation $\rho : \pi_1(S) \to \mathrm{PSL}_2(\mathbb{C})$, which is faithful and has discrete image. With these notations, we have:

Theorem [Mas2]. — *The group $\rho(\pi_1(S))$ is quasi-Fuchsian.*

Remark. — This last result is a particular case of a theorem of Maskit which asserts that any finitely generated subgroup of a geometrically finite group G (eventually with parabolic elements) with $\Omega(G) \neq \emptyset$, is geometrically finite ([**Mor**], [**OP**]). However, this weaker statement will be sufficient for us: its most important application will be to define *the skinning map* (cf.§2).

The Kleinian groups that will appear during the proof of the Hyperbolization theorem are constructed by induction. It follows from this construction that the Maskit theorem can be checked for these groups by the same induction.

Remark. — An essential hypothesis of Maskit's theorem is $\Omega(G) \neq \emptyset$. A finitely generated subgroup of an arbitrary geometrically finite group is not necessarily geometrically finite. The basic example to keep in mind is the fundamental group of the fiber of a hyperbolic manifold which fibers over the circle. This is precisely the reason why the proof of Thurston's hyperbolization theorem needs to consider separately the cases of fibered manifolds and non-fibered manifolds.

One corollary of Maskit's theorem is the next result which will be used in §2.

Proposition. — *Let G be a geometrically finite group. Then, for all $\eta > 0$, $\Omega(G)$ has only a finite number of components with a diameter bigger than η.*

Hyperbolic annuli.

We give some formulas for the hyperbolic metric on annuli, which will be used in §3 and §4.

For $r < r'$, we denote by $A_{e^r,e^{r'}}$ the annulus:

$$A_{e^r,e^{r'}} = \{z \in \mathbb{C}, \quad e^r < |z| < e^{r'}\}.$$

Any hyperbolic annulus $\mathbb{H}^2/\langle\gamma\rangle$, where γ is an hyperbolic isometry, is conformally equivalent to an annulus A_{e^{-R},e^R}. The hyperbolic metric on A_{e^{-R},e^R} is given by:

$$ds = \frac{\pi|dz|}{2R|z|\cos(\pi\frac{\log|z|}{2R})}.$$

The circle of radius 1 is the only embedded closed geodesic of this annulus. Its hyperbolic length is $\ell(\gamma) = \pi^2/R$.

For $r < R$, the two circles of radius e^r and e^{-r} are equidistant curves to the circle of radius 1 at distance D such that

$$\tanh D = \sin(\pi\frac{r}{2R}).$$

The hyperbolic length of these two circles is

$$\frac{\pi^2}{R\cos(\pi\frac{r}{2R})}.$$

The injectivity radius of the hyperbolic metric of A_{e^{-R},e^R} is constant on the circle of radius e^r: if it is bounded above, say by $\varepsilon(2)$, the injectivity radius on the circle of radius e^r is equivalent to the hyperbolic length of this circle, independently of R.

CHAPTER 2

The fixed point problem

Let N be a connected and orientable closed 3-manifold. Let S be a (not necessarily connected) closed, orientable, incompressible surface embedded in N. Suppose that the Euler characteristic of each component of S is strictly negative. Denote by M the complement in N of an open regular neighborhood of S. We say that M is obtained *by splitting* N *along* S (cf. §7). Since S is incompressible, M has incompressible boundary.

The boundary of M is made of the union of two copies S_1 and S_2 of S. There is an orientation reversing diffeomorphism $f : S_1 \to S_2$ such that N is diffeomorphic to the quotient of M by the relation: $x \equiv y$ if and only if $x \in S_1$, $y \in S_2$ and $y = f(x)$. Rather than to consider the diffeomorphism $f : S_1 \to S_2$, it is more convenient to introduce the map $\tau : \partial M \to \partial M$ defined by $\tau(x) = f(x)$ for $x \in S_1$ and $\tau(y) = f^{-1}(y)$ for $y \in S_2$. Then τ is an orientation reversing involution of ∂M which permutes the components by pairs. And N is diffeomorphic to M/τ, i.e. to the quotient space of M by the equivalence relation

$$x \simeq y \text{ if and only if } x \in \partial M \text{ and } y = \tau(x).$$

The core of the proof of Thurston's hyperbolization theorem is the following result.

Final gluing theorem.— *Let M be a hyperbolic manifold with incompressible boundary which is not an interval bundle. Let τ be an orientation reversing involution of ∂M which permutes the components by pairs. If M/τ is atoroidal, then it is hyperbolic.*

In this chapter, we prove that if a certain map from $\mathcal{T}(\partial M)$ to itself has a fixed point, then the conclusion of the Final gluing theorem holds (M is assumed to satisfy the hypothesis in the Final gluing theorem).

2.1 Maskit's combination theorem

We keep the same notations as in the introduction of this chapter. But we suppose that S is connected and we don't suppose anymore that N is closed. Then M has two or one components according to whether S separates N or not. *We suppose also that N is not an interval bundle (cf. §7).*

Suppose that M is hyperbolic. Then there exists a geometrically finite group G_1 (resp. two geometrically finite groups G_1 and G_2) such that we can identify M with $\overline{M}(G_1)$ (resp. with the disjoint union of $\overline{M}(G_1)$ and $\overline{M}(G_2)$) in the case that S does not separate N (resp. separates N). Maskit's combination theorem provides a sufficient condition on G_1 (resp. on G_1 and G_2) which implies that N is hyperbolic. Since $S' = S_1 \cup S_2$ is incompressible in M, Maskit's theorem (cf. §1) says that the images of $\pi_1(S_1)$ and $\pi_1(S_2)$ in G_1 (resp. in G_1 and in G_2) are quasi-Fuchsian groups. We still denote these images by $\pi_1(S_1)$, $\pi_1(S_2)$. For $i = 1, 2$, we set $N_i = M(\pi_1(S_i))$ and $\overline{N}_i = \overline{M}(\pi_1(S_i))$. Since $\pi_1(S_i)$ is quasi-Fuchsian, $\overline{N}_i$ is diffeomorphic to $S_i \times [0,1]$. Under this diffeomorphism, one component of $\partial\overline{N}_i$ gets identified with S_i and will be still denoted by S_i. The diffeomorphism f induces a homotopy equivalence $\tilde{f} : N_1 \to N_2$. With these notations we have:

Maskit's combination theorem [Mas1].— *Assume that $\tilde{f}$ is homotopic to an isometry $\mathcal{J} : N_1 \to N_2$ whose extension to $\overline{N}_i$ satisfies $\mathcal{J}(S_1) = \partial\overline{N}_2 - S_2$. Then N is hyperbolic.*

Remark.— This theorem would not hold if $\overline{M}(G_1)$ and $\overline{M}(G_2)$ were twisted interval bundles.

Proof.— We consider the case when S separates N, the proof in the other case being similar. When one of the manifolds $\overline{M}(G_1)$ and $\overline{M}(G_2)$ is diffeomorphic to a trivial product, there is nothing to prove. We will suppose during the proof that say, $\overline{M}(G_1)$ is not a twisted interval bundle over a closed surface. Hence, only $\overline{M}(G_2)$ might be an interval bundle over a closed surface. Recall that $\overline{M}(G_i)$ is an interval bundle if and only if $\pi_1(S_i)$ has index one or two in G_i, if and only if $\Omega(G_i)$ has two components, see §7.

For $i = 1, 2$, $\overline{M}(G_i)$ is diffeomorphic to $N_\delta(G_i)$. Consider the covering $p_i : N_i \to M(G_i)$. The map p_i extends continuously to a map $N_i \cup S_i \to M(G_i) \cup S_i$ whose restriction to S_i is an embedding.

Let f_i be the (unique) harmonic function on N_i such that $f_i(x)$ tends to 1 when x tends to S_i and to 0 when x tends to $\partial\overline{N}_i - S_i$ (the existence of such a function is obtained by solving the corresponding Dirichlet problem in $\mathbb{H}^3$).

Suppose that $\overline{M}(G_i))$ is not an interval bundle. Then f_i has the following two properties:

(i) the level surface $f_i^{-1}(1/2)$ maps injectively into $M(G_i)$ under the covering map p_i. This follows from the maximum principle for harmonic functions and from the hypothesis on the index of $\pi_1(S_i)$ in G_i.

(ii) $f_i^{-1}(1/2)$ is compact. The reason is that $f_i(x)$ tends to 0 or 1 as x tends to ∞ in $M(G_i)$.

(iii) for $\delta > 0$, $f_i^{-1}(1/2)$ is contained in the interior of $N_\delta(G_i)$. This follows from the fact that $f_i^{-1}(1/2)$ is contained in the Nielsen core $N(G_i)$, by the maximum principle again.

Therefore, since p_i is a local homeomorphism, for each regular value c of f_i which is sufficiently close to $1/2$, $\Sigma_i = f_i^{-1}(c)$ is a compact surface embedded in the interior of $N_\delta(G_i)$ such that $p_i|\Sigma_i$ is an embedding.

Also, if c is chosen sufficiently close to $1/2$, $\Sigma_2 = f_2^{-1}(1-c)$ is an embedded compact surface in the interior of $N_\delta(G_2)$ which satisfies $p_2|\Sigma_2$ is an embedding.

When $\overline{M}(G_2)$ is an interval bundle, then (ii) and (iii) still hold, but (i) does not anymore. Let t be the deck transformation of the cover $N_2 \to M(G_2)$. Then t leaves $f_2^{-1}(1/2)$ invariant (inducing a degree two cover) and exchanges $f_2^{-1}([0,1/2[)$ with $f_2^{-1}(]1/2,1])$. Therefore, if $c < 1/2$ is a regular value of f_2 sufficiently close to $1/2$, $\Sigma_2 = f_2^{-1}(1-c)$ is a compact surface embedded in the interior of $N_\delta(G_2)$ such that p_2 restricts to $f_2^{-1}([1-c,1[)$ as an embedding. Since by hypothesis, the index of $\pi_1(S_1)$ in G_1 is greater than 2, we may suppose for this choice of c, that $\Sigma_1 = f_2^{-1}(c)$ is contained in the interior of $N_\delta(G_1)$ and that $p_1|\Sigma_1$ is an embedding.

By the maximum principle, no sum of components of Σ_i can be homologous to 0 in N_i. Therefore since Σ_i separates the two components of $\partial\overline{N}_i$, it is connected. Denote by H_i the submanifold of N_i bounded by Σ_i and whose closure in $\overline{N}_i$ contains S_i. The covering map p_i is an embedding when restricted to Σ_i and to the end of H_i approaching S_i. When $\pi_1(S_i)$ has index greater than 2 in G_i, it follows that $p_i|H_i$ is an embedding. When $\overline{M}(G_2)$ is an interval bundle, the same conclusion holds by the choice of Σ_2.

The hypothesis of the theorem and the uniqueness of the functions f_1 and f_2 imply: $f_2 \circ \mathfrak{J} = 1 - f_1$. Therefore $\Sigma_2 = \mathfrak{J}(\Sigma_1)$ and $\mathfrak{J}$ induces an orientation reversing diffeomorphism from Σ_1 to Σ_2 (Σ_i is oriented as boundary of H_i).

Let us consider now the manifold N' obtained as the result of the gluing of $M_1' = \overline{N_\delta(G_1) - p_1(H_1)}$ and $M_2' = \overline{N_\delta(G_2) - p_2(H_2)}$ identifying $p_1(\Sigma_1)$ and $p_2(\Sigma_2)$ by the diffeomorphism $p_2 \circ \mathfrak{J} \circ p_1^{-1}$. Then N' admits an atlas with charts modelled on convex sets of $\mathbb{H}^3$ (since a neighborhood of $\partial N'$ is isometric to a neighborhood of certain components of the boundary of the disjoint union of $N_\delta(G_1)$ and $N_\delta(G_2)$) and with coordinate changes in $\mathrm{PSL}_2(\mathbb{C})$. Therefore N' is a hyperbolic manifold.

In order to prove the Maskit combination theorem, it remains to prove that N' is diffeomorphic to N. For this, we could invoke Waldhausen's theorem on homotopy equivalences between Haken manifolds (cf. **[Mor]**). But since only a small part of this theorem is necessary, we prefer to explain this point.

Suppose first that $p_1(\Sigma_1)$ is incompressible in N'. Then Σ_i is incompressible in N_i. Therefore, by a theorem of Stallings **[Stal]**, $H_i \cup S_i$ is diffeomorphic to $S_i \times [0,1]$. Since $p_i|H_i$ is an embedding, this implies that M_i' is diffeomorphic to $\overline{M}(G_i)$. Under this diffeomorphism, $p_2 \circ \mathfrak{J} \circ p_1^{-1} : \Sigma_1 \to \Sigma_2$ is homotopic to

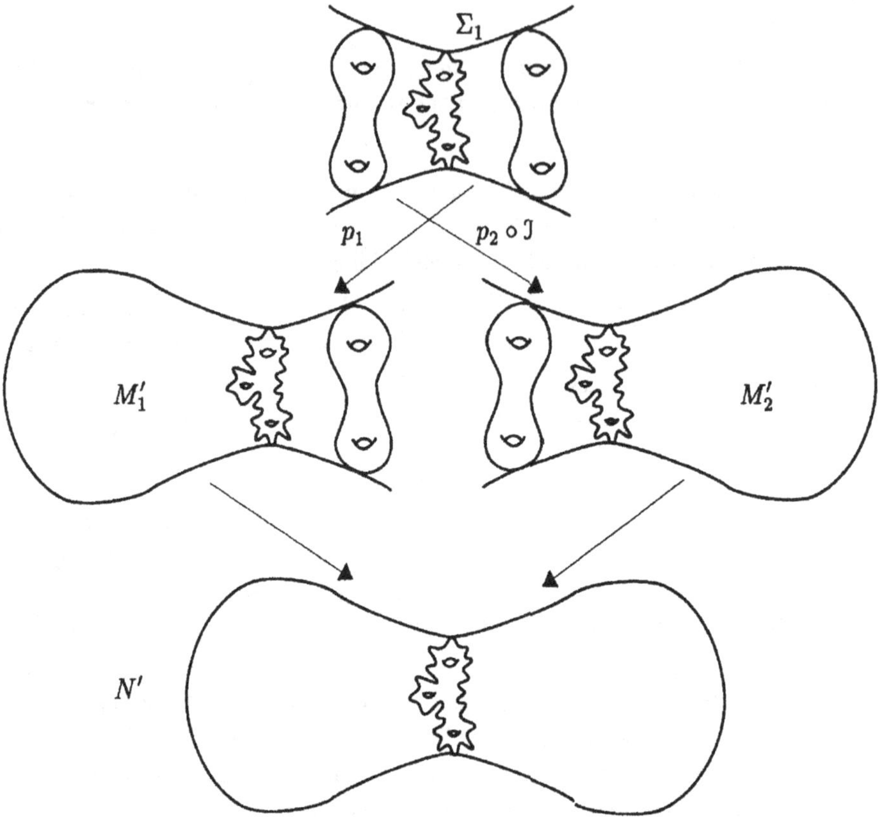

Figure 2.1

$f : S_1 \to S_2$. Thus, by a theorem of Nielsen [Nie], $p_2 \circ \mathfrak{J} \circ p_1^{-1}$ is isotopic to f. So N' is diffeomorphic to N.

When $p_1(\Sigma_1)$ is compressible in N', there exists, by Dehn's lemma, *a compression disc* for $p_1(\Sigma_1)$ in N', i.e. there is a disc D' embedded in N' which intersects $p_1(\Sigma_1)$ transversally and exactly along its boundary and such that the curve $\partial D'$ is not homotopic to 0 on $p_1(\Sigma_1)$ (cf. §7). Suppose that D' is contained, say in $M_1' \subset M(G_1)$. Then D' can be lifted isomorphically to a compression disc D for Σ_1 in N_1. Hence $p_1|(\Sigma_1 \cup D)$ is injective. Therefore, the restriction of p_1 to the union of H_1 and a regular neighborhood $N(D)$ of D is injective. The boundary of $H_1 \cup N(D)$ contains two or one components according to wether ∂D does or does not disconnect Σ_1. Let Σ_1' be the component of $\partial(H_1 \cup N(D))$ which is homologuous to Σ_1. If it exists, the other component, denoted by B_1, is homologuous to 0. It bounds therefore a compact (connected) manifold Z_1 in N_1. The surface Σ_1' cuts N_1 into two components, one of which, denoted by H_1', contains H_1. We prove that $p_1|H_1'$ is an embedding. Clearly, H_1' equals the union along B_1 of $H_1 \cup N(D)$ and Z_1. The surface $p_1(B_1)$ cuts $M(G_1)$ into two components: one is unbounded and contains $p_1(H_1)$, the other, denoted by T_1, is compact. The image $p_1(Z_1)$ is compact. Since p_1 is an open map, $p_1(Z_1)$ contains T_1. Since p_1 is an open map and since $M(G_1)$ is not compact, $p_1(Z_1)$ equals T_1. It follows that p_1 is an embedding when restricted to Z_1 and therefore also when restricted to H'_1. The surface $\Sigma'_2 = \mathfrak{I}(\Sigma'_1)$ cuts N_2 into two components, one of which, denoted by H'_2, is contained in H_2. Certainly p_2 restricts to H'_2 as an embedding since it does already on H_2. Also, for some constant $\delta' > 0$, $p_1(D)$ and $p_2(D)$ are contained respectively in $N_\delta(G_1)$ and $N_\delta(G_2)$. Therefore, up to replacing in the original description of N', $p_1(\Sigma_i)$ by $p_1(\Sigma'_i)$ and δ by δ', we obtain a hyperbolic manifold N'' (diffeomorphic to N').

The genus of $p_1(\Sigma'_1)$ is strictly smaller than the genus of $p_1(\Sigma_1)$ since ∂D is not homotopic to 0 on $p_1(\Sigma_1)$. Hence this process ends up, after a finite number of steps, with the case when the gluing surface is incompressible, bringing a hyperbolic manifold diffeomorphic to N. □

2.2 The skinning map

The skinning map, introduced by Thurston, allows us to formulate the hypothesis of Maskit's combination theorem as the existence of a fixed point for a certain map defined on a Teichmüller space. In this section, we define this map and we enunciate the Fixed point theorem.

We keep the same notations and hypothesis as for the definition of the Ahlfors-Bers map. For $1 \le i \le k$, Γ'_i is a quasi-Fuchsian group by Maskit's theorem (cf. §1). Thus $\Omega(\Gamma'_i)$ has two connected components, Ω_i and $\overline{\Omega}_i$. Let Γ_i be a Fuchsian group such that S_i is conformally equivalent to $\mathbb{H}^2/\Gamma_i$. For each i, there is a quasi-conformal homeomorphism f_i of $\overline{\mathbb{C}}$ such that:

(i) $f_i(\mathbb{H}^2) = \Omega_i$;

(ii) $f_i \circ \gamma \circ f_i^{-1} = \gamma'$, for all $\gamma \in \Gamma_i$.

Set $\mathcal{T}(\partial M) = \times_i \mathcal{T}(\Gamma_i)$ and $\mathcal{T}(\overline{\partial M}) = \times_i \mathcal{T}(\overline{\Gamma_i})$. Let $s \in \mathcal{T}(\partial M)$ and let $(\rho, \widetilde{\varphi})$ be the quasi-conformal deformation of G such that $s = \partial(\rho, \widetilde{\varphi})$. Then $\rho(\Gamma'_i)$ is conjugated to Γ_i by the quasi-conformal homeomorphism $\widetilde{\varphi} \circ f_i$. This defines a point in $\mathcal{GF}(\Gamma_i)$. By the Ahlfors-Bers theorem again, $\mathcal{GF}(\Gamma_i)$ is parametrized by the product $\mathcal{T}(\Gamma_i) \times \mathcal{T}(\overline{\Gamma_i})$. The first Ahlfors-Bers coordinate of $(\rho|\Gamma'_i, \widetilde{\varphi} \circ f_i)$ is s_i. We denote the second coordinate by s'_i.

Definition. — *The skinning map associated to* M is the map $\sigma : \mathcal{T}(\partial M) \to \mathcal{T}(\overline{\partial M})$ defined by

$$\sigma(s) = (s'_1, ... s'_k).$$

The Riemann surface s_i can be interpreted as the "outside" structure on the component S_i, whether s'_i is the "inside" structure, i.e. the one which appears when one takes off the "skin" of the manifold M. Another way is to consider the covering of M having fundamental group $\pi_1(S_i)$. This covering is homeomorphic to $S_i \times \mathbb{R}$. Any quasi-conformal deformation of G with Ahlfors-Bers parameters equals to $(s_1, \cdots, s_k)$ can be lifted to a quasi-conformal deformation of this covering. The parameters of this deformation are (s_i, s'_i).

Notation.— Let $s \in \mathcal{T}(\partial M)$. We will denote $\sigma(s)$ by $\sigma(\partial M^s)$, thinking of $\sigma(\partial M^s)$ as a hyperbolic metric on $\overline{\partial M}$ (in the same way as we think of s as a hyperbolic metric on ∂M, cf. §1).

A reformulation of Maskit's combination theorem.

We use now the skinning map to formulate in a different way the hypothesis of Maskit's combination theorem. We keep the same notations as in the statement of this theorem. Consider first the particular case when S separates N into two components.

Let σ_i be the skinning map associated to M_i. Let $(s_i, z_i) \in \mathcal{T}(\partial M_i)$ where s_i denotes the coordinate on the factor S_i and z_i denotes the coordinates on the components of ∂M_i others than S_i. Fix z_1 and z_2 and denote by $\sigma_i^1(s_i)$ the coordinate of $\sigma_i(s_i, z_i)$ on the factor $\overline{S_i}$. Consider the quasi-conformal deformation $M_i^{(s_i,z_i)}$ of M_i. Then N_i is deformed to the point with Ahlfors-Bers coordinates $(s_i, \sigma_i^1(s_i))$. The hypothesis of Maskit's combination means exactly that

$$(1) \qquad f^*(s_1) = \sigma_2^1(s_2) \quad \text{and} \quad f^* \circ \sigma_1^1(s_1) = s_2.$$

When (1) is satisfied, Maskit's combination theorem asserts that $M_1^{(s_1,z_1)}$ and $M_2^{(s_2,z_2)}$ can be "glued together", yielding a hyperbolic manifold diffeomorphic to N. (1) can be stated in a more symmetric way. The diffeomorphism τ from $S_1 \cup S_2$ to itself defined by $\tau(x,y) = (f^{-1}(y), f(x))$. induces a map $\tau^* : \mathcal{T}(\overline{S_1 \cup S_2}) \to \mathcal{T}(S_1 \cup S_2)$. It is straightforward to check that (i) means precisely that $\tau^*(\sigma_1^1(s_1), \sigma_2^1(s_2)) = (s_1, s_2)$, or in other terms, that (s_1, s_2) is a fixed point of the composition of τ^* with (σ_1^1, σ_2^1).

This particular case extends similarly to the case when S is connected but does not separate N.

Consider now the situation of the Final gluing theorem, i.e. when S is not necesarily connected. Then M is the disjoint union of hyperbolic manifolds with incompressible boundary $M_1, \cdots M_\ell, \cdots, M_p$.

Definition.— The Teichmüller space $\mathcal{T}(\partial M)$ (resp. $\mathcal{T}(\overline{\partial M})$) is defined as the product of the Teichmüller spaces $\mathcal{T}(\partial M_i)$ (resp. $\mathcal{T}(\overline{\partial M_i})$). *The skinning map* $\sigma : \mathcal{T}(\partial M) \to \mathcal{T}(\overline{\partial M})$ is the product of the skinning maps σ_i associated to M_i.

Let τ be an orientation reversing diffeomorphism of ∂M which permutes the components by pairs. Then τ induces a map $\tau^* : \mathcal{T}(\overline{\partial M}) \to \mathcal{T}(\partial M)$.

With these notations, the following theorem is merely an extension to this general case of the Maskit combination theorem in the formulation given above. Its proof follows from the original statement by induction on the number of components of S.

Theorem.— *If $\tau^* \circ \sigma$ has a fixed point, then M/τ is hyperbolic.*

Using that, the Final gluing theorem becomes equivalent to the following:

Thurston's fixed point theorem.— *Let M be a hyperbolic manifold with incompressible boundary which is not an interval bundle. Let τ be an orientation reversing* *involution of ∂M which permutes the components by pairs. If M/τ is atoroidal, then $\tau^* \circ \sigma$ has a fixed point.*

We conclude this section by computing the skinning map associated to a connected hyperbolic manifold M which is an interval bundle. There are only two possibilities for M up to diffeomorphism. It is either the product of a closed orientable surface S with the interval $[0,1]$ or *the* twisted interval bundle over a non-orientable surface T (cf. §7).

In the first case, we can identify S with the quotient $\mathbf{H}^2/\Gamma$ for some Fuchsian group Γ. Then $\mathcal{T}(\partial M) = \mathcal{T}(\Gamma) \times \mathcal{T}(\overline{\Gamma})$ and the skinning map is given by $\sigma((x,y)) = (y,x)$. In the second case, ∂M is identified with the orientation cover S of T. The deck transformation t of this cover reverses the orientation on S and induces therefore a map $t^* : \mathcal{T}(\Gamma) \to \mathcal{T}(\overline{\Gamma})$. By definition, we have: $\sigma = t^*$.

The computation in these two cases shows that σ is an isometry when $\mathcal{T}(\partial M)$ and $\mathcal{T}(\overline{\partial M})$ are endowed with their respective Teichmüller distances. Hence when M is diffeomorphic to an interval bundle, σ is an isometry. However if M is connected and is not an interval bundle, then σ is contracting. We will understand why it is so soon in this chapter, when we compute the derivative and the coderivative of σ. There is another way to prove this contraction property, by using the Teichmüller theorem which describes *the extremal quasi-conformal map* between two homeomorphic Riemann surfaces (cf. [**Mor**]).

2.3 The derivative and the coderivative of σ

McMullen's proof of Thurston's fixed point theorem entails a detailed analysis of the derivative of $\tau^* \circ \sigma$. The derivative and coderivative of σ at a point $s \in \mathcal{T}(\partial M)$ are expressed in terms of the geometry of the skinned surface $\sigma(\partial M^s)$. Like a leopard skin, $\sigma(\partial M^s)$ is covered by spots.

The leopard spots.

Same hypothesis and notations as for the definition of the skinning map. Let $s \in \mathcal{T}(\partial M)$.

Definition.— The image of a component $\widetilde{U}$ of $\Omega(G^s) \cap \widetilde{\varphi}(\widetilde{\Omega}_i)$ in the surface $\widetilde{\varphi}(\widetilde{\Omega}_i)/\rho(\Gamma'_i)$ is called a *spot.*

Notation.— Let $U \subset \sigma(\partial M^s)$ be a spot covered by a component $\widetilde{U}$ of $\Omega(G^s)$. Let $\Gamma'_{\widetilde{U}}$ be the stabilizer of $\widetilde{U}$ in G^s. Then U is conformally equivalent to the quotient $\widetilde{U}/\Gamma'_{\widetilde{U}} \cap \rho(\Gamma'_i)$. Hence U covers the component $\widetilde{U}/\Gamma'_{\widetilde{U}}$ of ∂M^s (cf. Figure 2.2). We denote this component by X_U.

Remark.— The quasi-conformal homeomorphism $\widetilde{\varphi}$ projects to a homeomorphism between $\sigma(\partial M)$ and $\sigma(\partial M^s)$ which maps the spots on $\sigma(\partial M)$ to the spots on $\sigma(\partial M^s)$. In particular, the topological configuration of the spots on $\sigma(\partial M^s)$ does not depend on s. Since $\widetilde{\varphi}$ conjugates G with G^s, the topological types of the covers $U \to X_U$ associated to the spots do not depend on s either.

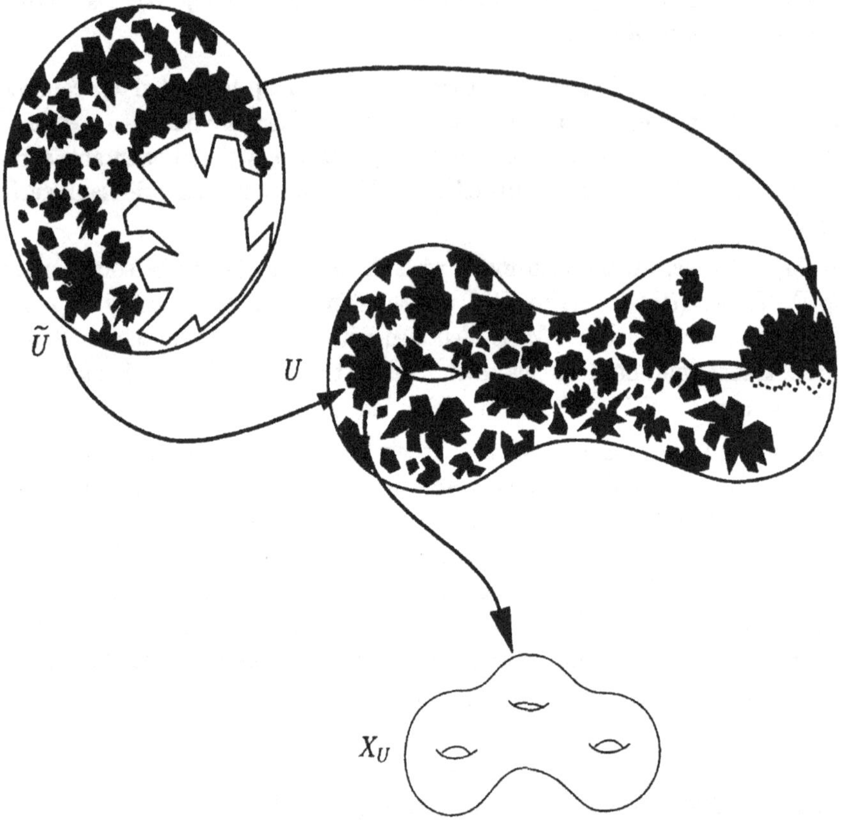

Figure 2.2

Since the components of $\Omega(G)$ are simply connected, the inclusion of each spot in $\sigma(\partial M)$ induces an injection on the fundamental group.

The coderivative of σ.

Notation.— For a spot U contained in $\sigma(\partial M^s)$, we denote by Θ_U the Theta operator Θ_{U/X_U} associated to the cover $U \to X_U$.

Let $\phi \in \mathcal{Q}(\sigma(\partial M^s))$. Since the restriction ϕ_U of ϕ to U is integrable, we can define a holomorphic quadratic differential $\Theta_U \phi_U$. It is an element of $\mathcal{Q}(\partial M^s)$ with norm less than or equal to $||\phi_U||$ (cf. §1). Thus, we can sum the differentials $\Theta_U \phi_U$ when U varies over all the spots contained in $\sigma(\partial M^s)$. This defines an element of $\mathcal{Q}(\partial M^s)$, denoted by $\sum_U \Theta_U \phi_U$.

The derivative of σ.

The tangent space to $\mathcal{T}(\partial M)$ (resp. to $\mathcal{T}(\sigma(\partial M))$) at s is isomorphic to the quotient $\mathcal{B}(\partial M^s)/\mathcal{N}(\partial M^s)$ (resp. to $\mathcal{B}(\sigma(\partial M^s))/\mathcal{N}(\sigma(\partial M^s))$. Let $\mu \in \mathcal{B}(\partial M^s)$. Let $\tilde{\mu}(z)\overline{dz}/dz$ be the pull-back of μ under the covering $\Omega(G^s) \to \partial M^s$. By setting

$\bar{\mu} \equiv 0$ on $L(\Omega(G^s))$, we obtain an element $\bar{\mu} \in L^\infty(\mathbb{C})$ such that

$$\bar{\mu}(\gamma(z))\frac{\overline{\gamma'(z)}}{\gamma'(z)} = \bar{\mu}(z),$$

for almost all $z \in \mathbb{C}$ and for all $\gamma \in G^s$. In particular for all i, the restriction of $\bar{\mu}$ to $\widetilde{\varphi}(\widetilde{\Omega}_i)$ is invariant under $\rho(\Gamma_i')$. It defines therefore a Beltrami form $\widetilde{\mu}_i$ on $\widetilde{\varphi}(\widetilde{\Omega}_i)/\rho(\Gamma_i')$. In this way, we obtain a Beltrami form $\widetilde{\mu} = (\widetilde{\mu}_i)$ on $\sigma(\partial M^s)$.

With the above notations, we have:

Proposition 2.1. —

(i) *$\mu \to \widetilde{\mu}$ induces a map from $\mathcal{B}(\partial M^s)/\mathcal{N}(\partial M^s)$ to $\mathcal{B}(\sigma(\partial M^s))/\mathcal{N}(\sigma(\partial M^s))$ which is the derivative of σ at s, and*

(ii) *the coderivative of σ at s is given by*

$$d_s^*\sigma\phi = \sum_U \Theta_U \phi_U.$$

Proof. — Let $\Pi : \mathcal{B}^1(\partial M) \to \mathcal{T}(\partial M)$, and $\overline{\Pi} : \mathcal{B}^1(\sigma(\partial M)) \to \mathcal{T}(\sigma(\partial M))$ be the projections defined in §1. Recall that Π (resp. $\overline{\Pi}$) induces an isomorphism between $\mathcal{B}(\partial M)/\mathcal{N}(\partial M)$ (resp. $\mathcal{B}(\sigma(\partial M))/\mathcal{N}(\sigma(\partial M))$) and the tangent space to $\mathcal{T}(\partial M)$ (resp. $\mathcal{T}(\sigma(\partial M))$) at ∂M (resp. $\sigma(\partial M)$). Choose a local differentiable section ℓ of Π, defined in an open neighborhood $\mathcal{V}$ of ∂M in $\mathcal{T}(\partial M)$ (cf. §1). It follows from the definition of σ that we have, for $s \in \mathcal{V}$: $\sigma(s) = \overline{\Pi} \circ \widetilde{\ell(s)}$. This implies that the map $\mu \to \widetilde{\mu}$ projects to a map from $\mathcal{B}(\partial M)/\mathcal{N}(\partial M)$ to $\mathcal{B}(\sigma(\partial M))/\mathcal{N}(\sigma(\partial M))$ which is the derivative of σ at ∂M. The differentiablity of σ at an arbitrary point follows from this special case, by using the naturality of the differentiable structure on $\mathcal{T}(\partial M)$ (cf. §1).

Let $\mu \in \mathcal{B}(\partial M^s)$. Denote by $\widetilde{\mu}_U$ the restriction of $\widetilde{\mu}$ to U, and by $\langle\cdot,\cdot\rangle_U$ the pairing between holomorphic quadratic differentials and Beltrami forms on U (cf. §1). By definition, $\widetilde{\mu}$ vanishes in the complement of the spots. Thus, for any $\phi \in \mathcal{Q}(\sigma(\partial M^s))$, we have

$$\langle\widetilde{\mu}, \phi\rangle = \sum_U \langle\widetilde{\mu}_U, \phi_U\rangle_U,$$

where the sum carries over all the spots $U \subset \sigma(\partial M^s)$. By definition, $\widetilde{\mu}_U$ is the pull-back of μ under the covering $U \to X_U$. Since the pull-back operator on Beltrami forms is the adjoint of the Theta operator, we have

$$\langle\widetilde{\mu}_U, \phi_U\rangle_U = \langle\mu, \Theta_U\phi_U\rangle_{X_U}.$$

Therefore

$$\langle\widetilde{\mu}, \phi\rangle = \langle\mu, \sum_U \Theta_U\phi_U\rangle.$$

This proves Proposition 2.1 (ii). □

Proposition 2.1 shows the relation between σ and the Theta operators associated to the spots. It implies that, for $s \in \mathcal{T}(\partial M)$ and $\phi \in \mathcal{Q}(\sigma(\partial M^s))$, we have

$$||d_s^* \sigma \phi|| \leq \sum_U ||\Theta_U|| ||\phi_U||.$$

In particular $||d_s^* \sigma|| \leq \sup_U ||\Theta_U||$. In order to study more precisely the contraction properties of σ, we will first establish related contraction properties for the operators Θ_U. We describe now the topological type of the covers associated to the spots.

The leopard spots and the topology of M.

If M is an interval bundle over a closed surface, each component of $\sigma(\partial M^s)$ is a spot and the cover associated to it is trivial: then $d^* \sigma$ is clearly an isometry (in that case, we had already noticed that σ was an isometry).

Therefore, we assume in the rest of this section that M is not an interval bundle. Since this topological type of the covers associated to the spots does not depend on s, we may suppose $M^s = M$.

Definition.— Let S be a compact surface contained in a connected surface X. We say that S is *incompressible* if it is connected and if $\pi_1(S)$ injects into $\pi_1(X)$. A cover $Y \to X$ between connected surfaces is *geometric* if $\pi_1(Y)$, viewed as a subgroup of $\pi_1(X)$, is equal to the fundamental group of a proper incompressible surface $S \subset X$. In this case we say that *the cover $Y \to X$ is associated to S*.

Examples.— The universal cover of any surface is geometric: it is associated to a disc. A non-trivial finite cover of a geometric cover is not geometric.

Proposition 2.2.—

(i) *Let $U \subset \sigma(\partial M)$ be a spot. Then the cover $U \to X_U$ is geometric.*

(ii) *All but finitely many spots in $\sigma(\partial M)$ are simply connected.*

Proof.— We keep the same notations that we used to define the spots. Let S_i be the component of ∂M such that U is contained in $\sigma(S_i)$. To simplify the notations we denote Ω_i by Ω, $\widetilde{\Omega}_i$ by $\widetilde{\Omega}$ and Γ_i' by Γ'. Then Γ' is a quasi-Fuchsian group and $\Omega(\Gamma') = \Omega \cup \widetilde{\Omega}$. Let $\widetilde{U} \subset \widetilde{\Omega}$ be a component of the preimage of U. Denote by Γ the stabilizer of $\widetilde{U}$ in G. Since $U = \widetilde{U}/\Gamma \cap \Gamma'$ and $X_U = \widetilde{U}/\Gamma$, the cover $U \to X$ corresponds to the subgroup $\Gamma \cap \Gamma'$ of $\pi_1(X_U) \simeq \Gamma$. Therefore, (i) is equivalent to say that $\Gamma \cap \Gamma'$ is the fundamental group of a proper incompressible surface of X_U.

Since Γ' (resp. Γ) is quasi-Fuchsian, the frontier of Ω (resp. $\widetilde{U}$) in $\overline{\mathbb{C}}$ equals the Jordan curve $L(\Gamma')$ (resp. $L(\Gamma)$). Consider the closed set $F = L(\Gamma) \cap L(\Gamma')$, which is invariant under $\Gamma' \cap \Gamma$. We note first that F is a proper subset of $L(\Gamma)$. For if F were equal to $L(\Gamma)$, then $\Omega(G)$ would be the disjoint union of Ω and $\widetilde{U}$. This would imply that Γ' is a subgroup of G of index at most 2, i.e. that M is an interval bundle (cf. §7).

Let $\gamma \in \Gamma$. We have $\gamma(\widetilde{\Omega}) \cap \widetilde{\Omega} = \emptyset$ if $\gamma \notin \Gamma \cap \Gamma'$ and $\gamma(\widetilde{\Omega}) = \widetilde{\Omega}$ if $\gamma \in \Gamma \cap \Gamma'$. This implies that F *and* $\gamma(F)$ *are not linked on* $L(\Gamma)$, i.e. that no pair of points in

F alternates on $L(\Gamma)$ with a pair of points in $\gamma(F)$. We consider now distinct cases according to the cardinality of F.

1) $\sharp F \geq 2$.

Let $C(F)$ denote the convex hull of F in $\widetilde{U} \cup L(\Gamma)$ for the hyperbolic metric on $\widetilde{U}$.

Lemma 2.3.— *Under the covering $\widetilde{U} \to X_U$, the frontier of $C(F)$ in $\widetilde{U}$ maps to a disjoint union of embedded closed geodesics.*

Proof.— Let $\widetilde{g}$ be a geodesic in the frontier of $C(F)$ in $\widetilde{U}$. Since, for $\gamma \in \Gamma$ the endpoints of $\widetilde{g}$ do not alternate on $L(\Gamma)$ with the endpoints of $\gamma(\widetilde{g})$, $\widetilde{g}$ maps to a geodesic g on X_U without transverse self-intersections. Since two distincts translates of F are not linked, two distincts translates of $C(F)$ intersect atmost along their frontiers; therefore, $\partial C(F)$ maps to a disjoint union of embedded geodesics. Suppose for a contradiction that for some geodesic $\widetilde{g} \subset \partial C(F)$, g is not a closed curve. Then g is not compact. Hence there is an infinite sequence (γ_k) of elements of Γ such that the geodesics $\gamma_k(\widetilde{g})$ are distincts and that their endpoints accumulate on two distincts points of $L(\Gamma)$. Since the endpoints of $\widetilde{g}$ are contained in the frontier of F in $L(\Gamma)$, the components $\gamma_k(\Omega)$ of $\Omega(G)$ are all distincts for sufficiently large k. Then $\Omega(G)$ has an infinity of components with diameter bounded from below by a non-zero constant. This is impossible (cf. §1). □

1a) $\sharp F = 2$.

Then $C(F)$ is a geodesic and, by Lemma 2.3 its projection is an embedded closed curve. Let $\gamma \in \Gamma$ be the element represented by this geodesic. Then γ leaves also Ω invariant. For, if this is not true, the components $(\gamma)^n(\Omega)$ of $\Omega(G)$ are all distinct and their diameter is bigger than the diameter of F. This is impossible (cf. §1). Thus, $\gamma \in \Gamma'$ and $\Gamma \cap \Gamma'$ is equal to the cyclic group generated by γ. This proves Proposition 2.2 (i) in this case.

1b) $\sharp F > 2$.

In this case, $C(F)$ has non-empty interior. Let Σ' be the projection of $C(F)$ to X_U. By Lemma 2.3 and since X_U is compact, the projection of $\partial C(F)$ to X_U is the disjoint union of a finite number of embedded closed geodesics. Therefore, Σ' is a compact connected surface with geodesic boundary (however, the projection of some components of $\partial C(F)$ is maybe contained in the interior of Σ'). The complement in Σ' of an open regular neighborhood of the projection of $\partial C(F)$ defines an incompressible surface S. Let $\widetilde{S}$ be the component of the preimage of S that is contained in $C(F)$. Since $\widetilde{S}$ is contained in the interior of $C(F)$, we have, for $\gamma \in \Gamma$: $\gamma(\widetilde{S}) = \widetilde{S}$ if $\gamma \in \Gamma \cap \Gamma'$ and $\gamma(\widetilde{S}) \cap \widetilde{S} = \emptyset$ if $\gamma \notin \Gamma \cap \Gamma'$. Therefore $\Gamma \cap \Gamma'$ equals $\pi_1(S)$ (up to conjugacy). This proves Proposition 2.2 (i) in this case.

2) $F = \emptyset$.

Then, for any non-zero $\gamma \in \Gamma'$, we have $\gamma(\widetilde{U}) \cap \widetilde{U} = \emptyset$. Hence $\widetilde{U}$ is homeomorphic to U which is therefore simply connected. Thus the cover $\widetilde{U} \to X_U$ is geometric.

3) $F = \{f\}$.

We show that this cannot happen in our case — when G does not contain parabolic elements (if G contained parabolics elements, it could happen that F were reduced to one point; a slight modification of the next argument could prove however Proposition 2.2 (i) in this situation as well).

Let $g \subset \widetilde{\Omega}/\Gamma'$ be the projection of the Jordan arc $L(\Gamma) - \{f\}$. Suppose that g is compact. Then g is a closed curve which is not homotopic to 0 since $L(\Gamma) - \{f\}$ is not compact. Thus g is homotopic to a closed geodesic, and any of its lifts to $\widetilde{\Omega}$ accumulates to the two fixed points of some hyperbolic element of Γ' (since G does not contain parabolics elements). This contradicts the fact that $L(\Gamma) - \{f\}$ accumulates to f.

Thus g is non-compact. Since $\widetilde{\Omega}/\Gamma'$ is compact, there exists a sequence of distincts elements $\gamma_k \in \Gamma'$ such that $\gamma_k(L(\Gamma) - \{f\})$ accumulates to a point $p \in \widetilde{\Omega}$. Then the domains $\gamma_k(\widetilde{U})$ are distinct components of $\Omega(G)$ and their diameter is bigger than the distance from p to $L(\Gamma')$. This excludes the case 3) and finishes the proof of Proposition 2.2 (i).

To prove (ii), recall that, for any spot $U \subset \widetilde{\Omega}/\Gamma'$, $\pi_1(U)$ maps injectively into Γ'. In particular, each spot is homeomorphic to the interior of a compact surface with boundary. Only finitely many spots can have strictly negative Euler characteristic, since such a spot contributes at least -1 to the Euler characteristic of $\widetilde{\Omega}/\Gamma'$. To exclude the presence of an infinity of spots which are homeomorphic to annuli, we argue by contradiction. Then there are also infinitely many spots $U_i \subset \widetilde{\Omega}/\Gamma'$ which are homotopic to the same simple closed curve c. These spots can be lifted in $\widetilde{\Omega}$ to distincts components of $\Omega(G)$ which have the same endpoints as some lift $\widetilde{c}$ of c. Since their diameter is bigger than the distance between the two endpoints of $\widetilde{c}$, we obtain a contradiction. This finishes the proof of (ii). □

The next result can be proven with the same arguments as Proposition 2.2.

Corollary 2.4.— *Let $\widetilde{\Omega}$ and $\widetilde{U}$ be components of $\Omega(G)$ with stabilizers Γ' and Γ respectively. Let $\gamma \in \Gamma$ be a hyperbolic element which has one fixed point in $L(\Gamma) \cap L(\Gamma')$. Then $\gamma \in \Gamma' \cap \Gamma$.* □

Acylindricity.

The hypothesis that all spots are simply connected will introduce an important dichotomy in the proof of Thurston's fixed point theorem. We now show that this situation reflects a topological property of M, namely that M *is acylindrical.*

Definition.— Let A denote the annulus $\mathbb{S}^1 \times [0,1]$. Let M be a compact 3-manifold. A continuous map $f : (A, \partial A) \to (M, \partial M)$ is *essential* if it induces an injective map on $\pi_1(A)$ and on $\pi_1(A, \partial A)$. The image $f(A)$ is *an essential annulus.*

We say that M is *acylindrical* if it does not contain any essential annulus.

Fact 2.5.— *The manifold $M = \overline{M}(G)$ is acylindrical if and only if all the spots contained in $\sigma(\partial M)$ are simply connected.*

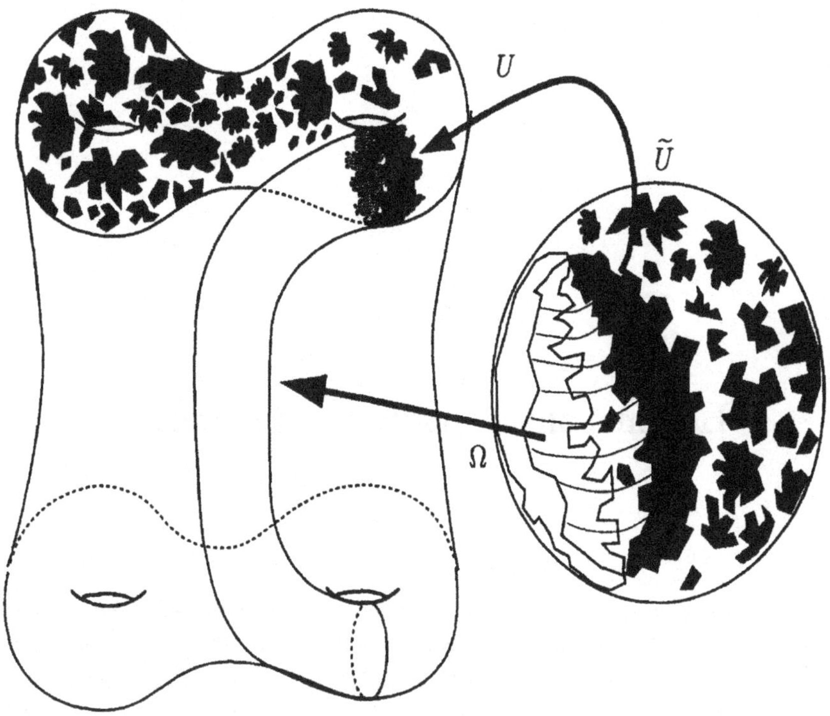

Figure 2.3

Proof. — Suppose that M is not acylindrical. Let $f : A \to M$ be an essential map. Let f^* be the map induced by f on the fundamental group. Then $f^*(\pi_1(A))$ is a cyclic group generated by a hyperbolic element g. Let $\widetilde{f} : \widetilde{A} \to \widetilde{M}$ be a lift of f to the universal cover. Since f induces an injection of $\pi_1(A, \partial A)$, the two components of $\widetilde{f}(\partial \widetilde{A}))$ are contained in distincts components Ω and Ω' of $\Omega(G)$. Then, the intersection of the closures of Ω and Ω' in $\overline{\mathbb{C}}$ contains the two fixed points of a conjugate of g. It follows from the proof of Proposition 2.2 that the image of Ω' on the component of $\sigma(\partial M)$ which is covered by $\widehat{\Omega} = \overline{\mathbb{C}} - \overline{\Omega}$ is a spot that is not simply connected.

Conversely let $U \subset \sigma(\partial M)$ be a spot which is not simply connected. Suppose that U is contained in the component $\sigma(S_i)$ of $\sigma(\partial M)$. Let $\widetilde{U} \subset \widehat{\Omega}_i$ be a component of the preimage of U. Since U is not simply connected, and since $\pi_1(U)$ maps injectively into $\pi_1(S_i) = \Gamma'_i$, $\widetilde{U}$ is invariant by a non-zero element of Γ'_i. Then the intersection of $\widetilde{U}$ with $L(\Gamma'_i)$ contains the two fixed points of this element and it is easy to construct an essential map from A into M (cf. Figure 2.3). □

CHAPTER 3

Holomorphic quadratic differentials

In this chapter, we study triples (X, x, ϕ) where X is a connected Riemann surface, $x \in X$ and $\phi \in \mathcal{Q}(X)$. We won't describe the topology on the set of these triples in all its generality. The reader is referred to **[McM1]** for the general definition. We will rather explain, in an elementary way, what it means for a sequence of such triples to converge, so that the proof of the main theorem, Theorem 3.1, can be reduced to classical compactness theorems on holomorphic functions.

3.1 Compactness properties of holomorphic quadratic differentials

Definition.— *A pointed Riemann surface* is a pair (X, x) where X is connected Riemann surface and $x \in X$.

We recall first how the space of pointed compact hyperbolic Riemann surfaces can be compactified (**[Thu1]**,**[Mu]**).

Limits of pointed Riemann surfaces.

Consider a sequence (X_i, x_i) of pointed compact hyperbolic Riemann surfaces with fixed topological type. The behaviour of this sequence, viewed as a sequence of pointed metric spaces where X_i is endowed with the hyperbolic metric depends on *the injectivity radius of X_i at x_i*. This is the largest radius of an open embedded hyperbolic ball centered at x_i. It is denoted by $\mathrm{inj}(x_i)$. Since the topological type of X_i is fixed, the hyperbolic volume of X_i is constant: in particular $\mathrm{inj}(x_i)$ is bounded from above by a constant depending only on $\chi(X_i)$. We distinguish two cases according to whether $\mathrm{inj}(x_i)$ is bounded from below by a non-zero constant or not.

a) $\mathrm{inj}(x_i)$ *is bounded from below away from* 0.

Let us identify X_i with the quotient of $\mathbb{D}^2$ by a Fuchsian group Γ_i in such a way that the origin $0 \in \mathbb{D}^2$ maps to x_i. By definition, *the Dirichlet domain of* Γ_i *with respect to* 0 is the set of points closer to 0 than to any of its translates by non-zero elements of Γ_i. It is a finite sided convex polygon $\mathcal{D}_i$. By an Euler characteristic argument, the number of sides of $\mathcal{D}_i$ is bounded independently of i. The hypothesis on $\mathrm{inj}(x_i)$ implies that $\mathcal{D}_i$ contains a ball centered at 0 of radius independent of i. Therefore, up to extracting a subsequence, $(\mathcal{D}_i)$ converges to a finite sided polygon (having perhaps some vertices on the circle at infinity) which is the Dirichlet domain of a discrete group Γ. This group has finite covolume and is called *a geometric limit of* Γ_i. Let $X = \mathbb{D}^2/\Gamma$ and let x be the projection of $0 \in \mathbb{D}^2$. We say that (X_i, x_i) *converges to* (X, x). One can prove that this convergence is equivalent to the convergence of Γ_i to Γ, for *the Chabauty topology*, i.e. for the Hausdorff topology on closed subsets of $\mathrm{PSL}_2(\mathbb{R})$ [CEG].

b) $\mathrm{inj}(x_i)$ *tends to* 0.

We use the formulas for the hyperbolic metric on an annulus (cf §1). Let $\varepsilon \leq \varepsilon(2)$. Suppose that x_i belongs to a component of $X_i^{]0,\varepsilon]}$ which is a Margulis tube around a geodesic g_i. Consider the geometric cover of X_i with fundamental group isomorphic to the cyclic group generated by g_i. This cover can be identified conformally with $A_{e^{-R_i}, e^{R_i}}$, for $R_i = \pi^2/\ell(g_i)$. By the Margulis lemma, the ε-thin part of $A_{e^{-R_i}, e^{R_i}}$ embeds under the covering $A_{e^{-R_i}, e^{R_i}} \to X_i$, and in particular, the annulus $A_{e^{-\rho_i}, e^{\rho_i}}$ for

$$\rho_i = \frac{2R_i}{\pi} \cos^{-1} \frac{\ell(g_i)}{\varepsilon}.$$

Let $\widetilde{x}_i$ be the lift of x_i which is contained in $A_{e^{-\rho_i}, e^{\rho_i}}$. Denote by Y_i the image of $A_{e^{-R_i}, e^{R_i}}$ by the homothety of ratio $1/\widetilde{x}_i$. Up to extracting a subsequence, Y_i converge to the annulus $A_{0,\infty} = \mathbb{C}^*$ and the hyperbolic metrics on Y_i, rescaled by the factor $1/\mathrm{inj}(x_i)$, converge to a flat complete metric on $\mathbb{C}^*$. We say that (X_i, x_i) *converges to* $(\mathbb{C}^*, 1)$.

This normalization allows us, by looking at an appropriate cover of X_i —which is either the universal cover $\mathbb{D}^2$ or the annulus Y_i— to compare charts around $x_i \in X_i$ when X_i varies.

Definition.— Let (X_i, x_i) be a sequence of pointed compact hyperbolic Riemann surfaces which converges to (X, x). Let $\phi_i \in \mathcal{Q}(X_i)$ and let ϕ be a holomorphic quadratic differential defined on X. If $\mathrm{inj}(x_i)$ does not tend to 0, let $\widetilde{\phi}_i(z)dz^2$ (resp. $\widetilde{\phi}(z)dz^2$) be the pull-back of ϕ_i (resp. ϕ) to $\mathbb{D}^2$. If $\mathrm{inj}(x_i)$ tends to 0, let Y_i be the covering of X_i associated to the fundamental group of that component and let $\widetilde{\phi}_i(z)dz^2$ (resp. $\widetilde{\phi}(z)dz^2$) be the pull-back of ϕ_i (resp. ϕ) to Y_i. We say that ϕ_i *converges uniformly to* ϕ if $\widetilde{\phi}_i$ converges to $\widetilde{\phi}$ uniformly over compact sets (in $\mathbb{D}^2$ or in $\mathbb{C}^*$).

Theorem 3.1.— *Let (X_i, x_i) be a sequence of pointed compact hyperbolic Riemann surfaces with a fixed topological type which converges to (X, x). Let $\phi_i \in \mathcal{Q}(X_i)$*

with $\phi_i \neq 0$. Then there exists constants c_i and a non-zero holomorphic quadratic differential ϕ on X such that $(c_i\phi_i)$ converges uniformly to ϕ up to extracting a subsequence.

Remark.— One important feature of this theorem is to produce a non-zero limit, merely by applying a homothety to ϕ_i. The existence of a limit for a sequence (ϕ_i), when $||\phi_i||$ is bounded can be shown by a more elementary argument. It follows from the precompacity of a sequence of holomorphic functions with bounded L^1-norm. However, this is not sufficient to guarantee that the limit is non-zero, even if $||\phi_i|| = 1$.

Remark.— The limit ϕ produced by Theorem 3.1 is not necessarily integrable. The basic example of a non-integrable limit appears in Lemma 3.4.

In order to prove Theorem 3.1, we construct first non-zero holomorphic quadratic differentials θ_i on X_i and θ on X such that (θ_i) converges uniformly to θ as i tends to infinity.

Proposition 3.2.— *Let (X_i, x_i) be a sequence of pointed compact hyperbolic Riemann surfaces with a fixed topological type which converges to (X, x). Then there exists non-zero $\theta_i \in \mathcal{Q}(X_i)$ and a non-zero holomorphic quadratic differential θ on X, such that (θ_i) converges uniformly to θ, up to extracting a subsequence.*

Proof.— We consider two cases according as X is a finite volume hyperbolic surface or is an annulus.

1) *X is a finite volume hyperbolic surface.*

We begin with the following lemma.

Lemma 3.3.— *Suppose that (X_i, x_i) converge to (X, x) where X is a finite volume hyperbolic surface. Then $\Theta_{\mathbb{D}^2/X_i}$ converge weakly to $\Theta_{\mathbb{D}^2/X}$, i.e. for any $P \in \mathcal{Q}(\mathbb{D}^2)$, $\Theta_{\mathbb{D}^2/X_i}P$ converges uniformly to $\Theta_{\mathbb{D}^2/X}P$.*

Proof.— Assume that X_i and X are uniformized by Fuchsian groups Γ_i and Γ acting on $\mathbb{D}^2$, in such a way that $0 \in \mathbb{D}^2$ projects to x_i and x respectively. Let $P \in \mathcal{Q}(\mathbb{D}^2)$. Fix a compact set $K \subset \mathbb{D}^2$. For each $r < 1$, there is a compact neighborhood C_r of $\mathrm{Id} \in \mathrm{PSL}_2(\mathbb{R})$ such that, for any $g \notin C_r$, $g(K)$ is contained outside of the disc Δ_r of radius r. We may assume that the frontier of C_r in $\mathrm{PSL}_2(\mathbb{R})$ is disjoint from Γ so that $\Gamma_i \cap C_r \to \Gamma \cap C_r$ as i tends to ∞. We can choose r so that the $|P|$-mass of $\mathbb{D}^2 - D_r$ is arbitrarily small. Therefore, for i sufficiently large, the difference $|\Theta_{\mathbb{D}^2/X}P(z) - \Theta_{\mathbb{D}^2/X_i}P(z)|$ can be made arbitrarily small over K: this follows directly from Cauchy's formula. □

We use now the fact that $\Theta_{\mathbb{D}^2/X}$ is surjective when X has finite volume. The proof can be sketched as follows. For a general Riemann surface X, the image of the restrictions of the polynomials to $\mathbb{D}^2$ is dense in $\mathcal{Q}(X)$, when $\mathcal{Q}(X)$ is endowed with the Weil-Peterson scalar product (cf. [Kr]). If the hyperbolic volume of X is finite, $\mathcal{Q}(X)$ is a finite dimensional vector space (cf. §1). It follows that $\Theta_{\mathbb{D}^2/X}$ is surjective in this case. For a composition of covers of Riemann surfaces $Z \to Y \to X$, we have: $\Theta_{Z/X} = \Theta_{Y/X} \circ \Theta_{Z/Y}$. Hence if X has finite volume, the operator $\Theta_{Y/X}$ is surjective.

Lemma 3.3 and the surjectivity of $\Theta_{\mathbb{D}^2/X}$ yield a proof of Proposition 3.2 as follows.

1a) *X is different from the thrice punctured sphere.*

In this situation, the vector space $\mathcal{Q}(X)$ is non-trivial. By the above, it contains a non-zero element of the form $\Theta_{\mathbb{D}^2/X}P$. By Lemma 3.3, $(\Theta_{\mathbb{D}^2/X_i}P)$ converges uniformly to $\theta = \Theta_{\mathbb{D}^2/X}P$. For all sufficiently large i, $\theta_i = \Theta_{\mathbb{D}^2/X_i}P$ is non-zero. This proves Proposition 3.2.

1b) *X is the thrice punctured sphere.*

Then the space of integrable holomorphic quadratic differentials is trivial and the reasonning above cannot be applied. The proof we will give could be extended with a minor modification to the case when X has finite volume but is not compact. However, we present it only when X is the thrice punctured sphere: any puncture on X corresponds to a closed geodesic $g_i \subset X_i$ whose length tends to 0, and since X is a thrice punctured sphere, we can choose such a geodesic g_i which gives rise to a single puncture on X.

The cover of X_i associated to the cyclic subgroup of Γ_i representing the curve g_i can be conformally identified with $A_{e^{-R_i},e^{R_i}}$ (cf. §1). Let $\varepsilon \leq \varepsilon(2)$. By Margulis lemma, the annulus $A_{e^{-\rho_i},e^{\rho_i}}$ embeds under the covering map into X_i, for

$$\rho_i = \frac{2R_i}{\pi} \cos^{-1}\left(\frac{\ell(g_i)}{\varepsilon}\right).$$

Without loss of generality we may assume, (up to applying an inversion through the unit circle) that the image of the circle of radius e^{ρ_i} remains at a bounded distance from x_i on X_i. Denote by Y_i the image of $A_{e^{-R_i},e^{R_i}}$ by the homothety of ratio $e^{-\rho_i}$, i.e. the annulus $A_{e^{-R_i-\rho_i},e^{R_i-\rho_i}}$ (the unit circle has now length ε). Since $\ell(g_i)$ tends to 0, R_i tends to infinity; therefore $R_i - \rho_i \simeq 2\pi/\varepsilon$. Up to extracting a subsequence, (Y_i) converges to the annulus $Y = A_{0,e^{2\pi/\varepsilon}}$, and the hyperbolic metrics on Y_i converge to the hyperbolic metric on Y. The annulus Y is also the covering of X corresponding to the puncture we have selected.

Consider the holomorphic quadratic differential $\phi = dz^2/z^2$ on $\mathbb{C}^*$. Since ϕ is integrable on Y_i, we can apply the operator Θ_{Y_i/X_i} to it, obtaining a differential $\theta_i = \Theta_{Y_i/X_i}\phi \in \mathcal{Q}(X_i)$. However ϕ is not integrable on Y and therefore $\Theta_{Y/X}\phi$ cannot be defined in the classical way. Nevertheless, ϕ is integrable in the complement of the unit disc. By the Margulis lemma, the punctured unit disc embeds under the covering $Y \to X$. Let B be a small ball contained in X. The preimage of B in Y consists of disjoint copies of B. At most one of them is contained in the punctured unit disc, all the others are in $A_{1,e^{2\pi/\varepsilon}}$. Therefore in the series defining $\Theta_{Y/X}\phi$, the sum of the terms coming from $A_{1,e^{2\pi/\varepsilon}}$ converges uniformly over B (by Cauchy's formula) and gives an integrable holomorphic quadratic differential. The term coming from the punctured unit disc contributes as a differential with a double pole at the puncture. This allows us to define $\theta = \Theta_{Y/X}\phi$. It is a holomorphic quadratic differential on X, which is non-zero because it has a pole of order 2 at the puncture. The following lemma implies Theorem 3.1 in this case.

Lemma 3.4.— *(θ_i) converges uniformly to θ.*

For this lemma to be true, it is necessary that g_i gives rise to a single puncture in X. In the case when g_i gives rise to two punctures, (θ_i) converges to the sum of the two differentials constructed in the same way as θ for each of these punctures.

Proof.— We will prove that the pull-back of θ_i to Y_i converges to the pull-back of θ to Y. This will imply the lemma. We denote by the same letter π the covering maps from Y to X or from Y_i to X_i. Let $K \subset Y$ be a compact set, which we choose to be a lift of a small ball $\pi(K)$ embedded in X. Since $A_{e^{-2\rho_i},1}$ embeds into X_i (as consequence of the Margulis lemma), at most one component of $\pi^{-1}(\pi(K))$ is contained in $A_{e^{-2\rho_i},1}$. Since the metrics on Y_i converge to the metric on Y, the distance on Y_i between K and the circle of radius 1 is bounded from above independently of i. Now, we analyze the contribution of the various components of $\pi^{-1}(\pi(K))$ to $\Theta_{Y_i/X_i}\phi$.

Fact 3.5.— *The hyperbolic distance on Y_i between the circle of radius $e^{-2\rho_i}$ and any component of $\pi^{-1}(\pi(K))$ contained in $A_{e^{-R_i-\rho_i},e^{-2\rho_i}}$ tends to ∞ with i.*

Proof.— If not, then the image on X_i of the circle of radius $e^{-2\rho_i}$ is at a bounded distance from $\pi(K)$ and thus at a bounded distance from x_i too. By the normalization of Y_i, the projection of the unit circle is at a finite distance from K. But since g_i gives rise to a single puncture in X, only one of the two boundary components of $A_{e^{-2\rho_i},1}$ can stay within a bounded distance of x_i. This contradiction proves Fact 3.5. □

The formula for the hyperbolic distance on Y_i between two circles shows that the components of $\pi^{-1}(\pi(K))$ contained in $A_{e^{-R_i-\rho_i},e^{-2\rho_i}}$ are confined inside an annulus $A_{e^{-R_i-\rho_i},e^{-r_i-\rho_i}}$ where $R_i - r_i$ tends to 0 when i tends to ∞. The ϕ-mass of this strip is equal to $2\pi(R_i - r_i)$ and so tends to 0. In particular, the ϕ-mass of $\pi^{-1}(\pi(K)) \cap A_{e^{-R_i-\rho_i},e^{-2\rho_i}}$ tends to 0 when i tends to ∞.

Since ϕ is integrable near the exterior end of Y, we can select, for all $\eta > 0$, a radius $r < e^{R_i-\rho_i}$ such that the ϕ-mass of $A_{r,e^{R_i-\rho_i}}$ is smaller than η for all i sufficiently large. In particular, the ϕ-mass of $\pi^{-1}(\pi(K)) \cap A_{r,e^{R_i-\rho_i}}$ is smaller than η, for all i sufficiently large.

The components of $\pi^{-1}(\pi(K)) \cap A_{e^{-2\rho_i},r}$ are K itself, and finitely many others which intersect $A_{1,r}$. As in Lemma 3.3, it follows now from Cauchy's formula that the pull-back of θ_i to Y_i converges to the pull-back of θ uniformly in the interior of K. This ends the proof of Lemma 3.4. □

2) $X = \mathbb{C}^*$.

We keep the same notations as for the description of the convergence of (X_i, x_i). The annulus Y_i is identified with an annulus $A_{e^{-2R_i+R'_i},e^{R'_i}}$ for a certain number R'_i, in order for the lift $\tilde{x}_i$ of x_i to be the point 1. As in the case of the thrice punctured sphere, we can apply Θ_{Y_i/X_i} to $\phi = dz^2/z^2$, and define a holomorphic quadratic differential θ_i on X_i.

Lemma 3.6.— *(θ_i) converges uniformly to $\phi = dz^2/z^2$.*

Proof.— Let $K \subset \mathbb{C}^*$ be a compact set. The length of the unit circle for the hyperbolic metric of Y_i of the unit circle is equivalent to $\mathrm{inj}(x_i)$. It follows from the formula for the length of the circles on Y_i that the injectivity radius tends to 0 uniformly over K. In particular, K is contained in $Y_i^{]0,\varepsilon]}$ for all i sufficiently large, and the hyperbolic distance between K and $\partial Y_i^{]0,\varepsilon]}$ tends to ∞ with i. By the Margulis lemma, $Y_i^{]0,\varepsilon]}$ maps injectively to X_i. Therefore K is the single component of $\pi^{-1}(\pi(K))$ contained in $Y_i^{]0,\varepsilon]}$, and the hyperbolic distance between $\partial Y_i^{]0,\varepsilon]}$ and the components of $\pi^{-1}(\pi(K))$ others than K tends to ∞ with i. It follows that the components of $\pi^{-1}(\pi(K))$ which are near the exterior end (say) are confined inside an annulus $A_{e^{R_i'-r_i}, e^{R_i'}}$ such that $R_i' - r_i$ tends to 0 when i tends to infinity. The same holds for the components which are near the interior end. Therefore, as in the case of the thrice punctured sphere, the pull-back of θ_i to the interior of K converges to ϕ uniformly in the interior of K. This proves Lemma 3.6. □

This completes the proof of Proposition 3.2. □

Proof of Theorem 3.1.— Let (θ_i) be the sequence constructed in Proposition 3.2. The ratio $f_i = \phi_i/\theta_i$ is a rational function $X_i \to \overline{\mathbb{C}}$. The degree d of f_i is independent of i, by the Hurwitz formula. The preimage $E_i = f_i^{-1}\{0, 1, \infty\}$ has cardinality less than $3d$. Denote by $\widetilde{E}_i$ the preimage of E_i in the universal cover $\mathbb{D}^2$ or in the annuli Y_i, according to the nature of $\mathrm{inj}(x_i)$. Let $\widetilde{f}_i$ be the lift of f_i to the corresponding cover $\mathbb{D}^2$ or Y_i. Let K be an arbitrary compact set contained in $\mathbb{D}^2$ if $\mathrm{inj}(x_i)$ is bounded away from 0, or in $\mathbb{C}^*$ if $\mathrm{inj}(x_i)$ tends to 0. If $K \subset \mathbb{D}^2$, the cardinality of $K \cap \widetilde{E}_i$ is bounded only in terms of d and of the "degree" of the restriction to K of the covering map $\pi : \mathbb{D}^2 \to X$, i.e. the maximal cardinality of $\pi^{-1}(z) \cap K$ of a point $z \in \pi(K)$. If $K \subset \mathbb{C}^*$, the covering map $\pi : Y_i \to X_i$ restricts to K as an embedding (cf. proof of Lemma 3.6); hence, the cardinality of $K \cap \widetilde{E}_i$ is bounded independently of i. Therefore, up to extracting a subsequence, the sets $\widetilde{E}_i$ converge to a discrete set $\widetilde{E}$ contained in $\mathbb{D}^2$ or in $\mathbb{C}^*$. By the Montel theorem, the rational functions $\widetilde{f}_i$ converge to a holomorphic function $\widetilde{f}$ uniformly on compact sets in $K - \widetilde{E}$, up to passing to a subsequence. The degree of $\widetilde{f}$ is finite because the degree of $\widetilde{f}_i$ over K is bounded independently of i. Hence by the Picard theorem, $\widetilde{f}$ extends across $\widetilde{E}$ to a meromorphic function. Thus, the functions $\widetilde{f}_i$ converge to a meromorphic function $\widetilde{f}$ uniformly on compact sets.

Up to scaling $\widetilde{f}_i$ by a constant, we can ensure that $\widetilde{f}$ is not identically 0 or ∞. To see this, consider a point $x_n \in X_n - E_n$ which tends to a point in $X - E$. Then the functions $f_n' = \widetilde{f}_n/f_n(x_n)$ converge to a meromorphic function uniformly over compact sets by the argument above. This function is not identically 0 nor ∞ by construction. Hence the functions $\widetilde{\phi}_n/\widetilde{f}_n(x_n)$ converge uniformly on compact sets in $\mathbb{D}^2$ (or in $\mathbb{C}^*$), to a non-zero holomorphic function $\widetilde{\phi}$.

When $\text{inj}(x_i)$ is bounded away from 0, $\widetilde{\phi}(z)dz^2$ is invariant under Γ by continuity. So $\widetilde{\phi}(z)dz^2$ induces a non-zero holomorphic quadratic differential ϕ on X. This concludes the proof of Theorem 3.1. □

3.2 Applications of Theorem 3.1

Let X be a hyperbolic Riemann surface. Let ϕ be a holomorphic quadratic differential. In a conformal chart around a point in X, the hyperbolic metric can be written $\lambda(z)|dz|$ and ϕ can be written $\phi(z)dz^2$. Then, the quantity

$$\langle\phi(z)\rangle = |\phi(z)|\lambda^{-2}(z)$$

is independent of the chart and it defines a function on X.

Definition.— We call $\langle\phi\rangle(z)$ *the hyperbolic norm of* ϕ *at the point* z.

The quantity $\langle\phi\rangle(z)$ can be viewed also as the Riemannian norm of the tensor ϕ on the hyperbolic surface X.

Definition.— *The systole of* X is the length of the shortest closed geodesic of X.

This is well defined since a hyperbolic surface X has finitely many closed geodesics shorter than $\varepsilon(2)$ (cf. §1). When X is compact, one can prove that a lower bound $\varepsilon > 0$ for the systole gives an upper bound for the diameter of X which is a function of ε and of $\chi(X)$.

Let Z be the set of zeroes of a non-zero $\phi \in \mathcal{Q}(X)$. If X has genus g, the cardinality of Z is smaller than $4g-4$ (cf. [Ga]). Denote by $Z(r)$ the neighborhood of radius r of Z on X and let

$$m(r) = \inf_{z\in X-Z(r)} \langle\phi(z)\rangle.$$

Proposition 3.7 [Di].— *Let* $\varepsilon > 0$. *Let* X *be a compact connected hyperbolic Riemann surface with systole bigger than* ε *and let* $\phi \in \mathcal{Q}(X)$ *with* $||\phi|| = 1$. *Let* $r < \varepsilon/2$. *Then,* $m(r) \geq m$ *where* $m > 0$ *is a function of* r, ε *and* $\chi(X)$.

Proof.— We argue by contradiction. Suppose that there exist compact hyperbolic Riemann surfaces X_i with fixed topological type whose systole is bigger than ε, and ϕ_i in $\mathcal{Q}(X_i)$ with $||\phi_i|| = 1$, such that

$$m_i(r) = \inf_{z\in X_i - Z_i(r)} \langle\phi_i(z)\rangle$$

tends to 0, when i tends to ∞.

We keep the same notations as for the description of the convergence of pointed Riemann surfaces. Let $x_i \in X_i$. Since $\text{inj}(x_i) \geq \varepsilon$, the pointed surfaces (X_i, x_i) converge to a hyperbolic surface (X, x), up to extracting a subsequence. Since the systole of X_i is bigger than ε, the diameter of X_i is bounded from above independently of i. Therefore the diameter of $\mathcal{D}_i$ is bounded independently of i and X is compact. By Theorem 3.1 there exist constants c_i and a non-zero

$\phi \in \mathcal{Q}(X)$ such that $(c_i\phi_i)$ converges uniformly to ϕ: this means that the functions $c_i\widetilde{\phi}_i(z)$ converge to $\widetilde{\phi}(z)$, where $c_i\widetilde{\phi}_i(z)dz^2$ (resp. $\widetilde{\phi}(z)dz^2$) is the pull-back of ϕ_i (resp. ϕ) to $\mathbb{D}^2$. Choose a ball $B \subset \mathbb{D}^2$ of finite radius which contains $\mathcal{D}$ in its interior. It meets only a finite number of translates of $\mathcal{D}$ by Γ. Hence, the number of the translates of $\mathcal{D}_i$ by elements of Γ_i that meet B is less than a constant C independent of i. Therefore, we have

$$\int_B |\widetilde{\phi}(z)||dz|^2 = \lim_{i\to\infty} \int_B |c_i\widetilde{\phi}_i(z)||dz|^2 \leq \lim_{i\to\infty} C|c_i|.$$

It follows that $|c_i|$ is bounded from below by a non-zero constant.

For all i sufficiently large, $\mathcal{D}_i$ is contained in B. Therefore

$$\int_B |\widetilde{\phi}||dz|^2 = \lim_{i\to\infty} \int_B |c_i\widetilde{\phi}_i||dz|^2 \geq \lim_{i\to\infty} |c_i|,$$

so that $|c_i|$ is bounded also from above. This means that we can choose all of the constants c_i equal to 1.

Since Z is discrete, we may suppose that ∂B does not intersect the set of zeroes $\widetilde{Z}$ of $\widetilde{\phi}$. Then $\widetilde{Z}_i \cap B$ converges to $\widetilde{Z} \cap B$, and $\widetilde{Z}_i(r) \cap B$ converges to $\widetilde{Z}(r) \cap B$. Therefore, if we write the Poincaré metric of $\mathbb{D}^2$ in the form $\lambda(z)|dz|$, we have

$$m_i(r) = \inf_{z\in B-\widetilde{Z}_i(r)} |\widetilde{\phi}_i(z)|\lambda(z)^{-2}.$$

So the uniform convergence of $\widetilde{\phi}_i$ to $\widetilde{\phi}$ over B implies that $m_i(r)$ tends to $m(r)$ as i tends to ∞. Since $m(r) \neq 0$, this is a contradiction. □

Notation. — Let X be a hyperbolic Riemann surface. We denote by $B(x,r)$ the hyperbolic ball of radius r centered at x.

Proposition 3.8 [McM2]. — *Let X be a connected compact hyperbolic Riemann surface and let $x \in X$. Let $\alpha \geq 1$. Then, for any non-zero $\phi \in \mathcal{Q}(X)$*

$$\frac{\int_{B(x,\alpha r)} |\phi|}{\int_{B(x,r)} |\phi|} \leq c(\alpha),$$

where the constant $c(\alpha) < \infty$ is a function of $\chi(X)$ and of α.

Proof. — Observe that multiplying the differential ϕ by a non-zero constant does not affect the ratio of the ϕ-masses that we are considering. To prove Proposition 3.8, we argue by contradiction. Then there is a sequence of pointed compact hyperbolic Riemann surfaces (X_i, x_i) with fixed topological type, non-zero $\phi_i \in \mathcal{Q}(X_i)$, and balls $B(x_i, r_i)$, constant such that the ratio of the ϕ_i-masses of $B(x_i, r_i)$ and $B(x_i, \alpha r_i)$ tends to ∞ with i. We keep the same notations as for the description of the convergence of pointed Riemann surfaces.

1) $\mathrm{inj}(x_i)$ *is bounded from below.*

Under the covering map $\mathbb{D}^2 \to X_i$, $B(x_i, r_i)$ is isomorphic to $B(0, r_i) \subset \mathbb{D}^2$ and $B(x_i, \alpha r_i)$ is the image of $B(0, s_i) \subset \mathbb{D}^2$. In particular, for any $\phi \in \mathcal{Q}(X_i)$, the ϕ-mass of $B(x_i, r_i)$ equals the $\tilde{\phi}$-mass of $B(0, r_i)$ and the ϕ-mass of $B(x_i, \alpha r_i)$ is less than or equal to the $\tilde{\phi}$-mass of $B(0, \alpha r_i)$, where $\tilde{\phi}$ denotes the pull-back of ϕ to $\mathbb{D}^2$.

By Theorem 3.1, up to passing to a subsequence and up to multiplying ϕ_i by a non-zero constant, we can suppose that (ϕ_i) converges uniformly to a holomorphic quadratic differential $\phi \neq 0$. Since $B(x_i, r_i)$ is embedded, and since the area of X_i is constant, r_i is bounded from above. If r_i admits a non-zero lower bound, we may suppose that it converges to $r > 0$. Then the $\tilde{\phi}_i$-mass of $B(0, r_i)$ and $B(0, \alpha r_i)$ converge respectively to the $\tilde{\phi}$-mass of $B(0, r)$ and $B(0, \alpha r)$. Both are non-zero, since $\tilde{\phi} \neq 0$. We obtain a contradiction in this case. When r_i tends to 0, suppose that $\tilde{\phi}$ vanishes exactly up to the order n at 0. Then after a change of variables, we find

$$\frac{\int_{B(0,\alpha r_i)} |\tilde{\phi}_i|}{\int_{B(0,r_i)} |\tilde{\phi}_i|} \to (\alpha)^{2n}.$$

This gives a contradiction.

2) $\mathrm{inj}(x_i)$ *tends to* 0.

In our normalizations, the rescaled hyperbolic metrics on Y_i tend to the flat metric on $\mathbb{C}^*$ and $1 \in \mathbb{C}^*$ maps to x_i. Denote by $B_i(1, r)$ the ball of radius r centered at 1 for the rescaled hyperbolic metric on Y_i and by $B_\infty(1, r)$ this ball for the limit flat metric. Then, under the covering $Y_i \to X_i$, $B_i(1, r_i/\mathrm{inj}(x_i))$ maps isomorphically to $B(x_i, r_i)$, and $B(x_i, \alpha r_i)$ is the image of $B_i(1, \alpha r_i/\mathrm{inj}(x_i))$. Since $B(x_i, r_i)$ is embedded, $r_i \leq \mathrm{inj}(x_i)$. As in 1), we are led to consider two cases according to whether the ratio $r_i/\mathrm{inj}(x_i)$ admits a non-zero lower bound or tends to 0. Observe that for any $R > 0$, there are non-zero constants a and b such that, for all $r \leq R$ each ball $B_\infty(1, r)$ is contained in the euclidean ball of radius ar and contains the euclidean ball of radius br. The same result holds for the balls $B_i(1, r)$ when i is sufficiently large because of the convergence of the rescaled metrics on Y_i. Using this property and applying the same arguments as in 1), we obtain a contradiction. □

Proposition 3.9 [McM2].— *Let* $0 < \varepsilon \leq \varepsilon(2)$. *Let* X *be a compact hyperbolic Riemann surface with systole smaller than* $\varepsilon/2$. *Let* $\phi \in \mathcal{Q}(X)$ *with* $||\phi|| = 1$. *Then the* ϕ*-mass of a component of* $X^{]0,\varepsilon]}$ *is bigger than a constant* $\zeta > 0$ *which is a function of* ε *and* $\chi(X)$.

Proof.— The number of components of $X^{]0,\varepsilon]}$ is less than $3g - 3$, where g is the genus of X. In order to prove Proposition 3.9 by contradiction, we may therefore suppose that there is a sequence of Riemann surfaces X_i with the same topological type, with systole smaller than $\varepsilon/2$ and $\phi_i \in \mathcal{Q}(X_i)$ with $||\phi_i|| = 1$ such that the ϕ_i-mass of $X_i^{]0,\varepsilon]}$ tends to 0 as i tends to ∞. The number of components of $X_i^{[\varepsilon,\infty[}$ is smaller than $|\chi(X_i)|$. Let Z_i be one of these components with the property that its ϕ_i-mass is bigger than some number $\nu > 0$, for all i sufficiently large. Let $x_i \in Z_i$. Since $\mathrm{inj}(x_i) \geq \varepsilon$ the sequence (X_i, x_i) converges, up to extracting a subsequence, to (X, x) where X is a hyperbolic Riemann surface. By assumption, $X^{]0,\varepsilon/2]}$ is non-empty. By Theorem 3.1, there exist constants c_i such that $(c_i\phi_i)$ converges uniformly to a non-zero holomorphic quadratic differential ϕ defined on X. As the ϕ_i-mass of Z_i is more than ν, c_i can be chosen equal to 1 (cf. the proof of Proposition 3.7). Since $\phi \neq 0$, the ϕ_i-mass of $X_i^{]0,\varepsilon]} - X_i^{]0,\varepsilon/2]}$ does not tend to 0. This is a contradiction. □

The next result asserts that the pairing $\langle\phi, \mu\rangle$ between a unit norm holomorphic quadratic differential and a unit norm Beltrami form can be estimated from a certain local data.

Definition.— Let X be a compact hyperbolic Riemann surface. Let E be a measurable set contained in X. For a non-zero $\phi \in \mathcal{Q}(X)$ and for a Beltrami form $\mu \in \mathcal{B}(X)$, define *the efficiency of the pairing between ϕ and μ over E* to be the ratio

$$e(E) = \frac{\langle\phi, \mu\rangle_E}{||\phi||_E},$$

where $\langle\phi, \mu\rangle_E = \Re(\int_E \phi(z)\mu(z)|dz^2|)$, and where $||\phi||_E$ is the ϕ-mass of E.

When $||\mu|| \leq 1$, we have $e(E) \leq 1$. Equality holds if and only if the restriction of μ to E equals $\overline{\phi}/|\phi|$ a. e.

Proposition 3.10 [McM2].— *Let X be a compact hyperbolic Riemann surface. Let $\phi \in \mathcal{Q}(X)$ with $||\phi|| = 1$ and let $\mu \in \mathcal{B}(X)$ with $||\mu|| \leq 1$ Let $E \subset X$ be a measurable subset of ϕ-mass bigger than m for some $m > 0$. Suppose that each point of E is the center of an embedded hyperbolic ball on which the efficiency of the pairing between ϕ and μ is less than $1-\alpha$, for some $\alpha > 0$. Then $\langle\phi, \mu\rangle \leq 1 - cm\alpha$, where $c > 0$ depends only on $\chi(X)$.*

Proof.— By a Vitali type argument, we can extract from the family of balls provided by the hypothesis, a family of disjoint balls $\{B_i\}$ such that the balls $5B_i$ cover E. Then by Proposition 3.8, we have:

$$(1) \qquad \int_E |\phi| \leq \int_{\sqcup 5B_i} |\phi| \leq c(5) \int_{\sqcup B_i} |\phi|.$$

Since the balls B_i are disjoint we have

$$\langle\phi, \mu\rangle = ||\phi||_{\complement\sqcup B_i} \frac{\langle\phi, \mu\rangle_{\complement\sqcup B_i}}{||\phi||_{\complement\sqcup B_i}} + \sum_i ||\phi||_{B_i} \frac{\langle\phi, \mu\rangle_{B_i}}{||\phi||_{B_i}}.$$

Using (1) we obtain

$$\langle\phi, \mu\rangle \leq 1 - ||\phi||_{\sqcup B_i} + (1-\alpha)||\phi||_{\sqcup B_i} \leq 1 - \alpha m/c(5).$$

□

CHAPTER 4

The volume form on open Riemann surfaces

Definition. — *An open Riemann surface* is a connected hyperbolic Riemann surface of finite topological type but of infinite hyperbolic volume.

Equivalently an open Riemann surface is conformally equivalent to $\mathbb{D}^2$, to an annulus, or to the quotient of $\mathbb{D}^2$ by a Fuchsian group Γ with $\Omega(\Gamma) \neq \emptyset$.

Let Y be a hyperbolic Riemann surface. *A 1-form of type* $(1,0)$ *on* Y is a smooth 1-form η which can be written in any complex chart on the form $\eta(z)dz$. In any complex chart, the hyperbolic metric can be written on the form $\lambda(z)|dz|$. So, when η is a 1-form of type $(1,0)$ on Y, the quantity $|\eta(z)|\lambda(z)^{-1}$ is independent of the chart.

Definition. — The function $\langle\eta\rangle : Y \to \mathbb{R}$ defined by $\langle\eta\rangle(z) = |\eta(z)|\lambda(z)^{-1}$ is called *the hyperbolic norm of* η.

Thus $\langle\eta\rangle$ is the norm of the 1-form η when Y is endowed with its hyperbolic metric. The norm of any tensor on Y can be similarly defined. For instance, when ω is a 2-form on Y, $\langle\omega\rangle$ equals the ratio ω/dv where dv is the hyperbolic volume form on Y.

Theorem 4.1 [Di]. — *Let* $\varepsilon > 0$. *Let* Y *be an open hyperbolic Riemann surface whose systole is bigger than* ε. *Then there is a 1-form* η *of type* $(1,0)$ *such that*

(i) $\overline{\partial}\eta = dv$, *and*

(ii) $||\langle\eta\rangle||_\infty$ *is finite and bounded by a constant* C *which is a function of* ε *and* $\chi(Y)$.

Remark. — This theorem applies to give an isoperimetric inequality for domains in Y. If $\mathcal{K} \subset Y$ is a compact domain with smooth boundary, a direct application of the Stokes formula gives

$$\text{Area}(\mathcal{K}) \leq \frac{1}{C}\ell(\partial\mathcal{K}).$$

This means that the hyperbolic metric of Y satisfies a linear isoperimetric inequality with a constant depending only on the systole of Y and on $\chi(Y)$.

Proof. — We consider only the case when Y has no cusps.

Suppose first that Y is conformally equivalent to a disk or to an annulus. On $\mathbb{D}^2$, the hyperbolic metric is given by

$$ds = \frac{2}{1-|z|^2}|dz|.$$

Hence the 1-form

$$\eta = \frac{-2i\bar{z}}{1-|z|^2}dz$$

satisfies $\bar{\partial}\eta = dv$ and $\langle\eta\rangle(z) = |z|$. Therefore $||\langle\eta\rangle||_\infty = 1$ and η satisfies the conclusions of Theorem 4.1.

Suppose that Y is conformally equivalent to

$$A_{e^{-R},e^R} = \{z \in \mathbb{C}, \quad e^{-R} < |z| < e^R\}.$$

The hyperbolic metric on this annulus is

$$ds = \frac{\pi}{2R|z|\cos(\pi\frac{\log|z|}{2R})}|dz|.$$

Hence the 1-form

$$\eta = \frac{-i\pi}{2Rz}\tan(\pi\frac{\log|z|}{2R})dz$$

satisfies $\bar{\partial}\eta = dv$. Moreover for all $z \in A_{e^{-R},e^R}$, we have

$$\langle\eta\rangle(z) = \sin(\pi\frac{\log|z|}{2R}) \leq 1.$$

Therefore η is the required differential. We note also that $||\langle\eta\rangle||_\infty = 1$ is independent of the systole of Y.

In the other cases, since Y has no cusps, the Nielsen core of Y is a compact surface Y_0 with geodesic boundary. The surface Y equals the union of Y_0 and a collection of half-infinite annuli. For an open Riemann surface such Y a lower bound on the systole does not guarantee that Y remains in a compact set of metrics, since the length of ∂Y_0 could tend to infinity. For instance, imagine a pair of pants tending closer and closer to a bikini. The strategy to prove Theorem 4.1 is to use the explicit solution constructed above in the half-infinite annuli which are components of $Y - Y_0$ and then to extend it over Y_0 using the Green's function on Y.

The Green's function on a Riemann surface X.

Definition. — Let X be a Riemann surface. *A Green's function of* X is a positive function $G(.,.)$ on $X \times X$ − diagonal which satisfies:

(i) $\Delta G(.,y) = \delta_y$, i.e. for x not equal to y, the function $x \to G(x,y)$ is harmonic and in a holomorphic chart around y, the function

$$x \to G(x,y) + \frac{1}{2\pi}\log|x-y|$$

is harmonic, and

(ii) G is minimal among all positive functions satisfying (iii).

In $\mathbb{D}^2$, the Green's function equals

$$-\frac{1}{2\pi}\log\tanh\left(\frac{d(x,y)}{2}\right),$$

where d is the hyperbolic distance. But there does not always exist a Green's function on a given Riemann surface (cf. **[Ah3]**, **[Nic]**, **[Ts]**). For instance on a hyperbolic Riemann surface of finite volume, it does not. But, when a Green's function exists, it is unique. Let Y be an open Riemann surface isomorphic to the quotient of $\mathbb{D}^2$ by a Fuchsian group Γ. Consider the positive function on $\mathbb{D}^2 \times \mathbb{D}^2$ – diagonal

$$\widetilde{G}(x,y) = -\frac{1}{2\pi}\sum_{\gamma\in\Gamma}\log\tanh\left(\frac{d(x,\gamma(y))}{2}\right). \tag{1}$$

This series is invariant under Γ and therefore induces a function G on $Y \times Y$ – diagonal.

Lemma 4.2.— *Let $Y = \mathbb{D}^2/\Gamma$ be an open Riemann surface. Then for all $\alpha > 0$, the series (1) converges uniformly for all pairs (x,y) such that $d(x,\Gamma(y)) \geq \alpha$. The function G is the Green's function of Y.*

Proof.— For $d(x,\gamma(y)) \geq \alpha$, we have:

$$-\log\tanh\left(\frac{d(x,\gamma(y))}{2}\right) \leq C(\alpha)e^{-d(x,\gamma(y))}$$

for some constant $C(\alpha)$. By applying the triangle inequality, we see that the convergence of $\sum_{\gamma\in\Gamma} e^{-d(0,\gamma(0))}$ implies the convergence of the series (1). From the formula of the hyperbolic metric in $\mathbb{D}^2$, we obtain that the general term in the second series is equivalent to $(1-|\gamma(0)|^2)$. A direct computation based on the fact that the Möbius transformations preserve the cross-ratio implies that for any $\theta \in \partial\mathbb{D}^2$ we have (cf. **[Nic]**)

$$|(\gamma^{-1})'(\theta)| = (1-|\gamma(0)|^2)|\theta - \gamma(0)|^{-2}.$$

Since Y is an open Riemann surface, $\Omega(\Gamma) \neq \emptyset$. Let $I \subset \Omega(\Gamma)$ be a small interval which is disjoint from all its translates by non-trivial elements of Γ. The total euclidean length of the union of the translates of I is less than the length of the circle. Therefore

$$\sum_{\gamma\in\Gamma}\int_I |\gamma'(\theta)|d\theta \leq 2\pi.$$

Since $|\gamma'(\theta)| \geq 1/4(1-|\gamma^{-1}(0)|^2)$, the convergence of the series (1) follows.

The function G defined by (1) is positive and satisfies property (i) of the Green's function. To check that it satisfies (ii) also, it suffices to prove that for any $y \in Y$, $G(x,y)$ tends to 0 as x tends to ∞ in Y. Let Y_0 be the Nielsen core of Y. Then Y_0 is a compact surface with geodesic boundary and its preimage $\widetilde{Y}_0$ in $\mathbb{D}^2$ is a closed convex set. Saying that x tends to ∞ in Y means that the distance $d(x,Y_0)$ tends to ∞. Let $\widetilde{x} \in \mathbb{D}^2$ be any point in the preimage of x. Denote by $\widetilde{x}_0$ the nearest point projection of $\widetilde{x}$ on $\widetilde{Y}_0$. Let $y \in Y$ and let $\widetilde{y} \in \mathbb{D}^2$ be a point in its preimage. Then an elementary distance estimate gives, for all $\gamma \in \Gamma$,

$$|d(\widetilde{x},\gamma(\widetilde{y})) - d(\widetilde{x},\widetilde{x}_0) - d(\widetilde{x}_0,\gamma(\widetilde{y}))| \leq c,$$

for a constant c independent of x and of γ. Therefore by the proof of (i), $G(x,y)$ tends to 0 as $d(x,Y_0) = d(\widetilde{x},\widetilde{x}_0)$ tends to ∞. □

The surface Y is the union along the boundary of Y_0 and a finite number of half-infinite annuli called $A'_1, \cdots, A'_k, \cdots, A'_\ell$. Denote by γ_k the geodesic $\partial Y_0 \cap A'_k$. Consider the covering $\pi_k : A_k \to Y$ with fundamental group equal to that of γ_k. The annulus A_k can be identified conformally with $A_{e^{-R_k},e^{R_k}}$, where R_k satisfies $\ell(\gamma_k) = \pi^2/R_k$ (cf. §1). In particular, the existence of a lower bound on $\ell(\gamma_k)$ is equivalent to the existence of an upper bound on R_k. The restriction of π_k to one of the two halves of A_k bounded by the unit circle is an embedding. Up to applying an automorphism of A_k we can assume that the identification of A_k with $A_{e^{-R_k},e^{R_k}}$ is such that π_k restricts to the "inner" half-annulus $A_{e^{-R_k},1}$ as an embedding. In this way A'_k gets identified with $A_{e^{-R_k},1}$.

Let η_k denote the restriction to A'_k of the 1-form constructed above that solves Theorem 4.1 for A_k. Choose a decreasing smooth function λ on $[-1,1]$ such that

(i) $\lambda(x) = 1$ for $x \leq -1/2$, and

(ii) $\lambda(x) = 0$ for $x \geq 0$.

Consider the 1-form η' on Y which vanishes on Y_0 and which is defined on A'_k by $\eta'|A'_k = \lambda_k \eta_k$, where

$$\lambda_k(z) = \lambda\left(\frac{\log|z|}{R_k}\right).$$

The 1-form η' is smooth and satisfies $\bar{\partial}\eta' = dv$ on the union of the annuli $A_{e^{-R_k},e^{-R_k/2}}$. Thus the 2-form

$$dv_0 = dv - \bar{\partial}\eta' = (1 - \sum \lambda_k)dv - \sum \bar{\partial}\lambda_k \wedge \eta_k$$

has compact support in Y. Therefore we can define a function h on Y by

$$h(z) = 4\int_Y G(z,w)dv_0(w). \tag{2}$$

The 2-form $\bar{\partial}\partial h$ equals dv_0. Therefore, $\eta = \eta' + \partial h$ is the required 1-form if we can show that its hyperbolic norm is bounded by a function of ε and $\chi(Y)$. Since the hyperbolic norm of η' on Y is easily seen to be bounded only in terms of λ, it suffices to study the L^∞-norm of $\langle \partial h \rangle$.

Lemma 4.3.— *There exist positive constants C_1 and C_2 such that if u is a C^2 function on $\mathbb{D}^2$, one has*

$$||\langle\partial u\rangle||_\infty \leq C_1||u||_\infty + C_2||\langle\bar\partial\partial u\rangle||_\infty.$$

Proof.— Using the invariance of $\langle.\rangle$ under conformal automorphisms of $\mathbb{D}^2$, it suffices to prove that, for appropriate constants C_1 and C_2, the right-hand side is bigger than $\langle\partial u\rangle(0)$. For z contained in the disc $1/2\mathbb{D}^2$ we have

$$|\Delta u| \leq \frac{4}{(1-|z|^2)^2}||\langle\bar\partial\partial u\rangle||_\infty \leq C||\langle\bar\partial\partial u\rangle||_\infty.$$

Green's formula on $1/2\mathbb{D}^2$ gives

$$u(z) = \int_{1/2\mathbb{D}^2} G(z,w)\Delta u(w)dw + \int_{|w|=1/2} \frac{\partial G}{\partial\nu}(z,w)u(w)dw.$$

where G denotes the Green's function on $1/2\mathbb{D}^2$. By differentiating this expression twice, we obtain Lemma 4.3. □

By Lemma 4.3 (applied to the lift of h to $\mathbb{D}^2$), in order to obtain the required bound for $||\langle\partial h\rangle||_\infty$, it suffices to bound $||\langle\bar\partial\partial h\rangle||_\infty$ and $||h||_\infty$. In fact it is easy to bound $||\langle\bar\partial\partial h\rangle||_\infty$. The computation of $\bar\partial\lambda_k$ shows $||\langle\bar\partial\lambda_k\wedge\eta_k\rangle||_\infty \leq c$, where c depends only on λ. It follows that $||\langle dv_0\rangle||_\infty \leq c+1$.

In order to estimate $||h||_\infty$, we split the formula (2) into the sum of three integrals:

$$(3)\quad \frac{1}{4}h(z) = \sum_k \int_{A'_k} G(z,w)(1-\lambda_k(w))dv(w)$$
$$- \sum_k \int_{A'_k} G(z,w)\bar\partial\lambda_k\wedge\eta_k + \int_{Y_0} G(z,w)dv(w).$$

To get a bound on each of these terms we use the following estimate on the circular averages of the lift of G to A_k (cf. **[Di]**).

Lemma 4.4.— *Let $\widetilde{G}_k$ be the lift of G to A_k. Let $\widetilde{z}\in A_k$. Then, for all r such that $-R_k < \log r < R_k$, we have*

$$\frac{1}{2\pi}\int \widetilde{G}_k(\widetilde{z}, re^{i\theta})d\theta \leq \alpha_k(\log r + R_k),$$

where the α_k are positive constants which satisfy $\sum\alpha_k = 1$.

Proof.— By definition $\widetilde{G}_k(\widetilde{z},\widetilde{w}) = G(\pi_k(\widetilde{z}),\pi_k(\widetilde{w}))$. Therefore $\widetilde{G}_k(\widetilde{z},\widetilde{w})$ is a harmonic function of $\widetilde{w}$ in the complement of the preimage of $\pi_k(\widetilde{z})$. This preimage is a discrete set in A_k which contains at most one point in A'_k.

Let u be a harmonic function on a circular annulus. It is well-known that the average of u over the circle of radius r is an affine function of $\log r$ (**[Ah3]**, p. 164). Let $\widetilde{z}\in A_k$ and set

$$\psi_k(r) = \frac{1}{2\pi}\int \widetilde{G}_k(\widetilde{z}, re^{i\theta})d\theta.$$

The minimality property of the Green's function implies that, for any $z \in Y$, $G(z,w)$ tends to 0 as w tends to ∞ in Y. Therefore since A'_k embeds under the covering $\pi_k : A_k \to Y$, $\widetilde{G}_k(\widetilde{z},\widetilde{w})$ tends to 0 as $\log|\widetilde{w}|$ tends to $-R_k$. Thus $\psi_k(r)$ tends to 0 when $\log r$ tends to $-R_k$. Set $z = \pi_k(\widetilde{z})$ and let $\widetilde{z}_k$ be the lift of z to A_k which has the smallest modulus. Then we have

$$\psi_k(r) = \alpha_k(\log r + R_k) \quad \text{for} \quad -R_k < \log r < \log|\widetilde{z}_k|.$$

Since $\widetilde{G}_k(\widetilde{z},.)$ is positive on A_k, we have $\alpha_k > 0$. There is an interpretation of α_k. Consider the 1-form $*d\widetilde{G}_k(\widetilde{z},.)$, which is the "complex conjugate" of $d\widetilde{G}_k(\widetilde{z},.)$ (**[Ah3]** p. 164). Then α_k is the period of $*d\widetilde{G}_k(\widetilde{z},.)$ along a circle of radius smaller than $|\widetilde{z}_k|$. The union of the projections to Y of those circles for all the annuli A_k is null-homologous on Y. In particular the sum of the periods of $*d\widetilde{G}_k(\widetilde{z},.)$ equals the period of $*dG(z,.)$ on a small circle around the single singularity of $G(z,.)$, i.e. the point z. In a conformal chart around z, $G(z,w)$ is equivalent to $-\frac{1}{2\pi}\log|z-w|$. Therefore this period is equal to 1. So we have

$$\sum \alpha_k = 1.$$

In particular each α_k is less than 1. Consider the change of ψ_k when $\log r$ crosses the value $\log|\widetilde{z}|$, corresponding to the first singularity of $\widetilde{G}_k(\widetilde{z},.)$. It is a continuous function of r. Just after crossing the circle of radius $|\widetilde{z}|$ it becomes again an affine function of $\log r$. The slope of this new function equals the period of $*d\widetilde{G}_k(\widetilde{z},.)$ along a circle of radius slightly bigger than $\log|\widetilde{z}|$. The same homological argument as before implies that this new period is equal to $\alpha'_k = \alpha_k - \nu$, where ν is the number of points of smallest modulus in $\pi_k^{-1}(z)$. Hence $\alpha'_k \le 0$. This argument can be used across the entire annulus A_k, showing that ψ_k is a piecewise affine function of $\log r$. The slope of ψ_k decreases by a positive integer each time the circle of radius $\log r$ contains a point in the preimage of z. This completes the proof of Lemma 4.4. □

Let us go back to the estimate of each term in (3). On the annulus $\{w \in A_k | -R_k/2 < \log|w| < R_k/2\}$, the hyperbolic volume element dv is bounded by

$$\frac{C}{R_k^2} dr d\theta$$

where C is a constant depending only on an upper bound on R_k, or equivalently on a lower bound on ε. Hence, each integral in the first term of (3) is dominated by

$$\frac{C}{R_k^2}\int_{e^{-R_k/2}}^{1} \widetilde{G}_k(\widetilde{z}, re^{i\theta}) dr d\theta, \tag{4}$$

where $\widetilde{z}$ is any point in the preimage of z. The estimate $\dfrac{1}{2\pi}\displaystyle\int \widetilde{G}_k(\widetilde{z}, re^{i\theta})d\theta \le$ $\log r + R_k$ (Lemma 4.4) implies that (4) is bounded above only in terms of R_k. We remarked earlier that there exists a constant c depending only on λ, such that $||\langle\bar{\partial}\lambda_k \wedge \eta_k\rangle||_\infty \le c$. Therefore, by the same argument, the second term of (3) is bounded from above. However the last term of (3) is of different nature. It is an integral over the Nielsen core Y_0. The following lemma allows us to replace it by an integral over a certain annulus.

Lemma 4.5.— *Let $\varepsilon > 0$. Let Y be an open Riemann surface whose systole is bigger than ε. Then, there exists a constant D which depends only on ε and on $\chi(Y)$ such that*

(i) *any point of Y_0 is at distance smaller than D from the longest component, say γ_0, of ∂Y_0;*

(ii) *for $\rho_0 = \frac{2R_0}{\pi} \sin^{-1}(\tanh D)$, the restriction of the covering map $\pi_0 : A_0 \to Y$ to $A_{1,e^{\rho_0}}$ is onto.*

Proof.— (i) follows essentially from the fact that the area of Y_0 is finite, depending only on $\chi(Y_0)$. Another way to think about this result is by considering the Margulis decomposition of the compact hyperbolic surface obtained by doubling Y_0 along its boundary. (ii) is a restatement of (i) using the formula for the metric in A_0. □

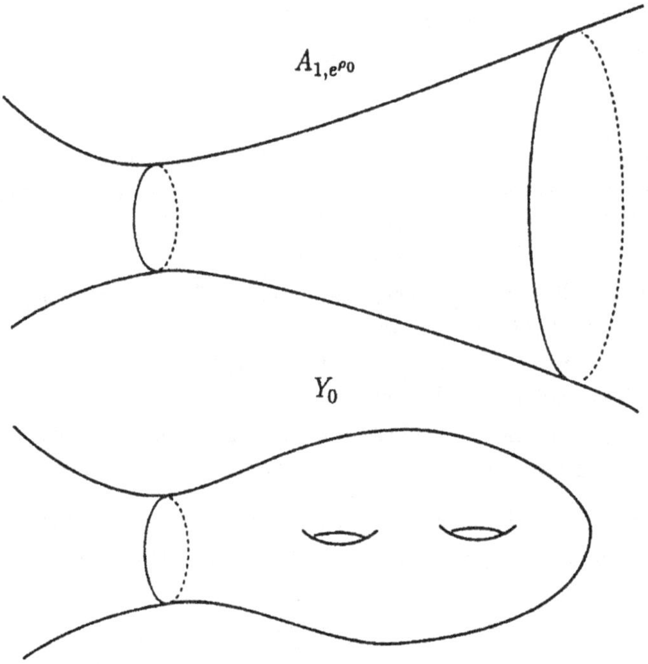

Figure 4.1

Using Lemma 4.5, the positivity of $\widetilde{G}_0$ and the fact that π_0 is a local isometry, we obtain

$$\int_{Y_0} G(z,w)dv(w) \leq \int_{A_{1,e^{\rho_0}}} \widetilde{G}_0(\widetilde{z},\widetilde{w})dv(\widetilde{w}),$$

for any point $\widetilde{z}$ in the preimage of z. On $A_{1,e^{\rho_0}}$ the hyperbolic volume element dv is bounded by $C/R_0^2 dr d\theta$ where C depends only on D, and therefore only on ε and $\chi(Y)$. So by Lemma 4.4 we have

$$\int_{A_{1,e^{\rho_0}}} \widetilde{G}(\widetilde{z},\widetilde{w})dv(\widetilde{w}) \leq \frac{C'}{R_0^2} \int_1^{e^{\rho_0}} (\log r + R_0)dr,$$

where C' depends only on ε and on $\chi(Y)$. An easy computation using the expression of ρ_0 gives that the integral is bounded from above in terms of R_0. This concludes the proof of Theorem 4.1. □

A primitive to the volume form in presence of short geodesics.

In the next chapter we will need a refinement of Theorem 4.1, which will allow us to deal with the case when Y contains geodesics shorter than the Margulis constant $\varepsilon(2)$. For $\varepsilon \leq \varepsilon(2)$, Y is the union of $Y^{]0,\varepsilon]}$ and $Y^{[\varepsilon,\infty[}$. Rather than finding solutions on Y to $\bar{\partial}\eta = dv$ which satisfy explicit bounds, we instead solve this equation on $Y^{[\varepsilon,\infty[}$. In fact, for the later applications, we only need to solve it on *the unbounded components of* $Y^{[\varepsilon,\infty[}$.

Definition.— Let Y be an open Riemann surface. *A component of* $Y^{[\varepsilon,\infty[}$ *is unbounded* if it is not compact (cf. Figure 4.2). Equivalently, a component of $Y^{[\varepsilon,\infty[}$ is unbounded if it is not entirely contained in Y_0.

Theorem 4.6.— *Let* $0 < \varepsilon \leq \varepsilon(2)$. *Let* Y *be an open Riemann surface. Then there exists a constant* C *depending only on* ε *and on* $\chi(Y)$, *such that on any unbounded component of* $Y^{[\varepsilon,\infty[}$ *there exists a 1-form* η *of type* $(1,0)$ *which satisfies*

(i) $\bar{\partial}\eta = dv$, *and*

(ii) $||\langle\eta\rangle||_\infty \leq C$.

Proof.— When Y is homeomorphic to a disc or to an annulus, the explicit solution described at the beginning of the proof of Theorem 4.1 has hyperbolic norm less than 1, independently of the length of the core geodesic.

In the other cases, let Y' be a component of $Y^{[\varepsilon,\infty[}$ which is unbounded. Suppose first that Y' is contained in a component of $Y - Y_0$, say A'_k. The 1-form constructed on A_k has norm less than 1. Since A'_k is isometrically embedded in A_k, the restriction of this form to Y' is a solution.

If Y' intersects Y_0, then each component of ∂Y_0 which intersects Y' is contained in Y' and so its length is bigger than ε. The surface Y' is the union of the compact surface $Y' \cap Y_0$ and a non-empty collection of half-infinite annuli, which are components of $Y - Y_0$. The proof of Theorem 4.6 follows exactly the same lines as that of Theorem 4.1. The restrictions of the 1-forms constructed on the annuli in $Y' - Y_0$ can be modified near the boundary of Y_0 to give a form η'. The 2-form $dv_0 = dv - \bar{\partial}\eta'$ has compact support in Y' so that we can define a function on Y' by the formula: $h(z) = 4\int_{Y'} G(z,w)dv(w)$, where G is the Green's function on Y. The proof that the form $\eta = \eta' + \partial h$ is the 1-form required in the theorem reduces to showing that h is bounded over Y'. The only difference between this proof and that of Theorem 4.1 appears in the estimate of the third integral in (3). The following lemma substitutes Lemma 4.5.

Lemma 4.7.— *Let* $\varepsilon > 0$. *Let* Y *be an open Riemann surface. Let* Y' *be a component of* $Y^{[\varepsilon,\infty[}$ *which intersects* Y_0. *Then there exists a constant* D' *which depends only on* ε *and on* $\chi(Y)$ *such that any point of* $Y' \cap Y_0$ *is at distance less than* D' *from the longest component of* ∂Y_0 *which is contained in* Y'.

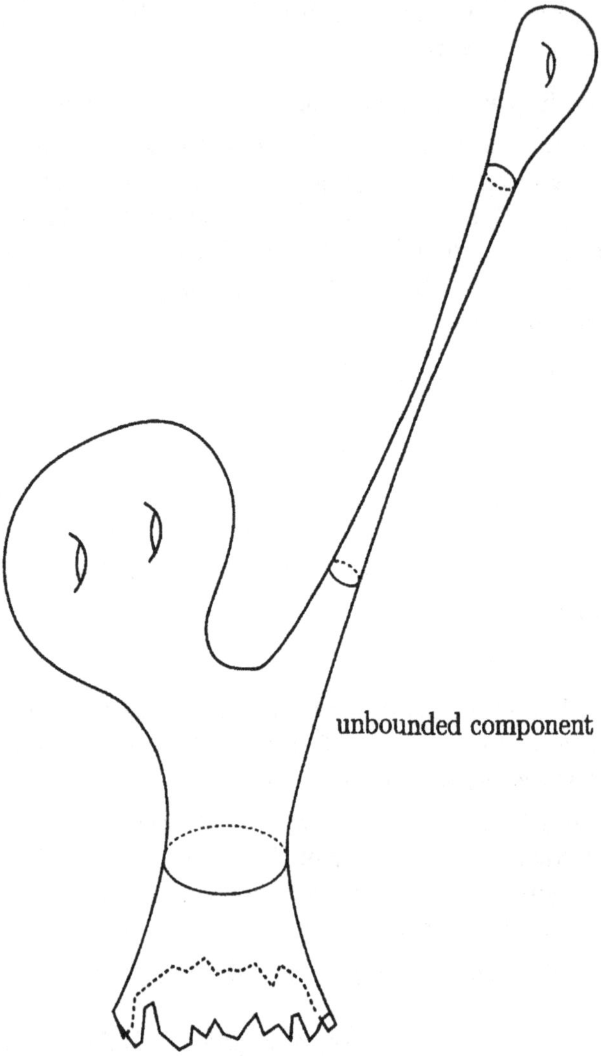

Figure 4.2

This concludes the proof of Theorem 4.6. □

CHAPITRE 5

Contraction properties of the Theta operator

Theorem 5.1. — *Let $\varepsilon > 0$. Let X be a compact hyperbolic Riemann surface whose systole is bigger than ε. Let $\pi : Y \to X$ be a geometric cover of X. There exists a constant $c = c(\varepsilon, \chi(X)) > 0$ such that $||\Theta_{Y/X}|| \leq 1 - c$.*

This theorem is a special case of a theorem of McMullen [**McM1**]. Here the hypothesis "geometric" has replaced the original hypothesis "non-amenable". We present the proof that Barett and Diller have given [**BD**] using the result of [**Di**] that was described in the previous chapter. Theorem 5.1 remains true when X has finite volume. It can be proved using the same techniques.

Averaging operators.

We need to extend the averaging procedure of Poincaré series to other types of tensors. For instance, if F is a function on Y we can sum it (in certain cases) over the sheets of the cover $\pi : Y \to X$ by the formula

$$\theta F(x) = \sum_{y \in \pi^{-1}(x)} F(y).$$

The Cauchy formula implies that this series converges uniformly when F is holomorphic and integrable with respect to the hyperbolic volume on Y. More generally, this will hold when F is not necessarily holomorphic, but has modulus less than that of an integrable holomorphic function. A special case is the following. For $\phi \in \Omega(Y)$, we defined in §3 a function $\langle\phi\rangle$: it is the density of the measure $|\phi(y)||dy^2|$ with respect to the hyperbolic volume measure on Y. Then $\theta\langle\phi\rangle$ can be defined using the formula above and $\theta\langle\phi\rangle(x)$ is the density with respect to the hyperbolic volume on X of a measure which has the same total mass as the measure $|\phi(y)||dy^2|$ on Y.

The same averaging procedure can be applied to 1-forms. Let η be a 1-form of type $(1,0)$ on Y. Let U be a conformal chart around a point $x \in X$. For each component U_i of $\pi^{-1}(U)$, let $s_i : U \to U_i$ be a holomorphic section of π. On U_i, η can be written as $\eta_i(z)dz$. Then the series

$$\sum_i \eta_i(s_i(z))s_i'(z)dz$$

defines, in certain cases, a 1-form of type $(1,0)$ on U. This happens for instance when the hyperbolic norm $\langle\eta\rangle(y)$ is bounded by the modulus of an holomorphic function which is integrable on Y with respect to the hyperbolic volume. These local expressions paste together defining a 1-form on Y denoted by $\theta\eta$. In the next paragraphs we will commute the exterior differentiation with the averaging operator θ. This will be allowed since the series that we consider converge uniformly.

Proof of Theorem 5.1.— We denote by Θ the classical operator $\Theta_{Y/X}$ acting on integrable holomorphic quadratic differentials, and by θ the averaging operator introduced above, acting on other types of tensors. Since the cover $Y \to X$ is geometric, Y is an open Riemann surface and the results of §4 can be applied.

Let $\phi \in \Omega(Y)$ with $||\phi|| = 1$. The pull-back $\pi^*(\Theta\phi)$ of $\Theta\phi$ to Y is a holomorphic quadratic differential (which, if not identically 0, is not integrable). Its zero set is exactly the preimage of the zero set Z of $\Theta\phi$. Assuming that $\Theta\phi$ does not vanish identically, we define a meromorphic function F on Y by $\phi = F\pi^*(\Theta\phi)$. Let $Z(r)$ denote the neighborhood of radius r of Z on X, and by $\widetilde{Z}(r)$ the preimage of $Z(r)$ in Y.

Lemma 5.2.— *For any $r > 0$, F is integrable in $Y - \widetilde{Z}(r)$ with respect to the hyperbolic volume.*

Proof.— Let $m(r) > 0$ be a lower bound for $\langle\Theta\phi\rangle(x)$ on $X - Z(r)$. On $Y - \widetilde{Z}(r)$, we have

$$|F(y)| \le \frac{\langle\phi\rangle(y)}{m(r)}.$$

Hence, $\int_{Y-\widetilde{Z}(r)} |F|dv \le 1/m(r)$. □

This observation means that we can define θF and $\theta|F|$ on $X - Z$. Since $\theta F \equiv 1$, $\theta|F| \ge 1$. As $||\phi|| = 1$, the integral of $\theta\langle\phi\rangle$ with respect to the hyperbolic volume is 1. Thus, we have

$$\begin{aligned} 1 - ||\Theta\phi|| &= \int_X \theta\langle\phi\rangle dv - \int_X |\Theta\phi| \\ &= \int_X (\theta|F| - 1)\langle\Theta\phi\rangle dv \\ &\ge \int_0^{r_0} m(r)\Big(\int_{\partial Z(r)} \theta(|F|ds) - \ell(\partial Z(r))\Big)dr. \end{aligned}$$

In the last line, we have bounded from below the integral over X of the positive function $(\theta|F| - 1)\langle\Theta\phi\rangle$ by its integral over $Z(r_0)$ for a sufficiently small constant r_0 which will be fixed later.

Let η be the 1-form on Y constructed in Theorem 4.1. Its hyperbolic norm is less than C. In particular, $\theta(F\eta)$ is well defined since F is integrable on Y. We have

$$\begin{aligned}\int_{\partial Z(r)} \theta(|F|ds) &\geq \frac{1}{C}\Big|\int_{\partial Z(r)} \theta(F\eta)\Big| \\ &= \frac{1}{C}\Big|\int_{X-Z(r)} \bar{\partial}\theta(F\eta)\Big| \quad \text{by Stokes formula} \\ &= \frac{1}{C}\Big|\int_{X-Z(r)} \theta(Fdv)\Big| \quad \text{since } F \text{ is holomorphic} \\ &= \frac{1}{C}\Big|\int_{X-Z(r)} dv\Big| \\ &= \frac{1}{C}\mathrm{Area}(X-Z(r)) \\ &= \frac{1}{C}(2\pi|\chi(X)| - \mathrm{Area}(Z(r))).\end{aligned}$$

The holomorphic quadratic differential $\Theta\phi$ has at most $4g-4$ zeroes, where g is the genus X. Hence, for $r \leq \varepsilon/2$, the area of $Z(r)$ is less than $c_1 r^2$ and the length of $\partial Z(r)$ is less than $c_2 r$ for constants c_1 and c_2 depending only on $\chi(X)$. It follows that for all r sufficiently small in terms of ε, $\int_{\partial Z_r} \theta(|F|ds) - \ell(\partial Z_r) \geq c$, where $c > 0$ depends only on ε and on $\chi(X)$.

If $||\Theta\phi|| \geq 1/2$, we know from Proposition 3.7 that $m(r)$ is bounded from below, independently of $\Theta\phi$, by a positive constant depending only on r, ε and $\chi(X)$. Recall that the constant C provided by Theorem 4.1 depends only on ε and $\chi(Y)$. Therefore, if $||\Theta\phi|| \geq 1/2$, $1 - ||\Theta\phi||$ is bounded from below by a constant which depends only on ε and $\chi(X)$.

When $||\Theta\phi|| \leq 1/2$, we obviously have $1 - ||\Theta\phi|| \geq 1/2$.

This concludes the proof of Theorem 5.1. □

Remark.— Geometric covers of a Riemann surface are examples of *non-amenable covers*, which were defined by McMullen's. Without going into the precise definition, let us just say that an example of amenable cover is given by a Galois cover whose automorphism group is amenable. The basic result of [**McM1**] is that the Theta operator associated to a non-amenable cover is strictly contracting. There is a link between McMullen's proof and the one of Barrett and Diller. The basic tool for the proof of Theorem 5.1 was the existence of a $\bar{\partial}$-primitive to the volume form, whose hyperbolic norm is bounded from above. As we noticed, this implied a linear isoperimetric inequality on Y. This is reminiscent of Fölner's criterium which characterizes non-amenable graphs as those which satisfy a linear isoperimetric inequality (cf. [**Gr**]).

Remark.— McMullen proved also a converse to his theorem, namely that the Theta operator associated to *an amenable cover* $Y \to X$, has norm 1 [**McM1**]: i.e. there is a sequence of elements ϕ_i in $\mathcal{Q}(Y)$ with $||\phi_i|| = 1$ such that $||\Theta\phi_i||$ tends to

1. However, one should notice that, for any infinite cover $\pi : Y \to X$ there does not exist a $\phi \in \mathcal{Q}(Y)$ such that $||\phi|| = 1$ and $||\Theta\phi|| = 1$. Let us justify this observation in the case of a Galois cover. This won't never be used as such in the sequel, but is behind the argument that we will use for proving Proposition 6.3. If, for some $\phi \in \mathcal{Q}(Y)$, $||\phi|| = ||\Theta\phi|| = 1$, then there exists $\mu \in \mathcal{B}(X)$ with $||\mu|| = 1$ such that $\langle\Theta\phi, \mu\rangle = \langle\phi, \pi^*\mu\rangle = 1$ (cf. §1). It follows easily from the definition of the pairing that

$$\pi^*\mu = \frac{\overline{\phi}}{|\phi|}. \tag{1}$$

Let g be an element of the automorphism group of the cover $Y \to X$. As $\pi^*\mu$ is invariant under g, (1) implies $\phi(gz)g'(z)^2 = k\phi(z)$ for a constant $k \neq 0$. If g has infinite order, this equality contradicts that $||\phi||$ is finite.

The norm of $\Theta_{Y/X}$ in presence of short geodesics.

Let $0 < \varepsilon \leq \varepsilon(2)$. The surface X can be decomposed as the union of the ε-thin part $X^{]0,\varepsilon]}$ and the ε-thick part $X^{[\varepsilon,\infty[}$. Recall that the cover $\pi : Y \to X$ we consider is associated to a proper incompressible surface $S \subset X$.

Definition.— A component of $X^{]0,\varepsilon]}$ or of $X^{[\varepsilon,\infty[}$ is *liftable* if it can be isotoped into S.

Let Z be a component of $X^{]0,\varepsilon]}$ or of $X^{[\varepsilon,\infty[}$. If Z is liftable, $\pi^{-1}(Z)$ consists of a single isomorphic copy of itself, which is called *the lift of* Z, and a disjoint union of copies of the universal cover of Z. Note that the latter are necessarily contained in the unbounded components of $Y^{[\varepsilon,\infty[}$. If Z is not liftable, any component of $\pi^{-1}(Z)$ is an infinite cover of Z. Therefore, it is entirely contained in an unbounded component of $Y^{[\varepsilon,\infty[}$. So the only components of the preimage of $X^{]0,\varepsilon]}$ or of $X^{[\varepsilon,\infty[}$ which are compact are the lifts of the liftable components. All the others are contained in the unbounded components of $Y^{[\varepsilon,\infty[}$.

Definition.— *The ε-amenable part of the cover $Y \to X$* is the union of the total preimage of $X^{]0,\varepsilon]}$ and of the lifts of the liftable components of $X^{[\varepsilon,\infty[}$. We denote it by $\mathcal{A}(X)^\varepsilon$.

The following result explains how the presence of short geodesics does influence the behaviour of of $||\Theta_{Y/X}||$.

Theorem 5.3.— *Let $\pi : Y \to X$ be a geometric cover of Riemann surfaces. Let $\varepsilon \leq \varepsilon(2)$. Let $\eta > 0$. There exists a $\delta > 0$ which depends only on η, ε and on $\chi(X)$, such that: for all $\phi \in \mathcal{Q}(Y)$ with $||\phi|| = 1$ and $||\Theta_{Y/X}\phi|| \geq 1 - \delta$, then the ϕ-mass of $\mathcal{A}(X)^\varepsilon$ is more than $1 - \eta$.*

Proof.— We argue by contradiction. Then, there is a sequence of isomorphic covers $Y_n \to X_n$, and $\phi_n \in \mathcal{Q}(Y_n)$ with unit norm, such that $||\Theta_{Y_n/X_n}\phi_n||$ tends to 1 and that the ϕ_n-mass of $\mathcal{A}(X_n)^\varepsilon$ is bounded away from 1. Observe that the topological type of a geometric cover associated to an incompressible surface $S \subset X$ only depends on the embedding of S into X; therefore we can assume that the covers $Y_n \to X_n$ have the same topological type. Denote Θ_{Y_n/X_n} by Θ_n. Since the number of components of $X_n^{[\varepsilon,\infty[}$ is bounded independently of n, there is some component K_n of $X_n^{[\varepsilon,\infty[}$ for which the ϕ_n-mass of the total union of the non-compact components of $\pi^{-1}(K_n)$ admits a lower bound $m > 0$. Let W_n denote the total union of the non-compact components of $\pi^{-1}(K_n)$. Let $x_n \in K_n$. Choose a uniformizing map $(\mathbb{D}^2, 0) \to (X_n, x_n)$. We can suppose, up to extracting a subsequence, that (X_n, x_n) converges to (K, x) where K is a hyperbolic surface of finite area, since $\mathrm{inj}(x_n) \geq \varepsilon$.

If K were compact, the injectivity radius of X_n would be bounded from below independently of n. Then, Theorem 5.1 would contradict the hypothesis of Theorem 5.3.

Therefore K is non-compact. Fix a positive number $\mu \leq \varepsilon$ smaller than the systole of K. For $\mu' \leq \mu$, let $K_n^{[\mu',\infty[}$ be the component of $X_n^{[\mu',\infty[}$ which contains K_n and let $W_n^{[\mu',\infty[}$ be the union of the non-compact components of $\pi^{-1}(K_n^{[\mu',\infty[})$. It is important to notice that $K_n^{[\mu',\infty[}$ is not necessarily diffeomorphic to K_n. However, since μ is smaller than the systole of K, for all $\mu' \leq \mu$, $K^{[\mu',\infty[}$ is diffeomorphic to $K^{[\mu,\infty[}$ and $K_n^{[\mu',\infty[}$ converges therefore to $K^{[\mu',\infty[}$.

In order to obtain a contradiction, we distinguish two cases according as $K_n^{[\mu,\infty[}$ does or does not lift to Y_n for sufficiently large n. When $K_n^{[\mu,\infty[}$ lifts, we will show that the ϕ_n-mass of $W_n^{[\mu,\infty[}$ tends to 0 when n tends to ∞. Since $W_n \subset W_n^{[\mu,\infty[}$, the ϕ_n-mass of W_n tends to 0 also. This contradicts our assumption that the ϕ_n-mass of W_n is bigger than m. When $K_n^{[\mu,\infty[}$ does not lift, we will contradict the hypothesis $||\Theta_n\phi_n|| \to 1$.

Let $0 < \mu' \leq \mu$ be a constant that will be fixed precisely later. As discussed already, we can associate to the cover $W_n^{[\mu',\infty[} \to K_n^{[\mu',\infty[}$ an averaging operator, which operates on integrable holomorphic quadratic differentials, and tensors of other types. We denote this operator by Θ'_n, when it is considered as acting on integrable holomorphic quadratic differentials and by θ'_n, when acting on tensors of other types.

1) *$K_n^{[\mu,\infty[}$ lifts to Y_n for sufficiently large n.*

Since $K_n^{[\mu',\infty[}$ is isotopic to $K_n^{[\mu,\infty[}$, it is liftable to Y_n also. The preimage of $K_n^{[\mu',\infty[}$ is the disjoint union of $W_n^{[\mu',\infty[}$ and the isomorphic lift $\widetilde{K}_n^{[\mu',\infty[}$ of $K_n^{[\mu',\infty[}$. Therefore in restriction to $K_n^{[\mu',\infty[}$, we have

$$(1) \qquad \Theta_n\phi_n = \phi_n|\widetilde{K}_n^{[\mu',\infty[} + \Theta'_n\phi_n,$$

where we have identified the restriction $\phi_n|\widetilde{K}_n^{[\mu',\infty[}$ with its projection on $K_n^{[\mu',\infty[}$.

For any measurable subset $U \subset X_n$, we have $\int_U |\Theta_n \phi_n| \leq \int_{\pi^{-1}(U)} |\phi_n|$ and therefore

$$\begin{aligned}
&1 - \|\Theta_n \phi_n\| \\
&= \int_{\pi^{-1}(K_n^{[\mu',\infty[})} |\phi_n| + \int_{Y_n - \pi^{-1}(K_n^{[\mu',\infty[})} |\phi_n| - \int_{K_n^{[\mu',\infty[}} |\Theta_n \phi_n| - \int_{X_n - K_n^{[\mu',\infty[}} |\Theta_n \phi_n| \\
&\geq \int_{\pi^{-1}(K_n^{[\mu',\infty[})} |\phi_n| - \int_{K_n^{[\mu',\infty[}} |\Theta_n \phi_n|.
\end{aligned}$$

To obtain a contradiction, it suffices to bound from below the last expression in the above inequality. Applying the triangle inequality and (1), we find that it is bounded from below by $\int_{W_n^{[\mu',\infty[}} |\phi_n| - \int_{K_n^{[\mu',\infty[}} |\Theta'_n \phi_n|$. The hypothesis that $K_n^{[\mu,\infty[}$ lifts to Y_n has the following topological consequence.

Fact 5.4.— *Let $\mu' \leq \mu$. Then for all sufficiently large n, $W_n^{[\mu',\infty[}$ is contained in $Y_n^{[\mu,\infty[}$.*

Proof.— If this is false, some component W' of $W_n^{[\mu',\infty[}$ intersects a component Y' of $Y_n^{]0,\mu]}$. Since the cover $\pi : Y_n \to X_n$ is geometric, the restriction of π to each component of $Y_n^{]0,\mu]}$ is a homeomorphism. In particular, $W' \cap Y'$ is homeomorphic to its image by π. Since $\pi(W') = K_n^{[\mu',\infty[}$, $\pi(W' \cap Y')$ is the intersection of $K_n^{[\mu',\infty[}$ with a component of $X_n^{]0,\mu]}$. By our choice of μ, any component of $K^{]0,\mu]}$ is a cusp. Such a cusp is approximated in $X_n^{]0,\mu]}$ by an annulus whose core geodesic has length tending to 0. Therefore, for all sufficiently large n, the intersection of $K_n^{[\mu',\infty[}$ with any component of $X_n^{]0,\mu]}$ is an annulus such that the injectivity radius equals μ on one boundary component and μ' on the other. It follows that the component of $W_n^{[\mu,\infty[}$ which is contained in W' contains a boundary curve of $Y_n^{]0,\mu]}$. This is impossible because all components of $W_n^{[\mu,\infty[}$ are simply connected since $K_n^{[\mu,\infty[}$ lifts. □

Since the differentials $\Theta'_n \phi_n$ have norm less than 1, they converge uniformly to a holomorphic quadratic differential ϕ defined on $K_n^{[\mu',\infty[}$, up to possibly passing to a subsequence (cf. the remark after the statement of Theorem 3.1). But this does not guarantee that ϕ is non-zero, in contrast to Theorem 3.1. Therefore, we consider two subcases.

1a) $\phi = 0$.

The diameter of $K_n^{[\mu',\infty[}$ is bounded from above independently of n. Hence there exists a ball of fixed radius in $\mathbb{D}^2$ centered at 0 whose projection to X_n contains $K_n^{[\varepsilon',\infty[}$, for all n. Therefore the uniform convergence of $\Theta'_n \phi_n$ to 0 implies

$$\lim_{n \to \infty} \int_{K_n^{[\mu',\infty[}} |\Theta'_n \phi_n| = 0.$$

So we obtain a non-zero lower bound on $\int_{W_n^{[\mu',\infty[}} |\phi_n| - \int_{K_n^{[\mu',\infty[}} |\Theta'_n \phi_n|$ for all sufficiently large n. This contradicts the hypothesis of Theorem 5.3.

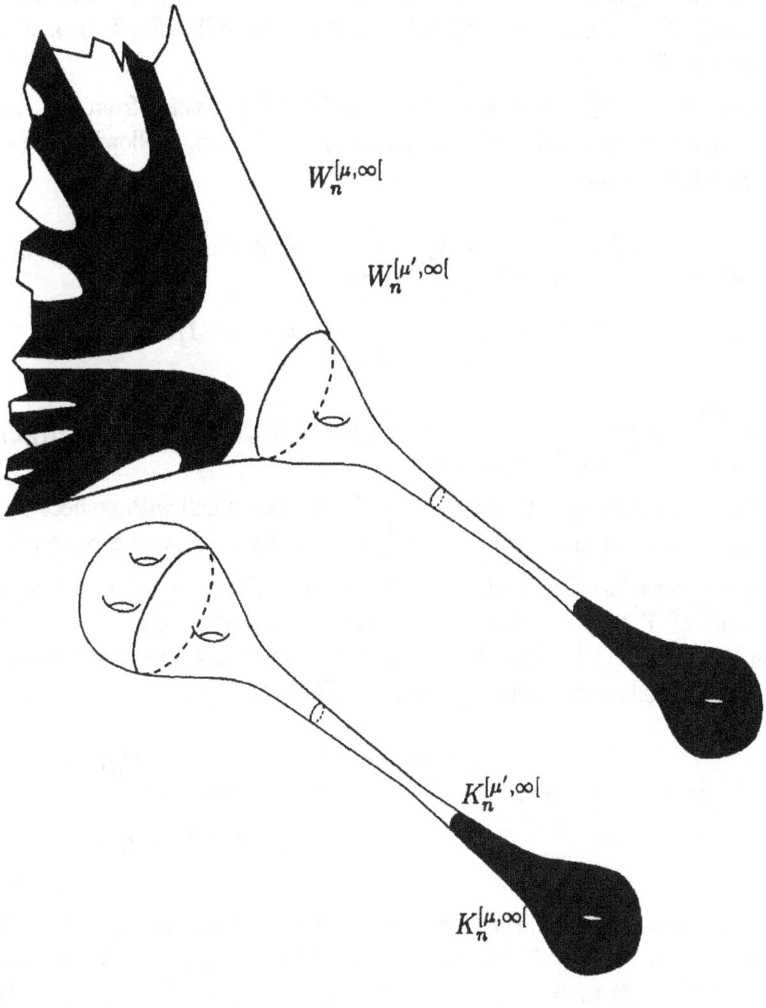

Figure 5.1

1b) $\phi \neq 0$.

The proof in this case is similar to that of Theorem 5.1. Denote by Z (resp. Z_n) the zero set of ϕ (resp. $\Theta'\phi_n$). Denote by $Z_n(r)$ the neighborhood of radius r of Z_n in $K_n^{[\mu',\infty[}$. Let $\langle\Theta'_n\phi_n\rangle$ be the hyperbolic norm of $\Theta'_n\phi_n$.

Fact 5.5. — *Let* $\mu' \leq \mu$. *For any* $r > 0$, *there is lower bound* $m(r) > 0$ *for* $\langle\Theta'_n\phi_n\rangle$ *over* $K_n^{[\mu',\infty[} - Z_n(r)$ *which is independent of* n.

This lower bound $m(r)$ depends on μ'.

Proof. — The proof is exactly the same as the one of Proposition 3.7: it follows directly from the convergence of $\Theta'_n\phi_n$ to ϕ. □

Let $\pi^*(\Theta'_n\phi_n)$ be the pull-back of $\Theta'_n\phi_n$ to $W_n^{[\mu',\infty[}$. The zero set of $\pi^*(\Theta'_n\phi_n)$ is the preimage $\tilde{Z}_n$ of Z_n. Let F_n be the meromorphic function on $W_n^{[\mu',\infty[}$ defined by $\phi_n = F_n\pi^*(\Theta'_n\phi_n)$. Then F_n is integrable on $W_n^{[\mu',\infty[} - \widetilde{Z}_n(r)$ (as in the proof of Lemma 5.2). Thus the averages $\theta'_n|F_n|$ and $\theta'_n F_n$ are well defined; also, $\theta'_n F_n \equiv 1$. Therefore, $\theta'_n|F_n| \geq 1$.

Suppose that $\mu' \leq \mu$ is chosen so that $\partial K^{[\mu',\infty[}$ is disjoint from Z. Then, for all n sufficiently large, $\partial K_n^{[\mu',\infty[}$ is disjoint from Z_n too. Following the proof of Theorem 5.1, we obtain

$$\int_{W_n^{[\mu',\infty[}} |\phi_n| - \int_{K_n^{[\mu',\infty[}} |\Theta'_n\phi_n| = \int_{K_n^{[\mu',\infty[}} \theta'_n\langle\phi_n\rangle dv - \int_{K_n^{[\mu',\infty[}} |\Theta'_n\phi_n|$$

$$= \int_{K_n^{[\mu',\infty[}} (\theta'_n|F_n| - 1)\langle\Theta'_n\phi_n\rangle dv$$

$$\geq \int_0^{r_0} m(r)\Big(\int_{\partial(K_n^{[\mu'+r,\infty[}-Z_n(r))} \theta'_n(|F_n|ds) - \ell(\partial(K_n^{[\mu'+r,\infty[} - Z_n(r)))\Big)dr.$$

In the formula above, $r_0 > 0$ is chosen for the moment small with respect to μ' and to the distance between Z and $\partial K^{[\mu',\infty[}$, but will be fixed more precisely later.

Consider the 1-form η_n that was constructed in Theorem 4.6 on the unbounded components of $Y_n^{[\mu,\infty[}$. Its hyperbolic norm is less than a constant C depending only on μ and on $\chi(Y)$. Since $W_n^{[\mu',\infty[} \subset Y_n^{[\mu,\infty[}$ (Fact 5.4), we obtain by imitating the argument used at the end of the proof of Theorem 5.1:

$$\int_{\partial(K_n^{[\mu'+r,\infty[}-Z_n(r))} \theta'_n(|F_n|ds) \geq \frac{1}{C}\int_{\partial(K_n^{[\mu'+r,\infty[}-Z_n(r))} \theta'_n|F_n\eta_n|$$

$$= \frac{1}{C}|\text{Area}(K_n^{[\mu'+r,\infty[} - Z_n(r))|.$$

If r_0 is sufficiently small, then for all $r \leq r_0$, the boundary of $K_n^{[\mu'+r,\infty[} - Z_n(r)$ is the disjoint union of $\partial K_n^{[\mu'+r,\infty[}$ and $\partial Z_n(r)$. When μ' and r tend to 0, the length of $\partial K_n^{[\mu'+r,\infty[}$ tends to 0. When r tends to 0, the length of $\partial Z_n(r)$ also tends to 0. However since the cardinality of $Z_n \cap K_n^{[\mu',\infty[}$ might increase when μ' tends to 0, we first fix a choice of μ'. The hyperbolic area of $K_n^{[\mu,\infty[}$ admits a lower bound $\nu > 0$ which depends only on $\chi(X)$. Choose $\mu' \leq \mu/2$ such that $\ell(\partial K_n^{[2\mu',\infty[}) < \nu/2$ and such that $\partial K^{[\mu',\infty[} \cap Z = \emptyset$. Then, for all $r \leq \mu'$, we have

$$\text{Area}(K_n^{[\mu'+r,\infty[}) - \ell(\partial K_n^{[\mu'+r,\infty[}) > \frac{\nu}{2}.$$

With this choice of μ', the number of zeroes of $\Theta'_n\phi_n$ that are contained in $K_n^{[\mu',\infty[}$ is bounded above independently of n because it converges (counting multiplicity) to the number of zeroes of ϕ contained in $K^{[\mu',\infty[}$. Hence the area of $Z_n(r)$ and the length of its boundary are bounded from above by $c_1 r^2$ and $c_2 r$ respectively for constants c_1 and c_2 which are independent of n. Therefore, there is some $r_0 \leq \mu'$ such that, for all $r \leq r_0$, and for all sufficiently large n, we have

$$\int_{\partial(K_n^{[\mu'+r,\infty[}-Z_n(r))} \theta'_n(|F_n|ds) - \ell(\partial(K_n^{[\mu'+r,\infty[} - Z_n(r))) \geq \frac{\nu}{2C}.$$

Therefore

$$\int_0^{r_0} m(r)\Big(\int_{\partial(K_n^{[\mu'+r,\infty[}-Z_n(r))} \theta'_n(|F_n|ds) - \ell(\partial(K_n^{[\mu'+r,\infty[} - Z_n(r)))\Big)dr$$

is bounded away from 0 independently of n by Fact 5.5. This leads to a contradiction as in 1a).

2) *$K_n^{[\mu,\infty[}$ does not lift to Y_n for sufficiently large n.*

An important subcase to keep in mind occurs when Y_n is the universal cover of X_n. However this situation could be handled with the same methods as above. The main difference between the first case and second case occurs when $K_n^{[\mu,\infty[}$, although not liftable to Y_n, contains some boundary components which are liftable. Because of such curves, certain components of $W_n^{[\mu',\infty[}$ intersect $Y_n^{]0,\mu]}$ for $\mu' \leq \mu$ and Fact 5.4 is no longer true. We need to argue differently.

Since $K_n^{[\mu,\infty[}$ is not liftable, $\pi^{-1}(K_n^{[\mu,\infty[}) = W_n^{[\mu,\infty[}$. Hence $\Theta_n = \Theta'_n$ and $\theta_n = \theta'_n$. Keeping the same notations as in 1), we assume that (X_n, x_n) converge to (K, x) where K is a hyperbolic surface of finite area. By Theorem 3.1, $(\Theta_n \phi_n)$ converges uniformly to a holomorphic quadratic differential ϕ on K. Since the $\Theta_n\phi_n$-mass of K_n is bounded away from 0, ϕ is non-zero (cf. 1a)). Let Z_n (resp. Z) be the zero set of $\Theta_n\phi_n$ (resp. ϕ). Let $\widetilde{Z}_n$ be the preimage of Z_n in Y_n and $\widetilde{Z}_n(r)$ be the neighborhood of radius r of $\widetilde{Z}_n$. Let F_n be the meromorphic function on Y_n defined by $\phi_n = F_n \pi^*(\Theta_n\phi_n)$. For any $r > 0$, F_n is integrable on $Y_n - \widetilde{Z}_n(r)$ with respect to the hyperbolic volume. Its L^1-norm is smaller than $1/m(r)$, where $m(r) > 0$ is a lower bound for $\langle \Theta_n\phi_n \rangle$ on $K_n - Z_n(r)$ (cf. Lemma 5.2).

Since Y_n covers X_n, its universal cover is naturally identified with $\mathbb{D}^2$. Let $\bar{F}_n$ be the lift of F_n to $\mathbb{D}^2$. Let $\bar{Z}$ (resp. $\bar{Z}_n$) be the preimage of Z (resp. Z_n) in $\mathbb{D}^2$.

Lemma 5.6. — *Up to extracting a subsequence, the functions $\bar{F}_n$ converge uniformly on compact subsets of $\mathbb{D}^2 - \bar{Z}$ to a holomorphic function $\bar{F}$.*

Proof. — Let $\mathcal{K} \subset \mathbb{D}^2$ be a compact set. Under the covering $\mathbb{D}^2 \to K$, the "degree" of the projection $\mathcal{K} \to K$ is finite, i.e. the cardinality of the preimage of any point of K which is contained in $\mathcal{K}$ is bounded, independently of that point. Hence since (X_n, x_n) tend to (K, x), the degree of the projections $\mathcal{K} \to X_n$ is bounded independently of n. The same property holds a fortiori for the projections $\mathcal{K} \to Y_n$. By the uniform convergence of $\Theta_n\phi_n$ to ϕ, $\bar{Z}_n$ converges to $\bar{Z}$ and $\bar{Z}_n(r)$ converges to $\bar{Z}(r)$ when n tends to ∞. It follows that for any $r > 0$, the L^1-norm of $\bar{F}_n|\mathcal{K} - \bar{Z}(r)$ is bounded independently of n. By Cauchy formula, this implies that $\bar{F}_n$ converges uniformly to a holomorphic function $\bar{F}$ on compact sets in $\mathcal{K} - \bar{Z}$, up to passing to a subsequence. □

2a) *$\bar{F}$ is not constant.*

Since $\bar{F}$ is holomorphic and not constant, there exists a point $\bar{q} \in \mathbb{D}^2 - \bar{Z}$ such that $\bar{F}(\bar{q})$ has a non-zero imaginary part. By the uniform convergence of $\bar{F}_n$ to $\bar{F}$, there exists positive numbers α, η and ρ so that for all q in the ball $B(\bar{q}, \rho)$ and for all sufficiently large n, we have

(i) $|\bar{F}_n(q)| > \alpha > 0$, and

(ii) $0 < \eta < \arg \bar{F}_n(q) < \pi - \eta$.

Suppose that $\bar{q}$ is near the origin in $\mathbb{D}^2$, so that the projection B_n of $B(\bar{q}, \rho)$ on X_n is contained in $K_n^{[\mu,\infty[}$. If $\rho \leq \mu/2$, B_n is embedded for all sufficiently large n. Since $\bar{q} \notin Z$, the ball $B(\bar{q}, \rho)$ is at distance bigger than ρ from $\bar{Z}_n$ for any sufficiently small ρ and for all sufficiently large n. In particular B_n is contained in $K_n - Z_n(\rho)$. Let q_n denote the image of $\bar{q}$ in Y_n. For sufficiently large n, $B(q_n, \rho)$ maps homeomorphically to B_n, under the covering $Y_n \to X_n$.

Let x and y be complex numbers such that $x + y = 1$, $|x| \geq \alpha > 0$ and $0 < \eta \leq \arg(x) \leq \pi - \eta$. Then, we have $|x| + |y| - 1 \geq c(\eta, \alpha) > 0$. Hence, for all $z \in B(q_n, \rho)$

$$|F_n(z)| + |1 - F_n(z)| - 1 \geq c(\eta, \alpha).$$

Thus, for all sufficiently large n

$$\begin{aligned}\int_{\pi^{-1}(B_n)} |\phi_n| - \int_{B_n} |\Theta_n \phi_n| &\geq \int_{B_n} (\theta_n |F_n| - 1)\langle \Theta_n \phi_n \rangle dv \\ &\geq \int_{B(q_n,\rho)} (|F_n(z)| + |1 - F_n(z)| - 1) m(\rho) dv \\ &\geq c(\eta, \alpha) m(\rho) \int_{B_n} dv,\end{aligned}$$

where $m(\rho)$ is a lower bound of $\langle \Theta_n \phi_n \rangle$ on $K_n - Z_n(\rho)$. This is impossible as $||\Theta_n \phi_n||$ tends to 1.

2b) *$\bar{F}$ is a non-zero constant.*

Let $y_n \in Y_n$ be the image of $0 \in \mathbb{D}^2$ under the covering map $\mathbb{D}^2 \to Y_n$. Since y_n projects to x_n, $\mathrm{inj}(y_n) \geq \varepsilon$. Let $B(y_n, R)$ denote the ball in Y_n of radius R centered at y_n.

Lemma 5.7.— *For any sufficiently small positive r, the hyperbolic area of $B(y_n, R) - \widetilde{Z}_n(r)$ tends to ∞ with R, uniformly in n.*

Proof.— Since $K_n^{[\mu,\infty[}$ does not lift to Y_n, y_n is contained in an unbounded component of $Y_n^{[\mu,\infty[}$. Denote $B(y_n, R)^{[\mu,\infty[} = B(y_n, R) \cap Y_n^{[\mu,\infty[}$.

Suppose that y_n belongs to the Nielsen core Y_n' of Y_n. By Lemma 4.7, there is a geodesic $\gamma_n \subset \partial Y_n'$ whose length $\ell(\gamma_n)$ is bigger than μ, such that $d(y_n, \gamma_n) \leq C(\mu, \chi(Y))$. Thus, $B(y_n, R)$ contains a ball of radius $R - C(\mu, \chi(Y))$ centered at some point on γ_n. Since $\ell(\gamma_n) \geq \mu$, the volume of $B(y_n, R)^{[\mu,\infty[}$ is (much) bigger than $(R - C(\mu, \chi(Y)))\mu$, for large R and for all n.

Suppose that y_n belongs to component of $Y_n - Y_n'$ (which is an annulus). Since $\mathrm{inj}(y_n) \geq \mu$ a short computation shows that the volume of $B(y_n, R)^{[\mu,\infty[}$ is (much) bigger than $R\mu$.

In both cases, the volume of $B(y_n, R)^{[\mu,\infty[}$ tends to infinity with R, uniformly in n.

Any ball of radius μ in Y_n which is contained in the preimage of $K_n^{[\mu,\infty[}$ embeds into X_n. Therefore the cardinality of the intersection of $\widetilde{Z}_n$ with any ball of radius μ is smaller than the number of zeroes of $\Theta_n\phi_n$ and so is smaller than $4g-4$. Thus, for all sufficiently small $r>0$, the volume of $B(y_n,R)^{[\mu,\infty[}-\widetilde{Z}_n(r)$ tends to ∞ with R uniformly in n. This implies Lemma 5.7. □

Fix $r>0$ so that the conclusions of Lemma 5.7 are satisfied. For any R, $|F_n|$ is bounded from below over $B(y_n,R)-\widetilde{Z}_n(r)$ by $|\bar{F}|/2$ for all sufficiently large n because $|\bar{F}_n|$ tends uniformly on $B(0,R)-\widetilde{Z}_n(r)$ to the non-zero constant $|\bar{F}|$. By Lemma 5.7 the volume of $B(y_n,R)-\widetilde{Z}_n(r)$ is bigger than $3/(m(r)|\bar{F}|)$, for a sufficiently large R independent of n. This is impossible since the L^1-norm of F_n on $Y_n-\widetilde{Z}_n(r)$ is less than $1/m(r)$.

2c) $\bar{F}=0$.

We will apply essentially the same argument as in 1). We introduce first some notations. Let $\mu'\le\mu$ to be fixed later. Recall that $K_n^{[\mu',\infty[}$ is isotopic to $K_n^{[\mu,\infty[}$, for all n sufficiently large. Restrict in what follows to those values of n. By hypothesis, $K_n^{[\mu',\infty[}$ cannot be lifted to Y_n. However, some components of $\partial K_n^{[\mu',\infty[}$ might be. We denote the union of these curves by ${}^S\partial K_n^{[\mu,\infty[}$.

For $\mu'<\mu$, each component of ${}^S\partial K_n^{[\mu,\infty[}$ cuts $K_n^{[\mu',\infty[}$ into two components, one of which is an annulus. We call ${}^SK_n^{[\mu,\mu']}$ the union of those annuli. Then the boundary of ${}^SK_n^{[\mu,\mu']}$ equals ${}^S\partial K_n^{[\mu,\infty[}\cup{}^S\partial K_n^{[\mu',\infty[}$. Since ${}^SK_n^{[\mu,\mu']}$ is isotopic into S, it can be lifted isomorphically to Y_n. The following is a generalization of Fact 5.4.

Fact 5.8.— *For all $\mu'<\mu$, $W_n^{[\mu',\infty[}$ is contained in the unbounded components of $Y_n^{[\mu,\infty[}$ except the isomorphic lift of ${}^SK_n^{[\mu,\mu']}$, which is contained in $Y_n^{]0,\mu]}$.*

Proof.— The preimage $W_n^{[\mu',\infty[}$ equals the union of $W_n^{[\mu,\infty[}$ and $\pi^{-1}(K_n^{[\mu',\mu]})$. Clearly, $W_n^{[\mu,\infty[}\subset Y_n^{[\mu,\infty[}$. Moreover since $K_n^{[\mu,\infty[}$ does not lift, $W_n^{[\mu,\infty[}$ is contained in the unbounded components of $Y_n^{[\mu,\infty[}$. Let A be a component of $K_n^{[\mu',\infty[}-K_n^{[\mu,\infty[}$. If $A\not\subset{}^S K_n^{[\mu,\mu']}$, it is not liftable to Y_n. Therefore $\pi^{-1}(A)$ is a disjoint union of discs and thus is disjoint from $Y_n^{]0,\mu]}$ (cf. Fact 5.4). If $A\subset{}^S K_n^{[\mu,\mu']}$, $\pi^{-1}(A)$ equals the isomorphic lift of A and a union of discs. Again only this isomorphic lift meets (is contained in) $Y_n^{]0,\mu]}$. This proves Fact 5.8. □

We need now to estimate $\int_{\partial(K_n^{[\mu',\infty[}-Z_n(r))}\theta_n(|F_n|ds)$. Let η_n be the 1-form on the unbounded components of $Y_n^{[\mu,\infty[}$ constructed in Theorem 4.6. Then $\langle\eta_n\rangle$ is bounded from above by a constant C independent of n. In contrary to 1), η_n is not defined over all $W_n^{[\mu',\infty[}$ but only on the complement of the isomorphic lift of ${}^SK_n[\mu',\mu]$. We need to modify the definition of θ_n to take these annuli into consideration.

A 1-form $\theta_n^S|F_n\eta_n|$ can be defined on $K_n^{[\mu',\infty[}$ by applying θ_n to the *discontinuous* 1-form defined by extending $|F_n\eta_n|$ by 0 in the complement of the unbounded

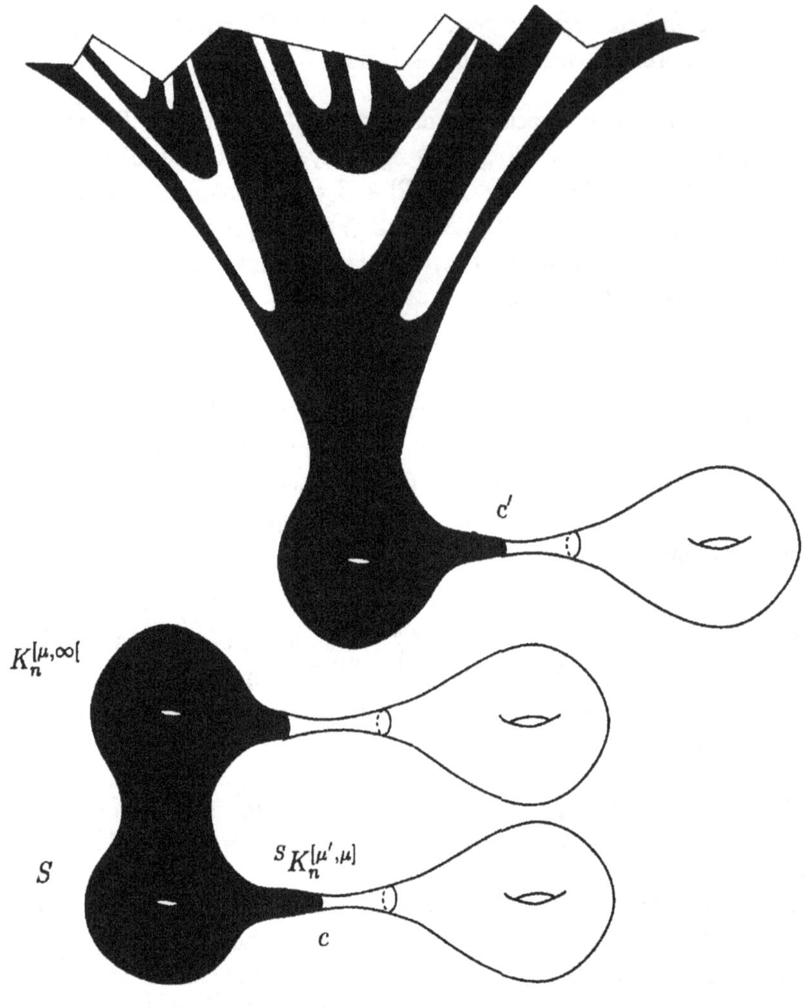

Figure 5.2

components of $Y_n^{[\mu,\infty[}$. Let c be a component of $\partial K_n^{[\mu',\infty[} - {}^S\partial K_n^{[\mu',\infty[}$. Since $\pi^{-1}(c)$ is contained in the unbounded components of $Y_n^{[\mu,\infty[}$, we have

$$\theta_n(|F_n|ds) \geq \frac{1}{C}\theta_n|F_n\eta_n| = \frac{1}{C}\theta_n^S|F_n\eta_n|.$$

More generally, if c is a component of $\partial K_n^{[\mu',\infty[}$, we have $\theta_n(|F_n|ds) \geq \frac{1}{C}\theta_n^S|F_n\eta_n|$.

A 1-form $\theta_n^S(F_n\eta_n)$ can be defined on $K_n^{[\mu',\infty[}$ using the same construction as for $\theta_n^S|F_n\eta_n|$. It is a form of type $(1,0)$ in the complement of ${}^S\partial K_n^{[\mu,\infty[}$. Its discontinuity along ${}^S\partial K_n^{[\mu,\infty[}$ can be described as follows. Each component A of ${}^S K_n^{[\mu',\mu]}$ lifts to Y_n. Over A the covering map $\pi : Y_n \to X_n$ has an inverse π^{-1}. We denote by f_n the function defined on ${}^S K_n^{[\mu',\mu]}$ which equals $F_n \circ \pi^{-1}$ on each of his components. Then the 1-form $\theta_n^S(F_n\eta_n)$ has a jump discontinuity along the curves ${}^S\partial K_n^{[\mu,\infty[}$, which equals $f_n\eta_n$. We have

$$\bar{\partial}\theta_n^S(F_n\eta_n) = dv, \quad \text{on} \quad K_n^{[\mu',\infty[} -{}^S K_n^{[\mu',\mu]}, \quad \text{and}$$
$$\bar{\partial}\theta_n^S(F_n\eta_n) = (1-f_n)dv, \quad \text{on} \quad {}^S K_n^{[\mu',\mu]}.$$

Suppose, up to increasing μ slightly, that $\partial K^{[\mu,\infty[}$ is disjoint from Z. Then, $\partial K_n^{[\mu,\infty[}$ is disjoint from Z_n for all sufficiently large n. For all $\mu' < \mu$ such that $\partial K^{[\mu',\infty[}$ is disjoint from Z, for all sufficiently large n and for all sufficiently small r, we have:

$$\left|\int_{\partial(K_n^{[\mu',\infty[}-Z_n(r))} \theta_n^S(F_n\eta_n)\right|$$
$$= \left|\int_{\partial(K_n^{[\mu',\infty[}-Z_n(r))} \theta_n^S(F_n\eta_n) - \int_{{}^S\partial K_n^{[\mu,\infty[}} \theta_n^S(F_n\eta_n) + \int_{{}^S\partial K_n^{[\mu,\infty[}} \theta_n^S(F_n\eta_n)\right|$$
$$\geq \left|\int_{{}^S K_n^{[\mu',\mu]}-Z_n(r)} (1-f_n)dv + \int_{K_n^{[\mu',\infty[}-{}^S K_n^{[\mu',\mu]}-Z_n(r)} dv + \int_{{}^S\partial K_n^{[\mu,\infty[}} f_n\eta_n\right|$$
$$= \left|\int_{K_n^{[\mu',\infty[}-Z_n(r)} dv - \int_{{}^S K_n^{[\mu',\mu]}-Z_n(r)} f_n dv + \int_{{}^S\partial K_n^{[\mu,\infty[}} f_n\eta_n\right|.$$

Fix a μ' as in 1). Since (f_n) tends uniformly to 0 over ${}^S K_n^{[\mu',\mu]} - Z_n(r)$ the same inequalities as in 1) hold for all n sufficiently large. This gives a contradiction which finishes the proof of Theorem 5.3. □

CHAPTER 6

McMullen's proof of the Fixed point theorem

In this chapter we prove, following [McM3], the Fixed point theorem.

Thurston's fixed point theorem.— *Let M be a hyperbolic manifold with incompressible boundary which is not an interval bundle. Let τ be an orientation reversing involution of ∂M which permutes the components by pairs. Suppose that M/τ is atoroidal. Then $\tau^* \circ \sigma$ has a fixed point.*

Through all this chapter, ε will denote a strictly positive constant smaller than $\varepsilon(3)$.

In the first two sections, we denote by M a connected hyperbolic manifold with incompressible boundary which is not an interval bundle, and by G be a geometrically finite group such that M is diffeomorphic to $\overline{M}(G)$.

6.1 Consequences of the contraction properties of the Theta operators for the skinning map

In this section, we deduce two corollaries for the norm of $d^*\sigma$ from the results of §5. An immediate consequence of Theorem 5.1 is the following.

Proposition 6.1.— *Let $s \in \mathcal{T}(\partial M)$. Then the norm of $d^*\sigma$ at s is less than or equal to a constant $k < 1$ which is a function of the systole of s and of $\chi(\partial M)$.*

Proof.— Let $s \in \mathcal{T}(\partial M)$. Let $d^*_s\sigma$ denote the coderivative of σ at s. We have

$$d^*_s\sigma\phi = \sum_U \Theta_U \phi_U,$$

where U varies over all the spots contained in $\sigma(\partial M^s)$ (Proposition 2.1). Then

$$(1) \qquad ||d^*_s\sigma|| \leq \sup_U ||\Theta_U||$$

where the supremum is taken over all the spots $U \subset \sigma(\partial M^s)$.

Since M is not an interval bundle, all the spots $U \subset \sigma(\partial M^s)$ are proper surfaces (cf. §2). By Proposition 2.2, all the covers $U \to X_U$ are geometric and those which are not isomorphic to the universal cover are finite in number. Therefore, by Theorem 5.1, the norms of the various operators Θ_U are all smaller than a constant $k < 1$ which is a function of the systole of s and of $\chi(\partial M)$. By (1), the same k is also an upper bound for the norm of $d_s^*\sigma$. This proves Proposition 6.1. □

By Proposition 6.1, in order to study the contraction properties of σ over all $\mathcal{T}(\partial M)$, we need to understand what happens when s contains short geodesics. Theorem 5.3 gave a precise information on Theta operators in that situation. In order to apply this theorem to the skinning map, we introduce first another definition.

Let $s \in \mathcal{T}(\partial M)$. We denote by $\mathcal{A}(\sigma(\partial M^s))^\varepsilon$ the part of $\sigma(\partial M^s)$ which equals the disjoint union of the ε-amenable parts $\mathcal{A}(X_U)^\varepsilon$ of the covers $U \to X_U$ over all the spots $U \subset \sigma(\partial M^s)$.

Proposition 6.2.— *Let $0 < \eta < 1$. There exists $\delta > 0$ such that: for any $s \in \mathcal{T}(\partial M)$ and for any $\phi \in \mathcal{Q}(\sigma(\partial M^s))$ with $||\phi|| = 1$ and $||d_s^*\sigma\phi|| \geq 1 - \delta$, the ϕ-mass of $\mathcal{A}(\sigma(\partial M^s))^\varepsilon$ is bigger than $1 - \eta$.*

Proof.— Let $0 < \delta < 1$. Assume that $||\phi|| = 1$ and $||d_s^*\sigma\phi|| \geq 1 - \delta$. Then, since $d_s^*\sigma\phi = \sum \Theta_U \phi_U$, the ϕ-mass of the union of the spots U for which $||\Theta_U\phi_U|| \geq (1-\sqrt{\delta})||\phi_U||$ is bigger than $1-\sqrt{\delta}$. For any spot U such that $||\Theta_U\phi_U|| \geq (1-\sqrt{\delta})||\phi_U||$, Theorem 5.3 asserts that the ϕ_U-mass of $\mathcal{A}(X_U)^\varepsilon$ is bigger than $(1 - c(\delta))||\phi_U||$ for a constant $c(\delta)$ which tends to 0 with δ and which depends only on ε and $\chi(X_U)$. Since there are only finitely many possibilities for the topology of the covers $U \to X_U$ (Proposition 2.2), $c(\delta)$ can be taken independent of U. Thus the ϕ-mass of $\mathcal{A}(\sigma(\partial M^s))^\varepsilon$ is bigger than $(1-\sqrt{\delta})(1-c(\delta))$. Proposition 6.2 directly from this. □

6.2 Inefficiency over the thin part

By Proposition 6.2, if $||d_s^*\sigma\phi||$ is near 1 for a unit norm ϕ, the ϕ-mass concentrates over the amenable part. In this section, we show that the ϕ-mass concentrates over *the ε-liftable part.*

Definition.— *The ε-liftable part* $\mathcal{L}(\sigma(\partial M^s))^\varepsilon$ is the union over the spots $U \subset \sigma(\partial M^s)$ of the lifts of the (liftable) components of $X_U^{]0,\varepsilon]}$ and of $X_U^{[\varepsilon,\infty[}$. Note that $\mathcal{L}(\sigma(\partial M^s))^\varepsilon$ is a compact surface that might be empty. We denote by $\mathcal{S}(\sigma(\partial M^s))^\varepsilon$ the union over all the spots $U \subset \sigma(\partial M^s)$ of the simply connected components in the preimage of $X_U^{]0,\varepsilon]}$. Hence $\mathcal{A}(\sigma(\partial M^s))^\varepsilon$ is the disjoint union of $\mathcal{L}(\sigma(\partial M^s))^\varepsilon$ and $\mathcal{S}(\sigma(\partial M^s))^\varepsilon$.

Proposition 6.3.— *There exists a constant $\varepsilon_M > 0$ depending only on $\chi(\partial M)$, such that for all $\varepsilon \leq \varepsilon_M$ and for all $\eta > 0$, there exists $\delta > 0$ such that: let $s \in \mathcal{T}(\partial M)$, let $\phi \in \mathcal{Q}(\sigma(\partial M^s))$ with $||\phi|| = 1$ and $||d_s^*\sigma\phi|| \geq 1 - \delta$, then the ϕ-mass of $\mathcal{L}(\sigma(\partial M^s))^\varepsilon$ is bigger than $1 - \eta$.*

Although this proposition looks very similar to Proposition 6.2, its proof is entirely different: it lies on the next result, whose proof is partially "3-dimensional".

Proposition 6.4.— *There exists a constant ε_M depending only on $\chi(\partial M)$ such that, for all $0 < \varepsilon \leq \varepsilon_M$, we have: let $s \in \mathcal{T}(\partial M)$, let $\phi \in \mathcal{Q}(\sigma(\partial M^s))$ with $||\phi|| = 1$, let $\mu \in \mathcal{B}(\partial M^s)$ with $||\mu|| = 1$, then every $x \in \mathcal{S}(\sigma(\partial M^s))^\varepsilon$ is the center of an embedded hyperbolic ball on which the efficiency of the pairing between ϕ and $d_s\sigma\mu$ is smaller than $1 - \alpha$, where the constant $\alpha > 0$ depends only on ε and on $\chi(\partial M)$.*

Proof.— We argue by contradiction. The contradiction will follow easily from Theorem 3.1 once we suitably normalize the limit set of G^s in $\overline{\mathbb{C}}$. Suppose that there exists a sequence (ε_i) tending to 0 and a sequence (s_i) in $\mathcal{T}(\partial M)$ such that the systole of s_i is smaller than ε_i, for which Proposition 6.4 fails. Denote to simplify $M^i = M^{s_i}$ and $G^i = G^{s_i}$. Then for each i, there is a point $x_i \in \mathcal{S}(\sigma(\partial M^i))^{\varepsilon_i}$ such that: there exists $\phi_i \in \mathcal{Q}(\sigma(\partial M^i))$ and $\mu_i \in \mathcal{B}(\partial M^i)$ with $||\phi_i|| = ||\mu_i|| = 1$, such that on any embedded hyperbolic ball $B_i \subset \sigma(\partial M^i)$ centered at x_i, the efficiency of the pairing between ϕ_i and $d_{s_i}\sigma\mu_i$ tends to 1.

For each i, choose a component Ω_i of $\Omega(G^i)$, with stabilizer Γ_i, such that x_i belongs to the component $\sigma(\Omega_i/\Gamma_i)$ of $\sigma(\partial M^i)$. Then $\Omega(\Gamma_i)$ is the disjoint union of $\Omega(\Gamma_i)$ and another component denoted by $\widetilde{\Omega}_i$. By definition, $\widetilde{\Omega}_i$ covers $\sigma(\Omega_i/\Gamma_i)$. Let $\widetilde{x}_i \in \widetilde{\Omega}_i$ be a point in the preimage of x_i. Let $\widetilde{U}_i$ be the component of $\Omega(G^i)$ which contains $\widetilde{x}_i$. The projection of $\widetilde{U}_i$ to $\sigma(\Omega_i/\Gamma_i)$ is the spot denoted by U_i which contains x_i. Since $x_i \in \mathcal{S}(\sigma(\partial M^i))^{\varepsilon_i}$, there is a non-zero element $\gamma_i \in G^i$ which stabilizes $\widetilde{U}_i$ and which moves $\widetilde{x}_i$ a distance smaller than ε_i for the hyperbolic metric on $\widetilde{U}_i$. Since the component of $\mathcal{S}(\sigma(\partial M^i))^{\varepsilon_i}$ which contains x_i is simply connected, $\gamma_i \notin \Gamma_i$ (cf. the proof of Proposition 2.2). Since G^i has no parabolic elements, γ_i is a hyperbolic isometry. Let α_i and ω_i be its two fixed points in $\partial\mathbb{D}^3 \simeq \overline{\mathbb{C}}$. With these notations, we have:

Lemma 6.5.— *Let $0 < \varepsilon_0 \leq \varepsilon(3)$ We can conjugate G^i inside* $\mathrm{PSL}_2(\mathbb{C})$ *so that, for all sufficiently large i, the following properties hold*

(i) $\alpha_i = 0$,

(ii) $\infty \in L(\Gamma_i)$ *and*

(iii) γ_i *moves the center of $\mathbb{D}^3$ a hyperbolic distance equal to ε_0.*

Proof.— By a conjugacy of G^i in $\mathrm{PSL}_2(\mathbb{C})$, it is easy to achieve (i) and (ii).

The element γ_i stabilizes $\widetilde{U}_i$. By Corollary 2.4, α_i (resp. ω_i) does not belong to $L(\Gamma_i)$ since $\gamma_i \notin \Gamma_i$. For the hyperbolic metric on $\widetilde{U}_i$, γ_i is a hyperbolic isometry with translation distance smaller than ε_i. By the Ahlfors lemma, the translation distance of γ_i in $\mathbb{H}^3$ is less than $2\varepsilon_i$. In particular, for all i sufficiently large, this translation distance is less than ε_0. The set of points in $\mathbb{D}^3$ which γ_i moves a distance exactly ε_0 is the boundary of the tube $n^{\varepsilon_0}(\gamma_i)$ (cf. §1). Since $\omega_i \neq \infty$, the boundary of $n^{\varepsilon_0}(\gamma_i)$ is pierced in exactly one point by the geodesic 0∞. Therefore, after conjugating G^i by a hyperbolic transformation fixing 0∞, we can achieve (iii).

□

Denote by Δ_R the disc of radius R centered at $0 \in \mathbb{C}$.

Lemma 6.6.— *Let $R > 0$. If ε_0 is smaller than a constant which depends only on $\chi(\partial M)$ and on R, then in the normalization provided by Lemma 6.5, $\widetilde{\Omega}_i$ contains Δ_R, for all sufficiently large i.*

Proof.— Recall that $\gamma_i \notin \Gamma_i$. The convex hull $C(\Gamma_i)$ of $L(\Gamma_i)$ is the universal cover of $N(\Gamma_i)$, the Nielsen core of $M(\Gamma_i)$ (cf. §1). With the induced path metric, $\partial N(\Gamma_i)$ is isometric to a hyperbolic surface (cf. §1). Since ∂M is compact, $\partial N(\Gamma_i)$ is compact and there are only a finite number of possibilities for its topological type. Thus the injectivity radius of the induced metric on $\partial N(\Gamma_i)$ is bounded from above by a constant d depending only on $\chi(\partial M)$. Therefore any point of $\partial C(\Gamma_i)$ is moved a distance smaller than $2d$ by a non-zero element of Γ_i.

For $\nu > 0$, $n^\nu(\gamma_i)$ contains the neighborhood of radius $2d$ of $n^{e^{-2d}\nu}(\gamma_i)$ (cf. §1). Recall that, by the Margulis lemma, for any $\nu \leq \varepsilon(3)$, the tubes $n^\nu(g)$ corresponding to elements $g \in G^i$ which are not contained in the same cyclic subgroup are disjoint. Using this, we prove by contradiction that for any $\varepsilon \leq \varepsilon(3)$, $n^{e^{-2d}\varepsilon}(\gamma_i)$ is disjoint from $C(\Gamma_i)$. Suppose this is not the case. Then, $n^{e^{-2d}\varepsilon}(\gamma_i)$ must intersect $\partial C(\Gamma_i)$, since $\gamma_i \notin \Gamma_i$. We saw above that any point in $\partial C(\Gamma_i)$ is translated a distance less than $2d$ by some non-trivial element $g_i \in \Gamma_i$. Thus the tubes $n^\varepsilon(\gamma_i)$ and $n^\varepsilon(g_i \circ \gamma_i \circ g_i^{-1})$ have non-empty intersection. By Margulis lemma, $g_i \circ \gamma_i \circ g_i^{-1}$ and γ_i are contained in the same cyclic group. Then γ_i and g_i must have the same fixed points. Therefore $\gamma_i \in \Gamma_i$. This provides the required contradiction.

In our normalization, γ_i moves $0 \in \mathbb{D}^3$ a distance equal to ε_0 for sufficiently large i. Thus for any $K > 0$, the hyperbolic ball of radius K centered at $0 \in \mathbb{D}^3$ is contained in $n^{e^K\varepsilon_0}(\gamma_i)$. By the last observation, if we choose $\varepsilon_0 = \varepsilon_0(K)$ so that $e^{2d+K}\varepsilon_0 \leq \varepsilon(3)$, then this ball is disjoint from $C(\Gamma_i)$. Since $\infty \in L(\Gamma_i)$, this implies that $\widetilde{\Omega}_i$ contains $\Delta_{R(K)}$, where $R(K)$ depends only on K and tends to ∞ with it. This implies Lemma 6.6. □

Lemma 6.7.— *If ε_0 is smaller than a constant which depends only on $\chi(\partial M)$, then, for all sufficiently large i, Δ_1 is contained in $\widetilde{\Omega}_i$ and embeds in $\widetilde{\Omega}_i/\Gamma_i$.*

Proof.— By Lemma 6.6, for all $R > 0$, we can choose $\varepsilon_0 = \varepsilon_0(R)$ such that $\Delta_R \subset \widetilde{\Omega}_i$, for all i sufficiently large. If $R > 1$, then $\Delta_1 \subset \widetilde{\Omega}_i$ for all i sufficiently large.

We show now by contradiction that Δ_1 embeds in $\widetilde{\Omega}_i/\Gamma_i$, for all i sufficiently large (in the normalization of Lemma 6.6). If Δ_1 does not embed into $\widetilde{\Omega}_i/\Gamma_i$, then there is a non-zero $\delta_i \in \Gamma_i$ such that $\delta_i(\Delta_1) \cap \Delta_1 \neq \emptyset$. The two fixed points of δ_i are contained in $L(\Gamma_i)$ and in particular they are outside from Δ_R. Thus, for the natural metric on $\mathrm{PSL}_2(\mathbb{C})$ δ_i is $C(R)$-close to a parabolic isometry δ which fixes ∞ and moves $0 \in \mathbb{C}$ an euclidean distance less than 2. In particular, δ moves then $0 \in \mathbb{D}^3$ a hyperbolic distance smaller than 2. When R tends to ∞, $C(R)$ tends to 0. Thus we can choose R sufficiently large such that δ_i moves $0 \in \mathbb{D}^3$ a hyperbolic distance smaller than 3, for all i sufficiently large. But γ_i moves $0 \in \mathbb{D}^3$ a distance smaller than $\varepsilon_0 = \varepsilon_0(R)$ (depending only on $\chi(\partial M)$). Therefore, if ε_0

also satisfies $\varepsilon_0 e^3 \leq \varepsilon(3)$, the tubes $n^{\varepsilon_0 e^3}(\gamma_i)$ and $n^{\varepsilon_0 e^3}(\delta_i \gamma_i \delta_i^{-1})$ must intersect. Since $\delta_i \gamma_i \delta_i^{-1}$ and γ_i do not belong to the same cyclic subgroup of G^i, this is impossible by Margulis lemma. This ends the proof of Lemma 6.7. □

In the sequel, we fix ε_0 such that Lemma 6.7 and Lemma 6.6 for $R = 2$ are satisfied. In order to obtain a contradiction, we need to find, for each i, a hyperbolic ball centered at x_i on which the efficiency of the pairing between ϕ_i and $d_{s_i}\sigma\mu_i$ is bounded away from 1, for any non-zero holomorphic quadratic differential ϕ_i and for any Beltrami form μ_i of unit norm.

Note that up to extracting a subsequence, (γ_i) tends in $\mathrm{PSL}_2(\mathbb{C})$ to a parabolic isometry fixing 0. Since each γ_i moves $0 \in \mathbb{D}^3$ a distance exactly equal to ε_0, some subsequence of (γ_i) converges to a non-trivial element $\gamma \in \mathrm{PSL}_2(\mathbb{C})$. Since the translation distance of γ_i tends to 0, γ is parabolic. By the normalization of Lemma 6.5, γ fixes $0 \in \mathbb{C}$.

Fact 6.8.— *The point $\widetilde{x}_i$ tends to 0 as i tends to ∞.*

Proof.— The hyperbolic metric on $\widetilde{U}_i$ can be written $\lambda_i(z)|dz|$. By the Koebe 1/4-lemma and since $0 \in \partial\widetilde{U}_i$, $\lambda_i(z) \geq 1/4d_{euc}(z, \partial\widetilde{U}_i) \geq 1/4|z|$. The hyperbolic distance between $\widetilde{x}_i$ and $\gamma_i(\widetilde{x}_i)$, for the hyperbolic metric on $\widetilde{U}_i$, tends to 0 as i tends to ∞. Therefore, if $\widetilde{x}_i$ would remain a bounded euclidean distance away from 0, the spherical distance between $\widetilde{x}_i$ and $\gamma_i(\widetilde{x}_i)$ would tend to 0. Then any accumulation point of $\widetilde{x}_i$ would be a fixed point of γ. Since γ is a parabolic isometry fixing $0 \in \mathbb{C}$, this is impossible. □

Let $\widetilde{B}_i$ denote the largest ball for the hyperbolic metric on $\widetilde{\Omega}_i$ which is centered at $\widetilde{x}_i$ and contained in the disc $\Delta = \Delta_1$.

Fact 6.9.— *Any limit of $\widetilde{B}_i$ for the Hausdorff topology on compact subsets of $\mathbb{C}$ contains a neighborhood of 0.*

Proof.— Let R_i be the radius of the largest Euclidean ball centered at 0 and contained in $\widetilde{\Omega}_i$. In our normalization, R_i is bigger than 2. By the Schwarz lemma, the conformal factor $\lambda_i(z)$ of the hyperbolic metric on $\widetilde{\Omega}_i$ is less than the one of the hyperbolic metric on Δ_{R_i} which equals

$$\frac{2R_i}{R_i^2 - |z|^2}.$$

As we noticed above, Koebe's $1/4$-theorem implies

$$\lambda_i(z) \geq \frac{1}{4d_{euc}(z, \partial\widetilde{\Omega}_i)}.$$

Therefore on Δ we have $\sup_\Delta \lambda_i \leq C \inf_\Delta \lambda_i$, for some constant C independent on i. Thus on Δ, the hyperbolic metric of $\widetilde{\Omega}_i$ is Lipschitz equivalent to the euclidean metric by a factor independent on i. Fact 6.9 follows from this and from Fact 6.8. □

Let B_i be the projection of $\widetilde{B}_i$ in $\widetilde{\Omega}_i/\Gamma_i$. By Lemma 6.7, the ball B_i is embedded. Suppose that the efficiency of the pairing between ϕ_i and $d_{s_i}\sigma\mu_i$ on B_i tends to 1. From (the proof of) Theorem 3.1 we know that, up to extracting a subsequence and up to multiplying the differentials ϕ_i by non-zero constants, their pull-back $\widetilde{\phi}_i(z)dz^2$ to $\widetilde{\Omega}_i$ converge uniformly over compact sets to a non-zero holomorphic quadratic differential $\widetilde{\phi}(z)dz^2$. Let $\widetilde{\mu}_i(z)\overline{dz}/dz$ be the pull-back of $d_{s_i}\sigma\mu_i$ to $\mathbb{C}$. We can suppose that $(\widetilde{\mu}_i)$ converges weakly to $\widetilde{\mu} \in L^\infty(\mathbb{C})$ with $||\widetilde{\mu}|| \leq 1$. The uniform convergence of $\widetilde{\phi}_i$ to $\widetilde{\phi}$ and the weak convergence of $\widetilde{\mu}_i$ to $\widetilde{\mu}$ imply that the efficiency of the pairing between $\widetilde{\phi}$ and $\widetilde{\mu}$ on any fixed ball which is contained in the limit of $\widetilde{B}_i$ equals 1. We observed already that, when i tends to ∞, γ_i tends to a parabolic isometry γ fixing 0. By continuity, $\widetilde{\mu}$ is invariant under γ. The transformation γ leaves a family of round balls containing 0 in their boundary invariant. By Fact 6.9, one of these balls, denoted by $\widetilde{B}$, is contained in the geometric limit of the balls $\widetilde{B}_i$. Then the efficiency of the pairing between $\widetilde{\phi}$ and $\widetilde{\mu}$ on $\widetilde{B}$ equals 1. Since $||\widetilde{\mu}|| \leq 1$, we have therefore on $\widetilde{B}$

$$\widetilde{\mu} = \frac{\overline{\widetilde{\phi}}}{|\widetilde{\phi}|}.$$

Since $\widetilde{\mu}$ is invariant under γ, $\widetilde{\phi}(\gamma(z)) = \kappa\widetilde{\phi}(z)\gamma'(z)^2$, for a constant $\kappa \neq 0$. But this is impossible since $\widetilde{\phi}$ is integrable on $\widetilde{B}$, as being holomorphic and defined on a larger domain. This concludes the proof of Proposition 6.4. □

Proof of Proposition 6.3.— Let ε_M be the constant provided by Proposition 6.4. Let $0 < \varepsilon \leq \varepsilon_M$. Let $\phi \in \mathcal{Q}(\sigma(\partial M^s))$ with $||\phi|| = 1$. Let m be the ϕ-mass of $\mathcal{S}(\sigma(\partial M^s))^\varepsilon$. From Proposition 6.4 and Proposition 3.10, we deduce that for any $\mu \in \mathcal{B}(\partial M^s)$ with $||\mu|| = 1$, $\langle \phi, d_s\sigma\mu\rangle \leq 1 - c\alpha m$, where $c > 0$ only depends on $\chi(\partial M)$. But $||d_s^*\sigma\phi|| = \sup_\mu \langle d_s^*\sigma\phi, \mu\rangle = \sup_\mu \langle \phi, d_s\sigma\mu\rangle$, where the supremum are both taken over all $\mu \in \mathcal{B}(\partial M^s)$ of unit norm. Thus $||d_s^*\sigma\phi|| \leq 1 - c\alpha m$. Therefore if $||d_s^*\sigma\phi|| \geq 1-\delta$, then the ϕ-mass of $\mathcal{S}(\sigma(\partial M^s))^\varepsilon$ is less than $\delta/c\alpha$. Since $\mathcal{A}(\sigma(\partial M^s))^\varepsilon$ is the disjoint union of $\mathcal{L}(\sigma(\partial M^s))^\varepsilon$ and $\mathcal{S}(\sigma(\partial M^s))^\varepsilon$, Proposition 6.3 follows from this estimate and from Proposition 6.2. □

6.3 Proof of Thurston's fixed point theorem

The proof of the Fixed point theorem amounts now essentially to translate Proposition 6.3 in terms of the topology of M.

In this section M is not necessarily connected. Denote by $M_1, ., M_\ell, ., M_p$ its components. For $s = (s_1, \cdots, s_\ell, \cdots, s_p) \in \mathcal{T}(\partial M)$, we denote by M^s the disjoint union of the manifolds $M_\ell^{s_\ell}$. Recall that σ equals the product of the skinning maps σ_ℓ associated to M_ℓ and that the norm on $\mathcal{Q}(\sigma(\partial M^s)) = \bigoplus_\ell \mathcal{Q}(\sigma(\partial M_\ell^s))$ is the sum of the norms on the spaces $\mathcal{Q}(\sigma(\partial M_\ell^{s_\ell}))$. Therefore, we have

$$(1) \qquad ||d_s^*\sigma|| = \sup_\ell ||d_{s_\ell}^*\sigma_\ell||.$$

Now, we distinguish two cases.

1) *M is acylindrical.*

Theorem 6.10.— *Let M be an acylindrical hyperbolic manifold with incompressible boundary. Then σ contracts the Teichmüller distance by a factor strictly less than 1.*

Proof.— Since σ is differentiable and since Teichmüller distance is induced by a Finsler metric, it suffices to prove that $||d_s\sigma||$ is bounded by some constant $k < 1$ independent of s. It is more convenient to prove the dual statement, namely that $||d_s^*\sigma||$ is bounded away from 1 independently of s. Since M is acylindrical, each of its components is acylindrical also. In view of (1), we may suppose that M is connected. Since M is acylindrical, each spot in $\sigma(\partial M^s)$ is simply connected. Therefore $\mathcal{L}(\sigma(\partial M^s))^\varepsilon$ is empty, for all s and for all ε. Therefore the norm of $d_s^*\sigma$ is bounded away from 1 independently of s. □

Proof of the Fixed point theorem in the acylindrical case.— Since τ^* is an isometry of the Teichmüller distance (cf. §1), Theorem 6.10 implies that $\tau^* \circ \sigma$ contracts the distance on $\mathcal{T}(\partial M)$ by a factor strictly less than 1. Since Teichmüller space is complete, $\tau^* \circ \sigma$ has a fixed point. □

As an application, we have the following.

Theorem 6.11.— *Let N be an acylindrical hyperbolic manifold with incompressible boundary. Then N carries a hyperbolic metric for which ∂N is totally geodesic.*

Proof.— This result is a consequence of the Fixed point theorem when M is the union of two copies of N having opposite orientations, and when τ is the natural identification between the two copies of ∂M. By Corollary 6.11, $\tau^* \circ \sigma$ has a fixed point $s \in \mathcal{T}(\partial M)$. Consider the hyperbolic manifold M^s corresponding to s via the Ahlfors-Bers isomorphism. Denote by $\mathcal{N}$ the Nielsen core of the component of M^s which has the same orientation as N. Let Γ' be the image in $\mathrm{PSL}_2(\mathbb{C})$ of the fundamental group of a component of $\partial\mathcal{N}$. Then Γ' is a quasi-Fuchsian group. Saying that s is a fixed point of $\tau^* \circ \sigma$ means precisely that the two components of $\Omega(\Gamma')$ are isometric by an orientation reversing map which extends to the identity on $L(\Gamma')$. Since the Ahlfors-Bers map is injective, Γ' is conjugated in $\mathrm{PSL}_2(\mathbb{C})$ to a Fuchsian group. It follows that Γ' leaves invariant a totally geodesic plane. Therefore $\partial\mathcal{N}$ is totally geodesic. Since $\mathcal{N}$ is diffeomorphic to N, this proves Theorem 6.11. □

2) *M is cylindrical.*

The hypothesis of Thurston's fixed point theorem is that M is not an interval bundle. Still some connected components of M can be cylindrical, like for instance interval bundles, some others may not. The hypothesis excludes only the case when *all components are interval bundles.* Let ε_M be the smallest of the constants ε_{M_ℓ} provided by applying Proposition 6.4 to the components M_ℓ of M which are not interval bundles. Recall that σ is an isometry when M is an interval bundle (§2).

Definition. — Denote by C the number of components of ∂M. Denote by S the maximal number of homotopy classes of disjoint simple closed curves in ∂M. Set $K = C + S$.

Theorem 6.12. — *Let M be a hyperbolic manifold which is not an interval bundle. Let τ be an orientation reversing involution of ∂M which exchanges the components by pairs. If $N = M/\tau$ is atoroidal, then $(\tau^* \circ \sigma)^K$ has a fixed point.*

Observe that this theorem does not claim that $(\tau^* \circ \sigma)^K$ contracts uniformly Teichmüller distance, like in the proof of the Fixed point theorem in the acylindrical case. However, it is sufficient.

Proof of Thurston's fixed point theorem. — Some component of M is not an interval bundle. By Proposition 6.1, the skinning map associated to this component contracts strictly Teichmüller distance. Since τ^* is an isometry and since K is bigger than the number of components of M, it follows that $(\tau^* \circ \sigma)^K$ contracts strictly Teichmüller distance. Therefore the fixed point provided by Theorem 6.12 is unique. Since $(\tau^* \circ \sigma)^K$ and $\tau^* \circ \sigma$ commute, this fixed point is also fixed by $\tau^* \circ \sigma$.

□

Proof of Theorem 6.12. — Choose an arbitrary point $s^0 \in \mathcal{T}(\partial M)$. Let L be the Teichmüller distance between s^0 and $\tau^* \circ \sigma(s^0)$. Denote by $\mathcal{T}(\partial M)_L \subset \mathcal{T}(\partial M)$ the set of points which are moved a Teichmüller distance smaller than L by $\tau^* \circ \sigma$. Then $\mathcal{T}(\partial M)_L$ is a non-empty closed subset of $\mathcal{T}(\partial M)$ which has the following properties:

(i) it is invariant under $\tau^* \circ \sigma$ (because $\tau^* \circ \sigma$ decreases the Teichmüller distance), and

(ii) for any $s \in \mathcal{T}(\partial M)_L$, the Teichmüller geodesic between s and $\tau^* \circ \sigma(s)$ is contained in $\mathcal{T}(\partial M)_L$ (this is a consequence of the triangular inequality).

In order to prove Theorem 6.12, it suffices to establish that the norm of the derivative of $(\tau^* \circ \sigma)^K$ is bounded over $\mathcal{T}(\partial M)_L$ by a constant $c < 1$. This will imply that the path which equals the union of all the positive iterates by $(\tau^* \circ \sigma)^K$ of the geodesic joining s^0 to $(\tau^* \circ \sigma)^K(s^0)$ has finite Teichmüller length. Therefore, this path accumulates on a fixed point of $(\tau^* \circ \sigma)^K$.

To prove that the norm of the derivative of $(\tau^* \circ \sigma)^K$ at any point in $\mathcal{T}(\partial M)_L$ is less than a uniform constant $c < 1$, we argue by contradiction. Let $\delta > 0$ a constant to be fixed later. Then there is $s \in \mathcal{T}(\partial M)_L$ such that the coderivative $d_s^*(\tau^* \circ \sigma)^K$ has norm bigger than $1/(1+\delta)$. For $0 \leq k \leq K$, set $s^k = (\tau^* \circ \sigma)^k(s)$ and denote to simplify $M^k = M^{s^k}$. Hence, for $0 \leq k \leq K$, there exists $\phi_k \in \mathcal{Q}(\partial M^k)$ such that

(i) $\phi_k = d_{s^k}^*(\tau^* \circ \sigma)\phi_{k+1}$, and

(ii) $1 \leq ||\phi_k|| \leq 1 + \delta$.

Notation. — If X is a union of components of ∂M^k, we denote by $||\phi_k||_X$ the ϕ_k-mass of X.

Choose $\varepsilon > 0$ such that $\varepsilon \leq \varepsilon_M$ and $\log(\varepsilon) + 2KL \leq \log(\varepsilon(2))$. By Proposition 2.2, $\sigma(\partial M)$ contains only finitely many spots which are not simply connected. Let S' be twice the number of those spots and set

$$\eta = \frac{\zeta}{C(2S')^K}.$$

For all $k < K$ and for any component M_ℓ^k of M^k, we have:

$$0 \leq ||d^*\tau^*\phi_{k+1}||_{\sigma(\partial M_\ell^k)} - ||\phi_k||_{\partial M_\ell^k} \leq ||d^*\tau^*\phi_{k+1}|| - ||\phi_k|| \leq \delta.$$

By Proposition 6.3 and since $\phi_k = d^*\sigma(d^*\tau^*\phi_{k+1})$, we can choose δ sufficiently small so that for any component M_L^k which is not an interval bundle and whose ϕ_k-mass is more than η, the $d^*\tau^*\phi_{k+1}$-mass of $\sigma(\partial M_L^k) - \mathcal{L}(\sigma(\partial M_L^k))^{\varepsilon/2}$ is less than η.

Notation.— Let X be a compact hyperbolic surface and let $\gamma \subset X$ be a closed curve homotopic to a geodesic shorter than $\varepsilon/2$. We denote by $X(\gamma)$ the component of $X^{]0,\varepsilon]}$ which contains this geodesic.

By Proposition 3.9, there is a constant $\zeta \neq 0$ depending only on $\chi(\partial M)$ such that, if X is a component of ∂M^k which contains a closed geodesic shorter than $\varepsilon/2$, then the ϕ_k-mass of $X(\alpha)$ is bigger than $\zeta||\phi_k||_X$ for some geodesic $\alpha \subset X$ shorter than $\varepsilon/2$.

Fact 6.13.— *Let $0 \leq k \leq K-1$. Let α_k be a closed curve contained in the boundary of a component M_ℓ^k of M^k and suppose that α_k is homotopic to a geodesic shorter than $\varepsilon/2$. Set $\mu = ||\phi_k||_{\partial M^k(\alpha_k)}$ and assume $\mu > \eta$. Then, $\partial M^k(\alpha_k)$ lifts to an annulus contained in $\sigma(\partial M_\ell^k)$ whose $d^*\tau^*\phi_{k+1}$-mass is bigger than $\mu' = (\mu - \eta)/S'$.*

Proof.— Since $\phi_k = d^*\sigma(d^*\tau^*\phi_{k+1})$ and since $d^*\sigma$ decreases the mass, the $d^*\tau^*\phi_{k+1}$-mass of the entire preimage of $\partial M_\ell^k(\alpha_k)$ in $\sigma(\partial M_\ell^k)$ is bigger than μ.

If M_ℓ^k is an interval bundle, this preimage consists of exactly one isomorphic lift of $\partial M_\ell^k(\alpha_k)$. The $d^*\tau^*\phi_{k+1}$-mass of this lift equals μ and thus is bigger than μ'.

If M_ℓ^k is not an interval bundle, the preimage of $\partial M_\ell^k(\alpha_k)$ equals the disjoint union of isomorphic lifts of $\partial M_\ell^k(\alpha_k)$ and simply connected components contained in $\mathcal{S}(\sigma(\partial M_\ell^k))^{\varepsilon/2}$. By our choice of δ, the $d^*\tau^*\phi_{k+1}$-mass of $\mathcal{S}(\sigma(\partial M_\ell^k))^{\varepsilon/2}$ is less than η. Thus, the $d^*\tau^*\phi_{k+1}$-mass of the union of the isomorphic lifts of $\partial M_\ell^k(\alpha_k)$ is bigger than $\mu - \eta$. Since the cover associated to any spot is geometric, each spot which is not simply connected contains at most two isomorphic lifts of $\partial M_\ell^k(\alpha_k)$ (cf. §2). Therefore the number of the isomorphic lifts of $\partial M_\ell^k(\alpha_k)$ is less than S' so that the $d^*\tau^*\phi_{k+1}$-mass of one of them is bigger than μ'. □

The geometric consequence of this result is the following.

Fact 6.14.— *Under the hypothesis of Fact 6.13, there is an essential annulus contained in M_ℓ^k which joins α_k to a curve $\gamma_k \subset \partial M_\ell^k$ which satisfies*

(i) *$\tau(\gamma_k)$ is homotopic to a geodesic shorter than $\varepsilon/2$ for the hyperbolic metric on ∂M^{k+1}, and*

(ii) *the* ϕ_{k+1}*-mass of* $\partial M^{k+1}(\tau(\gamma_k))$ *is bigger than* μ'.

Proof. — Let A_k be the isomorphic lift of $\partial M_\ell^k(\alpha_k)$ which is provided by Fact 6.14. The annulus A_k is isometric to $\partial M_\ell^k(\alpha_k)$ for the hyperbolic metric of the spot which contains it. Let $\beta_k \subset A_k$ be the lift of α_k. Since the Poincaré metric decreases under inclusion, β_k is shorter than $\varepsilon/2$ for the hyperbolic metric of $\sigma(\partial M_\ell^k)$. For the same reason, A_k is entirely contained in $\sigma(\partial M_\ell^k)(\beta_k)$. In particular, the $d^*\tau^*(\phi_{k+1})$-mass of $\sigma(\partial M_\ell^k)(\beta_k)$ is bigger than μ'.

Let Y be the component of ∂M_ℓ^k such that $\beta_k \subset \sigma(Y)$. Let $\mathcal{Y}$ be the covering space of M_ℓ^k with fundamental group $\pi_1(Y)$. Then $\mathcal{Y}$ is naturally contained in the manifold $\overline{M}(\pi_1(Y))$. Since $\pi_1(Y)$ is a quasi-Fuchsian group, $\overline{M}(\pi_1(Y))$ has two boundary components and these are identified with Y and $\sigma(Y)$ respectively. One component of $\partial\mathcal{Y}$ is the canonical lift of Y, whereas the others components are the spots contained in $\sigma(Y)$. From the proof of Fact 2.5 (see also Figure 2.3) β_k can be homotoped inside $\mathcal{Y}$ to a curve $\gamma_k' \subset Y$. Let f be the composition of this homotopy with the covering map $\mathcal{Y} \to M_\ell^k$. Then f is a map from the annulus into M_ℓ^k whose image joins the image of β_k —i.e. α_k— to the image of γ_k' — a curve contained in Y which we denote by γ_k. The map f is essential. It is injective on the fundamental group since α_k is not homotopic to zero. It is injective on the relative fundamental group since the two components of its lift to $\mathcal{Y}$ are contained in distinct boundary components.

Now, the length of the geodesic homotopic to $\tau(\gamma_k)$ for the hyperbolic metric on $\partial M^{k+1} = \tau^* \circ \sigma(\partial M^k)$ equals the length of the geodesic homotopic to γ_k for the hyperbolic metric on $\sigma(\partial M^k)$. This is also the length of the geodesic homotopic to β_k for the metric on $\sigma(Y)$. Therefore $\tau(\gamma_k)$ is homotopic to a geodesic shorter than $\varepsilon/2$ for the metric of ∂M^{k+1}.

Similarly the ϕ_{k+1}-mass of $\partial M^{k+1}(\tau(\gamma_k))$ equals the $d^*\tau^*\phi_{k+1}$-mass of $\sigma(\partial M^k)(\beta_k)$, and so is bigger than μ'. This proves Fact 6.14. □

In order to obtain a contradiction, we will construct by induction a finite sequence of essential annuli contained in M. The boundary of these annuli will match up under τ to produce a π_1-injective map from a torus into M/τ. This will contradict atoroidality of M/τ.

The construction begins with the following result.

Fact 6.15. — *There exists* $0 \leq k \leq C-1$, *such that* ∂M^k *contains a closed geodesic* α *shorter than* $\varepsilon/2$ *and such that the* ϕ_k*-mass of* $\partial M^k(\alpha)$ *is bigger than* ζ/C.

Proof. — The ϕ_0-mass of some component X of ∂M^0 is at least $1/C$. Denote by V the component of M^0 which contains X.

Suppose that V is not an interval bundle over a closed surface. Since $\phi_0 = d^*\sigma(d^*\tau^*\phi_1)$ and since $d^*\sigma$ decreases the mass, the $d^*\tau^*\phi_1$-mass of the union of the spots on $\sigma(\partial M^0)$ which cover X is more than $||\phi_0||_X$. By the choice of δ and since $||\phi_0||_X \geq 1/C \geq \eta$, one of these spots must (intersect and therefore) contain a component of $\mathcal{L}(\sigma(\partial M^0))^{\varepsilon/2}$. Therefore ∂M^0 contains a closed geodesic shorter than $\varepsilon/2$. Fact 6.15 follows then fro

Suppose that V is an interval bundle. Since σ is an isometry in that case, the $d^*\tau^*\phi_1$-mass of $\sigma(X)$ equals $||\phi_0||_X$. The differential $d^*\tau^*\phi_1$ is defined on a boundary component of V endowed with the reversed orientation. This component is identified by τ with a component X^1 of ∂M^1. Since τ^* is an isometry, $||\phi_1||_{X^1} = ||\phi_0||_X \geq 1/C$. If X^1 lies in the boundary of a component of M^1 wich is not an interval bundle, the reasoning of the previous case can be applied. Then Fact 6.15 holds with $k = 1$. If X^1 lies in the boundary of a component of M^1 which is an interval bundle, we can repeat the same argument. Since M has less than C components and is not an interval bundle, there exists $k \leq C-1$ and a component X^k of ∂M^k such that

(i) $||\phi_k||_{X^k} \geq 1/C$ and

(ii) X^k is not contained in the boundary of a component of M^k which is an interval bundle.

The reasoning of the first case concludes then the proof of Fact 6.15. □

We are now ready for constructing essential annuli $A_i \subset M/\tau$. We start with the curve α_k provided by Fact 6.15, for some $0 \leq k \leq C-1$. Up to shifting the indices, we suppose that $k = 0$. Since $K = C + S$, we have now a sequence $M^0, \cdots, M^i, \cdots, M^R$ with $R \geq S+1$.

Recall that each M^i is identified with M by a diffeomorphism well defined up to isotopy. In what follows we use implicitely this identification.

By Fact 6.14, there exists an essential annulus $A_0 \subset M$ with $\partial A_0 = \alpha_0 \cup \gamma_0$. The curve α_0 is shorter than $\varepsilon/2$ for the hyperbolic metric ∂M^0 and $\alpha_1 = \tau(\gamma_0)$ is homotopic to a geodesic shorter than $\varepsilon/2$ for the hyperbolic metric ∂M^1. The ϕ_1-mass of $\partial M^1(\alpha_1)$ (and in particular the ϕ_1-mass of the boundary of the component of M^1 which contains α_1) is bigger than

$$\mu' = \frac{\zeta/C - \eta}{S'} \geq \frac{\zeta}{C(2S')^2}.$$

Since $K > 2$, Fact 6.14 can be applied again. On this way, we define by induction a sequence of curves (α_i) on ∂M, such that

(i) α_i is homotopic to a geodesic shorter than $\varepsilon/2$ for the hyperbolic metric ∂M^i,

(ii) there is an essential annulus $A_i \subset M$ whose boundary components are α_i and a curve γ_i,

(iii) for $i > 0$, $\alpha_i = \tau(\gamma_{i-1})$, and

(iv) the ϕ_i-mass of $\partial M^i(\alpha_i)$ is bigger than $\zeta/C(2S')^{i+1}$.

By (i) and (iv), the sequence (α_i) can be defined as long as $i+1 \leq K$ and $i \leq R$. Since $K \geq S+1$ and $R \geq S$, it is defined at least for all $i \leq S$. The length of α_i is less than $\varepsilon/2$ for the metric ∂M^i. By the triangular inequality, the Teichmüller distance between ∂M^i and ∂M^0 is smaller than KL. Thus, our choice of ε implies that the length of α_i for the metric ∂M^0 is smaller than $\varepsilon(2)$ (cf. §1). Therefore, by the Margulis lemma, the curves α_i can be homotoped to be disjoint. Hence, by the definition of S, their homotopy classes form a set of cardinality smaller than S. Therefore, two among the curves α_i, say α_ℓ and α_m must be homotopic on ∂M. This means that the result of gluing with τ the annuli A_i along their boundaries for $\ell \leq i < m$ can be closed up with an homotopy between α_ℓ and α_m to create a map from the torus T^2 into M. This map is π_1-injective. This follows from the fact that each annulus A_i is essential. This contradiction with the hypothesis that M/τ is atoroidal concludes the proof of Theorem 6.13. □

CHAPTER 7

Manifolds-with-corners

In this chapter, we describe the topological tools necessary to reduce the proof of Thurston's hyperbolization theorem to the Final gluing theorem.

Let M be a compact orientable 3-manifold (of class C^1).

Definition.— We say that M is *irreducible* if any 2-sphere embedded in M bounds a 3-ball. We say that M is *atoroidal* if the fundamental group of any component of M does not contain subgroups isomorphic to $\mathbb{Z}+\mathbb{Z}$.

In what follows, a surface is always a compact orientable surface.

Definition.— A surface $S \subset M$ is *properly embedded* when it is a submanifold of M and when $\partial S = \partial M \cap S$.

Definition.— Suppose that $S \subset M$ is either a properly embedded surface or the disjoint union of components of ∂M. We say that S is *incompressible* when any component S' of S satisfies

(i) if $\chi(S') \leq 0$, then $\pi_1(S')$ maps injectively into $\pi_1(M)$ and $\pi_1(S', \partial S')$ maps injectively into $\pi_1(M, \partial M)$,

(ii) if $\chi(S') = 1$, then S' is *an essential disc*, i.e. a disc whose boundary is not homotopic to 0 on ∂M.

Dehn's lemma allows to replace (i) above by the following more geometric conditions (cf. **[He1]**, **[Ja]**)

(i) if $\gamma \subset S'$ is a curve which bounds *a compression disc*, i.e. a disc embedded in M whose interior is disjoint from S', then γ also bounds a disc in S',

(ii) if $k \subset S'$ and $k' \subset \partial M$ are properly embedded arcs such that $k \cup k'$ bounds *a* ∂*-compression disc* i.e. a disc embedded in M whose interior is disjoint from S', then there is an arc $k'' \subset \partial S'$ such that $k \cup k''$ bounds a disc in S'.

Definition.— We say that $S \subset M$ is *a splitting surface* if S is an incompressible surface and if no component of S can be isotoped into ∂M.

Haken manifolds.

Definition. — A compact connected manifold M is *Haken* if it is irreducible and if it contains a splitting surface.

When M is Haken, it contains a splitting surface S and a new manifold M' can be formed by *splitting M along S*. This manifold M' is defined as the complement in M of an open regular neighborhood of S. It follows from Dehn's lemma that M' is irreducible. The boundary of any component of M' is non-empty. Therefore, M' is a 3-ball or is again Haken (cf. [**He1**], [**Ja**]).

We need now to distinguish a particular class of 3-manifolds.

Definition. — A manifold M is called *an handlebody* if it is diffeomorphic to the manifold obtained from the 3-ball by attaching g *1-handles* $I \times \mathbb{B}^2$ along $\partial I \times \mathbb{B}^2$. The integer $g = g(M)$ is called *the genus of M*.

Let M be an handlebody with $g(M) \geq 1$. Then M is Haken since it contains an essential disc, for instance $1/2 \times \mathbb{B}^2$. Furthermore the manifold obtained by splitting M along the surface which is the disjoint union of the g discs $1/2 \times \mathbb{B}^2$ contained in each 1-handle is diffeomorphic to $\mathbb{B}^3$.

Definition. — Let M be an handlebody. *A system of meridians for M* is a disjoint union of essential discs properly embedded in M such that the manifold obtained by splitting M along m is diffeomorphic to $\mathbb{B}^3$.

Clearly, each system of meridians for M contains g essential discs, if g is the genus of M.

Handlebodies are the most docile of all Haken manifolds: they can be immediatly reduced to the 3-ball. If M is a Haken manifold which is not an handlebody, then it contains a connected splitting surface which is not an essential disc (cf. [**Ja**, p. 59]). Furthermore, if $\partial M \neq \emptyset$, such a splitting surface exists which has non-empty boundary ([**Ja**, p. 35]). A connected splitting surface with these two properties is called *a special splitting surface.*

Definition. — Let M be a compact connected Haken manifold. *A partial hierarchy of length n for M* is a finite sequence $M_0, \cdots, M_n$ such that

(i) $M_0 = M$, and

(ii) for $k \leq n-1$, there is a special splitting surface $S_k \subset M_k$, such that M_{k+1} is obtained by splitting M_k along S_k.

If M is a Haken manifold which is not an handlebody, it admits a partial hierarchy of length at least 1.

Definition. — We define *the length of M*, as the largest integer n such that M admits a partial hierarchy of length n. We denote it by $\ell(M)$.

It is a basic observation due to Wolfgang Haken that $\ell(M)$ is always finite ([**Hak**], [**Ja**, p. 61]). Also, $\ell(M) = 0$ if and only if M is an handlebody. Another important property which follows from the defini   if M is not an handlebody and if M' is a component of the manifold obtained by splitting M along an *arbitrary* special splitting surface, then $\ell(M') < \ell(M)$. So, if M is a Haken manifold, $\ell(M)$ measures the distance from M to handlebodies. This measure of the complexity of M simply with an integer goes back to Haken. It will be used during the proof of the Hyperbolization theorem for manifolds-with-corners.

Definition.— Let $S \subset M$ be a properly embedded orientable surface. Let $D \subset M$ be a compression disc for S and let $\gamma = \partial D$. Let $N(D)$ be an open regular neighborhood of (D, γ) in (M, S): the boundary of $N(D)$ consists of an annulus contained in S and two parallel copies of D. Then we define a new surface S' properly embedded in M as the union of $S - N(D)$ and these two discs. Similarly, let D be a ∂-compression disc for S and set $k = D \cap S$, $k' = D \cap \partial M$. Let $N(D)$ be an open regular neighborhood of (D, k) in (M, S). We define a new surface S' as the union of $S - N(D)$ and the two parralel copies of D contained in $N(D)$. We say (in both cases) that S' is obtained from S by *surgery along* D.

Fact 7.1.— *Let S be a splitting surface for M. Let D be a ∂-compression disc for S. Let S' be the surface obtained from S by surgery along D. If S is a special splitting surface, then one component of S' also. If M is an handlebody and if S is a system of meridians for M, then S' contains a system of meridians.*

Proof.— When S is a special splitting surface, then one component of S' is a disc and the other component, denoted by Σ, is homeomorphic to S. It follows from standard technics in 3-dimensional topology that Σ is a splitting surface for M. Since Σ is homeomorphic to S, it is a special splitting surface.

Let M be a handlebody and let S be a system of meridians for M. Let m be the component of S which intersects D. Then the surface obtained from m by surgery along D is the disjoint union of two discs. Since the union of these two discs is homologuous to m, one of them, called m', is non-separating. It follows that $S - \{m\} \cup \{m'\}$ is a system of meridians for M. □

Interval bundles.

Definition.— *An interval bundle* is an orientable manifold which fibers over a closed surface with fiber diffeomorphic to $[0,1]$.

Up to diffeomorphism, there are two types of interval bundles: *the trivial product* $S \times [0,1]$ of a closed connected orientable surface S with the interval, and *the twisted product* $S \tilde{\times} [0,1]$ of S with the interval, i.e. the quotient of $S \times [0,1]$ by the relation: $(x,0) \simeq (y,0)$ if and only if $x = \tau(y)$, where τ is an orientation reversing fixed point free involution of S.

The following characterization of interval bundles was used in §2 (cf. [**He1**]).

Theorem [Sta1].— *Let M be a connected irreducible manifold with non-empty boundary and let S be a component of M. If the index of $\pi_1(S)$ in $\pi_1(M)$ equals 1, then the inclusion of S into M extends to a diffeomorphism between $S \times [0,1]$ and M. If the index of $\pi_1(S)$ in $\pi_1(M)$ equals 2, then the inclusion of S extends to a diffeomorphism between $S \tilde{\times} [0,1]$ and M.*

7.1 Manifolds-with-corners

Definition. — A *manifold-with-corners* is a triple $(M, \mathcal{G}, \partial^0 M)$ such that

(i) M is a compact 3-manifold,

(ii) $\mathcal{G}$ is a trivalent graph contained in ∂M,

(iii) each component of $\partial M - \mathcal{G}$ equals the interior of its closure, and

(iv) each component of $\partial^0 M$ equals the closure of a component of $\partial M - \mathcal{G}$.

The closure of a component of $\partial M - \mathcal{G}$ which is not contained in $\partial^0 M$ is called *a mirror of* $(M, \mathcal{G}, \partial^0 M)$. The surface $\partial^0 M$ is called *the boundary of* $(M, \mathcal{G}, \partial^0 M)$.

From a differentiable structure on M, we can construct *a differentiable structure with corners on M associated to* $(M, \mathcal{G}, \partial^0 M)$, i.e. there is an atlas of class C^1 on M such that

(i) each point of M has a neighborhood isomorphic to an open set in $(\mathbb{R}^+)^3$, and

(ii) the points on edges (resp. the vertices) of $\mathcal{G}$ are exactly the points which have a neighborhood isomorphic to the neighborhood of a point of $(\mathbb{R}^+)^3$ with 2 (resp. 3) coordinates equal to 0.

When we are given a manifold-with-corners $(M, \mathcal{G}, \partial^0 M)$, we always think that M has such a differentiable structure with corners, i.e. that corners are really corners. This structure is unique. Any homeomorphism of M which preserves $\mathcal{G}$ is isotopic to a diffeomorphism: this follows essentially from [**C**].

Remark. — The definition of a manifold-with-corners can be made in any dimension. For instance *a surface-with-corners* having empty boundary is a pair (S, P) where S is a compact surface and $P \subset \partial S$ a finite set such that any component of $\partial S - P$ equals the interior of its closure. This is clearly equivalent to say that each component of ∂S either is disjoint from P or contains at least 2 points of P. In particular, let $(M, \mathcal{G}, \partial^0 M)$ be a manifold-with-corners and let P be the set of vertices of $\mathcal{G}$ contained in (the boundary of) $\partial^0 M$. Then $(\partial^0 M, P)$ is a surface-with-corners.

Definition. — Let $(M, \mathcal{G})$ be a manifold-with-corners having empty boundary. Let $S' \subset \partial M$ be a surface which is a union of mirrors of $(M, \mathcal{G})$. Let $\mathcal{G}'$ the trivalent graph obtained by erasing the edges of $\mathcal{G}$ which intersect the interior of S'. Then $(M, \mathcal{G}', S')$ is a manifold-with-corners. We say that $(M, \mathcal{G}', S')$ is obtained from $(M, \mathcal{G})$ by *erasing the mirrors contained in* S'.

Given a differentiable structure with corners on M associated to $(M, \mathcal{G})$, one can define a differentiable structure on M associated to $(M, \mathcal{G}', S')$ by "rounding the corners" contained in the interior of S'. This differentiable structure is unique up to diffeomorphism [**Do**].

Let $(M, \mathcal{G}, \partial^0 M)$ be a manifold-with-corners.

Definition.— Let $\gamma \subset \partial M$ be an embedded closed curve. We say that γ *intersects* $\mathcal{G}$ *tranversally* if γ is disjoint from the vertices of $\mathcal{G}$ and if γ intersects the edges of $\mathcal{G}$ transversally.

Definition.— Let $(M, \mathcal{G}, \partial^0 M)$ be a manifold-with-corners. We say that $(M, \mathcal{G}, \partial^0 M)$ has *incompressible boundary* when, if $\gamma \subset \partial M$ is an embedded closed curve which bounds a disc in M and satisfies

(i) $\gamma \cap \partial^0 M \neq \emptyset$, and

(ii) γ intersects $\mathcal{G}$ transversally in at most 3 points,

then γ bounds a disc in ∂M which has one of the forms described in Figure 7.1.

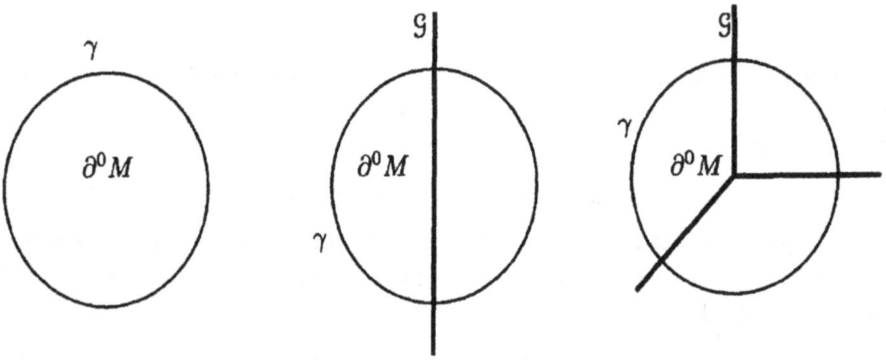

Figure 7.1

Definition.— Let $(M, \mathcal{G})$ be a manifold-with-corners having empty boundary. We say that $(M, \mathcal{G})$ *is irreducible* and *atoroidal* if the following conditions are satisfied:

(i) M is irreducible and atoroidal,

(ii) if A is an annulus properly embedded in M with $\partial A \cap \mathcal{G} = \emptyset$, and such that $\pi_1(A)$ maps injectively into $\pi_1(M)$, then A is parallel to an annulus contained in a mirror of $(M, \mathcal{G})$, and

(iii) if $\gamma \subset \partial M$ is an embedded closed curve which intersects $\mathcal{G}$ transversally in at most 4 points and which bounds a disc in M, then γ bounds a disc in ∂M whose intersection with $\mathcal{G}$ has one of the forms described in Figure 7.2.

Remark.— Let $(M, \mathcal{G})$ be an irreducible and atoroidal manifold-with-corners having empty boundary. Then it follows from the definition that any mirror of $(M, \mathcal{G})$ has one of the following shapes:

(i) a n-gon with $n \geq 5$,

(ii) an annulus with at least one vertex of $\mathcal{G}$ in its boundary, or

(iii) a surface with strictly negative Euler characteristic.

Remark.— The present notion of a manifold-with-corners essentially coincides with that of an orbifold that happens to be modelled on the quotient of $\mathbb{R}^3$ by the group of eight elements generated by sign reversal of one of the 3 coordinates.

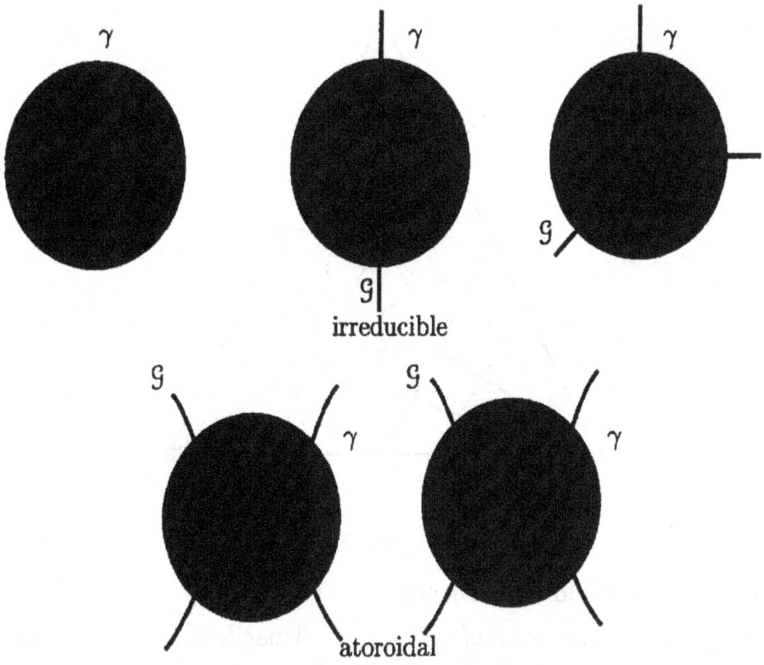

Figure 7.2

A manifold-with-corners is irreducible and atoroidal if and only if the corresponding orbifold is irreducible and atoroidal, i.e. does not contain spherical nor euclidean suborbifolds (cf. [Th1]).

Proposition 7.2. — *Let M be an irreducible and atoroidal 3-manifold. Then there is a graph $\mathcal{G} \subset \partial M$ such that $(M, \mathcal{G})$ is an irreducible and atoroidal manifold-with-corners having empty boundary.*

Proof. — We can suppose $\partial M \neq \emptyset$. Let $\mathcal{T}$ be a triangulation of ∂M. Refine $\mathcal{T}$ to a triangulation $\mathcal{T}'$ by modifying it inside each triangle as described on Figure 7.3 (this triangulation was shown to us by Emmanuel Giroux). Let $\mathcal{G}$ be 1-skeletton of the cellulation dual to $\mathcal{T}'$. Then $(M, \mathcal{G})$ defines a manifold-with-corners having empty boundary. Each mirror of $(M, \mathcal{G})$ is homeomorphic to a disc. To prove that $(M, \mathcal{G})$ is irreducible and atoroidal, it suffices to check property (iii). In fact, we will prove that any closed curve $\gamma \subset \partial M$ which intersects $\mathcal{G}$ in at most 4 points is as described on Figure 7.3. Such a curve γ gives rise to a closed path $\bar{\gamma}$ contained in the 1-skeletton of $\mathcal{T}'$ which follows at most 4 edges. From the way $\mathcal{T}'$ was defined, $\bar{\gamma}$ is contained in the union of at most 2 triangles of $\mathcal{T}$ which have a common edge or a common vertex. Then a new look at Figure 7.3 shows the existence of a disc bounded by γ which intersects $\mathcal{G}$ like on Figure 7.2. □

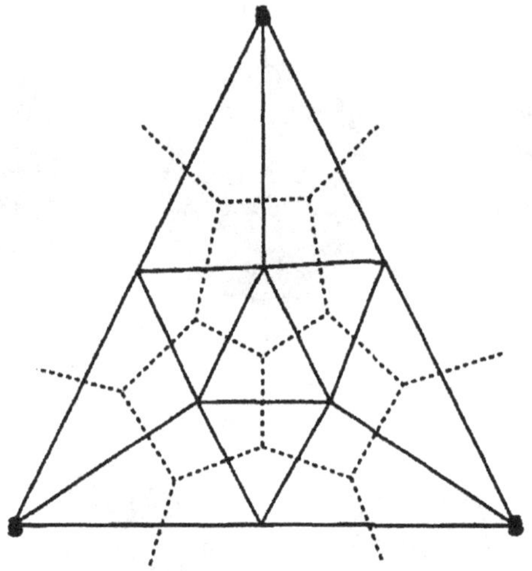

Figure 7.3

Splitting of a manifold-with-corners.

We now explain how the notion of splitting of a 3-manifold along a splitting surface can be extended to the context of manifolds-with-corners.

Let $(M, \mathcal{G})$ be a manifold-with-corners having empty boundary. Let S be a splitting surface for M such that ∂S intersects $\mathcal{G}$ transversally. .

Definition.— Let M_S be the manifold obtained by splitting M along S. Denote by S^+, S^- the two copies of S which are contained in ∂M_S and set $S' = S^+ \cup S^-$. There is an orientation reversing involution τ' of S' such that M is diffeomorphic to the quotient space M_S/τ' i.e. to the space of equivalence classes of the relation

$$x \simeq y \text{ if and only if } x \in S' \text{ and } y = \tau'(x).$$

Denote the quotient map by $\pi' : M_S \to M$. Let $\mathcal{G}_S \subset \partial M_S$ be the graph equal to $(\pi')^{-1}(\mathcal{G} \cup \partial S)$. Each vertex of $\mathcal{G}_S$ has valence 3. The closure of any component of $\partial M_S - \mathcal{G}_S$ equals either a component of S' or the closure of a component of $f - (f \cap \partial S)$ where f is a mirror of $(M, \mathcal{G})$. Thus $(M_S, \mathcal{G}_S, S')$ is a manifold-with-corners. We say that $(M_S, \mathcal{G}_S, S')$ is obtained by *splitting* $(M, \mathcal{G})$ *along* S.

Let $\mathcal{P}$ be the set of vertices of $\mathcal{G}_S$ which are contained in $\partial S'$. Then $(S', \mathcal{P})$ is a surface-with-corners having empty boundary. By construction, τ' preserves $\mathcal{P}$. Therefore τ' can be considered as an orientation reversing involution of $(S', \mathcal{P})$. Then, $(M, \mathcal{G})$ can be reconstructed from the data of $(M_S, \mathcal{G}_S)$ and τ'. The manifold M is diffeomorphic to the quotient M_S/τ' and $\mathcal{G}$ equals the complement in $\mathcal{G}_S/\tau'$ of the interior of the edges of $\mathcal{G}_S \cap \partial S'$.

Definition.— Let $(M, \mathcal{G})$ be a connected manifold-with-corners having empty boundary. *A good splitting surface for* $(M, \mathcal{G})$ is a splitting surface for M such that

(i) S is a system of meridians if M is an handlebody, and S is a special splitting surface if not,

(ii) ∂S intersects $\mathcal{G}$ transversally, and

(iii) $(M_S, \mathcal{G}_S, S')$ has incompressible boundary.

Proposition 7.3.— *Let $(M, \mathcal{G})$ be an irreducible and atoroidal manifold-with-corners having empty boundary. Suppose that M is Haken. Then there is a good splitting surface for $(M, \mathcal{G})$.*

Proof.— Since M is Haken, it contains a splitting surface S which is a system of meridians if M is an handlebody, or a special splitting surface if it is not. We can assume furthermore that S intersects $\mathcal{G}$ transversally and minimizes $\sharp(\partial\Sigma \cap \mathcal{G})$ among all the splitting surfaces Σ such that

(i) Σ is a special splitting surface for M if M is not an handlebody, or is a system of meridians if M is an handlebody, and

(ii) Σ intersects $\mathcal{G}$ transversally.

We show now that $(M_S, \mathcal{G}_S, S')$ has incompressible boundary.

Let $\gamma \subset \partial M_S$ be a closed curve such that $\gamma \cap S' \neq \emptyset$, which intersects $\mathcal{G}_S$ transversally in at most 3 points and which bounds a disc D embedded in M_S. Observe that $\gamma \cap \partial S'$ contains an even number of points.

1) $\gamma \cap \partial S' = \emptyset$.

Then $\gamma \subset S'$ and γ bounds a disc in S' since S is incompressible in M.

2) $\gamma \cap \partial S' \neq \emptyset$.

If $\sharp(\gamma \cap \mathcal{G}_S) = 2$ (resp. 3), γ is the union of one arc $k \subset S'$ and one arc k' contained in a mirror f' of $(M_S, \mathcal{G}_S, S')$ (resp two arcs k'_1 and k'_2 contained in mirrors f'_1, f'_2 of $(M_S, \mathcal{G}_S, S')$). Then $\pi'(D)$ is a disc embedded in M and its boundary is the union of the two arcs $\pi'(k) \subset S$ and $\pi'(k') \subset \partial M$ (resp. $\pi'(k'_1 \cup k'_2)$). By Fact 7.1, the surface obtained from S by surgery along $\pi'(D)$ contains a splitting surface for M which is a system of meridians if M is a handlebody or a special splitting surface if not. We observe that $\partial\Sigma$ is obtained from ∂S by replacing an arc $\kappa \subset \partial D'$ by $\pi'(k')$ (resp. $\pi'(k'_1 \cup k'_2)$) and that $\kappa \cup \pi'(k')$ (resp. $\kappa \cup \pi'(k'_1 \cup k'_2)$) is a closed curve embedded in ∂M which bounds a disc in M.

2a) $\sharp(\gamma \cap \mathcal{G}_S) = 2$.

The mirror f' is obtained by splitting a mirror f of $(M, \mathcal{G})$ and $\pi'(k')$ is contained in f. If $\kappa \cap \mathcal{G} = \emptyset$, then $\kappa \cup \pi'(k')$ is contained in f. Since $(M, \mathcal{G})$ is irreducible and atoroidal, $\kappa \cup \pi'(k')$ bounds a disc contained in f. It follows that γ bounds a disc on ∂M_S which is as described in Figure 7.1. If $\kappa \cap \mathcal{G} \neq \emptyset$, then $\sharp(\partial\Sigma \cap \mathcal{G}) < \sharp(\partial S \cap \mathcal{G})$. This is impossible by the choice of S.

2b) $\sharp(\gamma \cap \mathcal{G}_S) = 3$.

Then f'_1 and f'_2 intersect along an edge of $\mathcal{G}_S$. Each mirror f'_i is obtained by splitting a mirror f_i of $(M, \mathcal{G})$ so that f_1 and f_2 share an edge e of $\mathcal{G}$ in common. We have $\pi'(k'_1 \cup k'_2) \subset f_1 \cup f_2$ and $\pi'(k'_1 \cup k'_2)$ intersects e in exactly one point. If $\kappa \cap \mathcal{G} = 0$ or 1 point, then $\kappa \cup \pi'(k'_1 \cup k'_2)$ intersects $\mathcal{G}$ in 1 or 2 points. Since $(M, \mathcal{G})$

is irreducible and atoroidal, $\kappa \cup \pi'(k_1' \cup k_2')$ bounds a disc contained in $f_1 \cup f_2$. It follows that γ bounds a disc contained in ∂M_S which intersects $\mathcal{G}_S$ like on Figure 7.1. If $\sharp(\kappa \cap \mathcal{G}) \geq 2$, then $\sharp(\partial\Sigma \cap \mathcal{G}) < \sharp(\partial S \cap \mathcal{G})$. This contradicts our choice of S. □

Proposition 7.4.— *Let $(M, \mathcal{G})$ be an irreducible and atoroidal manifold-with-corners having empty boundary. Let S be a good splitting surface for $(M, \mathcal{G})$. Then there is a trivalent graph $\mathcal{G}_S' \subset \partial M_S$ such that*

(i) *$(M_S, \mathcal{G}_S')$ is an irreducible and atoroidal manifold-with-corners having empty boundary,*

(ii) *S' is a union of mirrors of $(M_S, \mathcal{G}_S')$, and*

(iii) *$(M_S, \mathcal{G}_S, S')$ is obtained from $(M_S, \mathcal{G}_S')$ by erasing the mirrors contained in S'.*

The meaning of this lemma is that we can add to $\mathcal{G}_S$ edges which are contained in S' to obtain a graph $\mathcal{G}_S'$ such that $(M_S, \mathcal{G}_S')$ is an irreducible and atoroidal manifold-with-corners having empty boundary.

Proof.— When $\partial S \neq \emptyset$, we begin by adding finitely many points to the set of vertices of $\mathcal{G}_S$ which sit on $\partial S'$, so that this new set of points forms the 0-skeletton of a triangulation of $\partial S'$ (i.e. so that there are at least 3 points per components). Denote by $\mathcal{V}$ this set of vertices. Choose now a triangulation $\mathcal{T}$ of S' which extends this triangulation of $\partial S'$. Refine then $\mathcal{T}$ to a new triangulation $\mathcal{T}'$ by modifying it inside each triangle as described on Figure 7.3. Let $\mathcal{E}$ be the 1-skeletton of the dual cellulation of $\mathcal{T}'$. Observe that each edge of $\mathcal{E}$ which is contained in $\partial S'$ contains at most one point of $\mathcal{V}$ in its interior, and in particular at most one vertex of $\mathcal{G}_S$. The union of the edges of $\mathcal{E}$ which intersect the interior of S' form a graph $\mathcal{E}'$. The vertices of $\mathcal{E}'$ which are contained in the interior of S' have valence 3 and those which are contained in $\partial S'$ have valence 1.

Define $\mathcal{G}_S' = \mathcal{G}_S \cup \mathcal{E}'$. By construction, $(M_S, \mathcal{G}_S')$ is a manifold-with-corners which satisfies Proposition 7.4 (ii), (iii). Observe also that each mirror of $(M_S, \mathcal{G}_S')$ which is contained in S' intersects $\partial S'$ in at most one edge, and that this edge contains at most one point of $\mathcal{G}_S$ in its interior.

We show now that $(M_S, \mathcal{G}_S')$ is irreducible and atoroidal. If A is an annulus properly embedded in M_S such that $\pi_1(A)$ maps injectively into $\pi_1(M_S)$, then $\pi_1(A)$ maps injectively into $\pi_1(M)$, by Van Kampen and since S is incompressible. Therefore, there is a parallelism in M between A and an annulus contained in a mirror of $(M, \mathcal{G})$, since $(M, \mathcal{G})$ is irreducible and atoroidal. By incompressibility, S is disjoint from this parallelism. Hence, A is parallel in M_S to an annulus contained in a mirror of $(M_S, \mathcal{G}_S)$.

Let $\gamma \subset \partial M_S$ be a closed curve which bounds a disc embedded in M_S and which intersects $\mathcal{G}_S'$ transversally in at most 4 points. We consider distinct cases, according to the cardinality of $\gamma \cap \partial S'$.

1) $\gamma \subset S'$.

Then γ gives rise to a closed path contained in the 1-skeletton of $\mathcal{T}'$ which follows at most 4 edges. One deduces from the way the triangulation $\mathcal{T}'$ was defined that γ bounds a disc on ∂M_S which intersects $\mathcal{G}'_S$ (i.e. $\mathcal{E}'$) like on Figure 7.2.

2) $\gamma \cap S' = \emptyset$.

Then $\pi'(\gamma)$ is a closed curve which intersects $\mathcal{G}$ transversally in at most 4 points and which bounds a disc embedded in M. Since $(M, \mathcal{G})$ is irreducible and atoroidal, $\pi'(\gamma)$ bounds a disc on ∂M which intersects $\mathcal{G}$ like on Figure 7.2. Since S is a splitting surface, this disc is disjoint from ∂S. It follows that γ bounds a disc on ∂M_S which intersects $\mathcal{G}'_S$ (i.e. $\mathcal{G}_S$) like on Figure 7.2 (cf. Proposition 7.2).

3) $\sharp(\gamma \cap \partial S') = 2$.

Then $\gamma \cap S'$ is a single arc that we denote by k. We distinguish two subcases.

3a) $k \cap \mathcal{E}' = \emptyset$.

Then k is contained in a mirror f of $(M_S, \mathcal{G}'_S)$. By the construction of $\mathcal{E}$, f is simply connected and $e = f \cap \partial S'$ is a single 1-cell of $\mathcal{E}$. Therefore, k can be isotoped relatively to ∂k to an arc $\kappa \subset e$. By construction, e contains at most one vertex of $\mathcal{G}_S$. Therefore if γ' denotes the curve obtained from γ by replacing k by κ and slightly perturbed to be disjoint from S', then $\pi'(\gamma')$ is a closed curve which intersects $\mathcal{G}_S$ transversally in at most 3 points and which bounds a disc embedded in M. Since $(M, \mathcal{G})$ is irreducible and atoroidal, $\pi'(\gamma')$ bounds a disc on ∂M which intersects $\mathcal{G}$ like on Figure 7.2. This disc is disjoint from S. Therefore γ bounds also a disc contained in ∂M_S which intersects $\mathcal{G}'_S$ like on Figure 7.2.

3b) $k \cap \mathcal{E}' \neq \emptyset$.

Then $\sharp(\gamma \cap \mathcal{G}_S) \leq 3$. Since $(M_S, \mathcal{G}_S, S')$ has incompressible boundary, γ bounds a disc $\Delta \subset \partial M_S$ which intersects $\mathcal{G}_S$ like on Figure 7.1. Since $\sharp(k \cap \mathcal{G}'_S) \leq 4$, k gives rise to a path contained in the 1-skeletton of $\mathcal{T}'$ which follows at most 2 1-cells. From the way $\mathcal{T}'$ was defined, we deduce that Δ intersects $\mathcal{G}'_S$ as described on Figure 7.2.

4) $\sharp(\gamma \cap \partial S') = 4$.

Then $\gamma \cap S'$ is the union of two disjoints arcs k_1 and k_2 which are contained in mirrors of $(M_S, \mathcal{G}'_S)$. By the same reasoning as in 3a), k_i can be isotoped relatively to ∂k_i into an arc κ_i contained in an edge e_i of $\mathcal{E}$. Also, κ_i contains at most one vertex of $\mathcal{G}_S$ in its interior. Let γ' be the curve obtained from γ by replacing k_i by κ_i for $i = 1, 2$ and perturbed to be disjoint from S'. Then $\pi'(\gamma')$ is a closed curve which bounds a disc in M and which intersects $\mathcal{G}_S$ in at most 2 points. Since $(M, \mathcal{G})$ is irreducible and atoroidal, $\pi'(\gamma')$ bounds a disc on ∂M which intersects $\mathcal{G}$ like on Figure 7.2. It follows that γ bounds a disc contained in ∂M_S which intersects $\mathcal{G}'_S$ like in Figure 7.2. □

7.2 The mirrored manifold

Let $(M, \mathcal{G}, \partial^0 M)$ be a manifold-with-corners. We explain now a canonical procedure for associating to $(M, \mathcal{G}, \partial^0 M)$ a compact 3-manifold. This 3-manifold will carry an action by diffeomorphisms of a finite group which encodes the data of $(M, \mathcal{G}, \partial^0 M)$.

For a set $\mathcal{S}$, we denote by $(\mathbb{Z}/2\mathbb{Z})^{\mathcal{S}}$ the set of maps from $\mathcal{S}$ to $\mathbb{Z}/2\mathbb{Z}$.

Definition.— Let $\mathcal{M}$ be the set of mirrors of $(M, \mathcal{G}, \partial^0 M)$. *The mirrored manifold* $\widehat{M}$ is the quotient of $M \times (\mathbb{Z}/2\mathbb{Z})^{\mathcal{M}}$ by the equivalence relation $\mathcal{R}$ generated by

$(x, i) \sim (x, j)$ when x is contained in the mirror $f \in \mathcal{M}$ and the two maps i and j satisfy $i(f) = j(f) + 1$ and $i(f') = j(f')$ for $f' \neq f$.

By looking at these identifications in the neighborhood of point which is contained in a mirror of $(M, \mathcal{G})$ or in an edge of $\mathcal{G}$ or which is a vertex of $\mathcal{G}$, one checks that $\widehat{M}$ is a manifold with boundary (cf. Figure 7.4). Clearly the action of $H = (\mathbb{Z}/2\mathbb{Z})^{\mathcal{M}}$ on $M \times (\mathbb{Z}/2\mathbb{Z})^{\mathcal{M}}$ by translation on the second factor preserves the equivalence classes of $\mathcal{R}$. It projects therefore to an action by homeomorphisms of H on $\widehat{M}$. The group H is *the mirror group of* $\widehat{M}$. When $(M, \mathcal{G}, \partial^0 M)$ carries a differentiable structure with corners, then $\widehat{M}$ inherits a differentiable structure and the mirror group acts by diffeomorphisms of class C^1.

Remark.— If M is connected, then $\widehat{M}$ is connected. This is due to the fact that for any two maps i and j in $(\mathbb{Z}/2\mathbb{Z})^{\mathcal{M}}$, there is a sequence $(i_k)_{k=1,\cdots,p}$ such that

(i) $i_1 = i$, $i_p = j$ and

(ii) for $k \leq p-1$ the maps i_k and i_{k+1} agree except on a single mirror.

It follows that if M is connected, the two copies $M \times i$ and $M \times j$ map to the same component of $\widehat{M}$ under the equivalence relation $\mathcal{R}$.

An important feature of the construction of $\widehat{M}$ is that one can recover $(M, \mathcal{G}, \partial^0 M)$ from the action of H on $\widehat{M}$. First the quotient space $\widehat{M}/H$ is clearly homeomorphic to M and $\partial\widehat{M}$ maps to a surface $\partial\widehat{M}/H$ contained in $\partial(\widehat{M}/H)$. The isotropy group of a point in $\widehat{M}$ under the action of H is isomorphic to $\mathbb{Z}/2\mathbb{Z}$, $(\mathbb{Z}/2\mathbb{Z})^2$ or $(\mathbb{Z}/2\mathbb{Z})^3$. The set of points in $\partial(\widehat{M}/H)$ whose isotropy group is isomorphic to $(\mathbb{Z}/2\mathbb{Z})^2$ or $(\mathbb{Z}/2\mathbb{Z})^3$ form a trivalent graph $\mathcal{G}' \subset \partial(\widehat{M}/H)$. Each mirror of $(\widehat{M}/H, \mathcal{G}', \partial\widehat{M}/H)$ can be characterized as the closure of the set of points whose isotropy group is a given non-zero element in H. Therefore $(\widehat{M}/H, \mathcal{G}', \partial\widehat{M}/H)$ is a manifold-with-corners isomorphic to $(M, \mathcal{G}, \partial^0 M)$.

Definition.— Let f be a mirror of $(M, \mathcal{G}, \partial^0 M)$. Let h_f be the element of H which has the f-coordinate as its single non-zero coordinate. Then h_f acts as an involution on $\widehat{M}$. Its fixed point set is exactly the image under $\mathcal{R}$ of $f \times (\mathbb{Z}/2\mathbb{Z})^{\mathcal{M}}$. This fixed point set can be described as follows. Let $\mathcal{S}$ be the set of mirrors of $(M, \mathcal{G}, \partial^0 M)$ which intersect f. Consider the natural inclusion of $(\mathbb{Z}/2\mathbb{Z})^{\mathcal{S}}$ into H where $i : \mathcal{S} \to \mathbb{Z}/2\mathbb{Z}$ is extended to a map $\mathcal{M} \to \mathbb{Z}/2\mathbb{Z}$ by the constant map 0 on $\mathcal{M} - \mathcal{S}$. Then the image of $f \times (\mathbb{Z}/2\mathbb{Z})^{\mathcal{M}}$ under the equivalence relation $\mathcal{R}$ is made of disjoint homeomorphic copies of the image of $f \times (\mathbb{Z}/2\mathbb{Z})^{S}$. This image is an embedded surface denoted by $\mathcal{F}$ and called *the surface in* $\widehat{M}$ *above* f.

Remark.— Suppose that $(M, \mathcal{G})$ is irreducible and atoroidal. From the description of the possible shapes for f, it follows that f is a hyperbolic surface-with-corners, i.e. that f has a hyperbolic metric such that the edges of $\mathcal{G}$ contained in ∂f are totally geodesic and that adjacent edges meet orthogonally. This implies that $\mathcal{F}$ is a hyperbolic surface and therefore that $\chi(\mathcal{F}) < 0$.

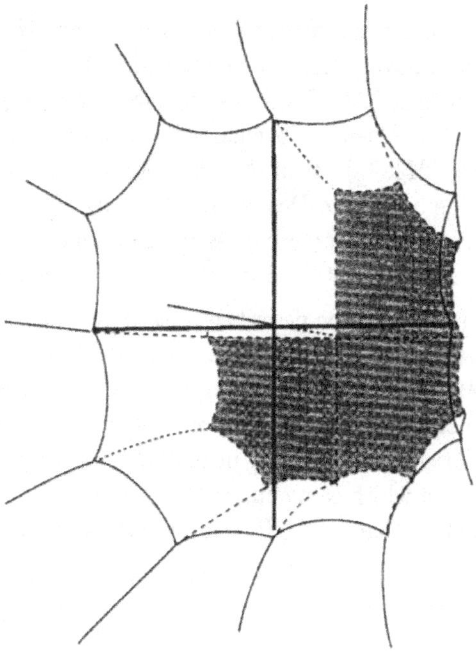

Figure 7.4

We describe now a relative version of the construction of the mirrored manifold.

The partially mirrored manifold.

Let $(M, \mathcal{G})$ be a manifold-with-corners having empty boundary. Let $S \subset \partial M$ be a splitting surface for $(M, \mathcal{G})$. Let $(M_S, \mathcal{G}_S, S')$ be the manifold-with-corners obtained by splitting $(M, \mathcal{G})$ along S. Then each mirror of $(M_S, \mathcal{G}_S, S')$ is contained in a unique mirror of $(M, \mathcal{G})$.

Definition.— We denote by $\widehat{M_S}$ the quotient space of $M_S \times (\mathbb{Z}/2\mathbb{Z})^{\mathcal{M}}$ by the equivalence relation $\mathcal{R}_S$ generated by the relation $(x, i) \sim (x, j)$ if and only if

(i) x belongs to a mirror f' of $(M_S, \mathcal{G}_S, S')$ which is contained in the mirror $f \in \mathcal{M}$, and

(ii) the maps i and j satisfy $i(f) = j(f) + 1$ and $i(f'') = j(f'')$ for $f'' \neq f$.

$\widehat{M_S}$ is called *the partially mirrored manifold of* $(M_S, \mathcal{G}_S, S')$.

Then $\widehat{M}_S$ is a compact 3-manifold on which the group $H = (\mathbb{Z}/2\mathbb{Z})^{\mathcal{M}}$ acts by homeomorphisms. The boundary of $\widehat{M}_S$ is the image of $S' \times (\mathbb{Z}/2\mathbb{Z})^{\mathcal{M}}$ under the equivalence relation $\mathcal{R}_S$. Like in the case of $\widehat{M}$, $\widehat{M}_S$ is connected once M_S is. If we put a differentiable structure with corners on M associated to $(M, \mathcal{G})$, then M_S has a differentiable structure with corners associated to $(M_S, \mathcal{G}_S, S')$. Then $\widehat{M}_S$ is manifold with boundary of class C^1 on which H acts by diffeomorphisms, and, like for the case of the mirrored manifold, the manifold-with-corners $(M_S, \mathcal{G}_S, S')$ can be reconstructed from the action of H on $\widehat{M}_S$.

We will consider too other differentiable structures on $\widehat{M}_S$. Let $(M_S, \mathcal{G}'_S)$ be a manifold-with-corners having empty boundary such that S' is a union of mirrors of $(M_S, \mathcal{G}'_S)$. Suppose that $(M_S, \mathcal{G}_S, S')$ is obtained from $(M_S, \mathcal{G}'_S)$ by erasing the mirrors contained in S'. When M is endowed with a differentiable structure with corners associated to $(M_S, \mathcal{G}'_S)$, then $\widehat{M}_S$ inherits a differentiable structure with (maybe) corners on the boundary. Denote by $\widehat{M}'_S$ this differentiable manifold-with-corners. Then H acts by diffeomorphisms on $\widehat{M}'_S$ and $\widehat{M}_S$ is H-equivariantly diffeomorphic to the manifold obtained from $\widehat{M}'_S$ by rounding the corners.

Recall now that $(M, \mathcal{G})$ can be described as the quotient space $(M_S, \mathcal{G}_S, S')/\tau'$. In a similar way, we can describe $\widehat{M}$ as a quotient of $\widehat{M}_S$. Define for that an involution τ'' of $\partial M_S \times (\mathbb{Z}/2\mathbb{Z})^{\mathcal{M}}$ by $\tau''(x, i) = (\tau'(x), i)$. Then τ'' preserves the classes of the equivalence relation $\mathcal{R}_S$ and commutes with each element in H. It induces therefore an orientation reversing involution τ of $\partial\widehat{M}_S$ which exchanges the components by pairs and which commutes with the action of H. It follows from the definitions of $\widehat{M}_S$ and τ, that $\widehat{M}$ is H-equivariantly diffeomorphic to $\widehat{M}_S/\tau$.

We justify now the terminology irreducible and atoroidal for a manifold-with-corners.

Proposition 7.5.— *Let $(M, \mathcal{G})$ be an irreducible and atoroidal manifold-with-corners having empty boundary. Then $\widehat{M}$ is irreducible and atoroidal.*

Proof.— This proposition could be proven using elementary technics from 3-dimensional topology. But to simplify the exposition we will use "the Equivariant Dehn's lemma", "the Equivariant sphere theorem" and "the Equivariant torus theorem". Recall that $\widehat{M}$ carries an action of $H = (\mathbb{Z}/2\mathbb{Z})^{\mathcal{M}}$.

Suppose for a contradiction that $\widehat{M}$ is not irreducible. Then $\widehat{M}$ contains *an essential sphere*, i.e. an embedded sphere which does not bound a ball. By the Equivariant sphere theorem ([**MY**], [**Du**]) there is an essential sphere Σ which is disjoint or equal to any of its translates by elements of H. Let $H(\Sigma)$ denote the stabilizer of Σ in H. Then $H(\Sigma)$ is isomorphic to the trivial group, to $\mathbb{Z}/2\mathbb{Z}$, $(\mathbb{Z}/2\mathbb{Z})^2$ or to $(\mathbb{Z}/2\mathbb{Z})^3$. A fundamental domain for the action of $H(\Sigma)$ on Σ is naturally contained in a copy $M \times i$ of M sitting in $\widehat{M}$. This fundamental domain is Σ or a disc whose boundary intersects $\mathcal{G}$ transversally in 0, 2 or 3 points. Since $(M, \mathcal{G})$ is an irreducible manifold-with-corners, this contradicts that Σ is essential.

In order to show that $\widehat{M}$ is atoroidal, we first prove that it is Haken. To see this, select a mirror $f \in \mathcal{M}$. Consider the involution h_f of $\widehat{M}$ which is associated to f and consider the surface $\mathcal{F}$ above f. The quotient $\widehat{M}/h_f$ is a manifold with boundary which naturally embeds in $\widehat{M}$. The group $H' = H/\langle h_f\rangle$ acts by diffeomorphism on $\widehat{M}/h_f$ and $\partial\widehat{M}$ equals the H'-orbit of $\mathcal{F}$. Suppose for a contradiction that $\mathcal{F}$ is not incompressible in $\widehat{M}$. Then by Van Kampen's theorem $\mathcal{F}$ is not incompressible in $\widehat{M}/h_f$ and by Dehn's lemma, there is an essential disc properly embedded in $\widehat{M}/h_f$. By the Equivariant Dehn's lemma **[MY]**, there is such a disc D which is either disjoint or equal to any of its translates by H'. Let $H'(D)$ be the stabilizer of D' in H'. Then $H'(D)$ is isomorphic to the trivial group, to $\mathbb{Z}/2\mathbb{Z}$ or to $(\mathbb{Z}/2\mathbb{Z})^2$. A fundamental domain for the action of $H'(D)$ on D is naturally contained in a copy $M \times i$ of M sitting in $\widehat{M}/h_f$. This fundamental domain is isomorphic to D or to a disc whose boundary intersects $\mathcal{G}$ transversally in 2 or 3 points. It follows from the irreducibility and the atoroidality of $(M, \mathcal{G})$ that D cannot be an essential disc. Therefore $\mathcal{F}$ is an incompressible surface in $\widehat{M}$ (which has a strictly negative Euler characteristic). Thus $\widehat{M}$ is Haken.

Let us show by contradiction that $\widehat{M}$ is atoroidal. Suppose that $\pi_1(M)$ contains a subgroup isomorphic to $\mathbb{Z}+\mathbb{Z}$. Since $\widehat{M}$ is Haken, the Torus theorem asserts that $\widehat{M}$ contains an incompressible torus (**[F]**, **[JS]**, **[Joh]**). By the Equivariant torus theorem (**[BS]**, **[JR]**, **[MS]**), there exists such a torus T which is disjoint or equal to any of its translates by elements of H. Let $H(T)$ be the stabilizer of T in H. Then $H(T)$ is either the trivial group or isomorphic to $\mathbb{Z}/2\mathbb{Z}$ or to $(\mathbb{Z}/2\mathbb{Z})^4$. A fundamental domain for the action of $H(T)$ on T is naturally contained in a copy of M sitting in $\widehat{M}$. This fundamental domain is either a torus, or an annulus with boundary disjoint from $\mathcal{G}$, or a disc whose boundary intersects $\mathcal{G}$ transversally in 4 points. Again, the irreducibility and the atoroidality of $(M, \mathcal{G})$ imply that T cannot be incompressible. □

The next proposition shows an important property of a good splitting surface.

Definition.— We say that a 3-manifold M is *fibered* if it contains a splitting surface S such that the manifold obtained by splitting M along S is an interval bundle over a (not necessarily connected) closed surface.

Note that a fibered manifold is not necessarily fibered over the circle.

Let $(M, \mathcal{G})$ be a connected irreducible and atoroidal manifold-with-corners with empty boundary. Let S be a good splitting surface for $(M, \mathcal{G})$. Let $(M_S, \mathcal{G}_S, S')$ be the manifold-with-corners obtained by splitting $(M, \mathcal{G})$ along S. Let $\widehat{M}_S$ be the partially mirrored manifold of $(M_S, \mathcal{G}_S, S')$.

Proposition 7.6.— *With these notations, we have:*

(i) *the boundary of $\widehat{M}_S$ is incompressible;*

(ii) *if M is not fibered over the circle or if $\partial M \neq \emptyset$, then $\widehat{M}_S$ is not an interval bundle over a closed surface.*

Proof.— The proof of (i) is similar to that of Proposition 7.5.

To prove (ii) suppose first that $\partial M = \emptyset$. Then since M is not fibered, the manifold obtained by splitting M along any embedded surface is not an interval bundle.

Suppose now that $\partial M \neq \emptyset$. Then $\partial S \neq \emptyset$, since S is a good splitting surface. Let f be a mirror of $(M, \mathcal{G})$ such that $f \cap \partial S \neq \emptyset$. Let $\mathcal{F}$ be the surface above f. Since $(M, \mathcal{G})$ is irreducible and atoroidal, the Euler characteristic of $\mathcal{F}$ is strictly negative. Let f' be the union of the mirrors of $(M_S, \mathcal{G}_S)$ which are contained in f. Let $\mathcal{S}$ be the set of mirrors of $(M, \mathcal{G})$ which intersect f. Let $\mathcal{F}'$ be the image of $f' \times (\mathbb{Z}/2\mathbb{Z})^{\mathcal{S}}$ under the equivalence relation $\mathcal{R}_S$. Then $\mathcal{F}'$ is homeomorphic to the complement in $\mathcal{F}$ of a non-empty disjoint union of annuli. In particular $\chi(\mathcal{F}') < 0$ and $\partial \mathcal{F}' \neq \emptyset$. The arguments used for proving the incompressibility of $\mathcal{F}$ and the fact that $(M_S, \mathcal{G}_S, S')$ has incompressible boundary show that $\mathcal{F}'$ is incompressible in $\widehat{M_S}$. Suppose for a contradiction that $\widehat{M_S}$ were an interval bundle. Then, by a theorem of Stallings [Stal], each component of $\mathcal{F}'$ would be an annulus. This is impossible since $\chi(\mathcal{F}') < 0$. □

7.3 Hyperbolic manifolds-with-corners

Let $(M, \mathcal{G}, \partial^0 M)$ be a manifold-with-corners. The definition of a hyperbolic metric on a manifold (cf. §1) can be made in exactly the same way if M has a differentiable structure with corners. For $(M, \mathcal{G}, \partial^0 M)$ to be hyperbolic, we require two more properties.

Definition.— We say that $(M, \mathcal{G}, \partial^0 M)$ *is hyperbolic* if M (with its differentiable structure with corners) has a hyperbolic metric such that

(i) each mirror of $(M, \mathcal{G}, \partial^0 M)$ is totally geodesic,

(ii) any two distincts mirrors or components of $\partial^0 M$ which share a common edge meet orthogonally.

Observe that along the components of ∂M which are not mirrors M is locally convex (from the definition of a hyperbolic metric).

We have:

Fact 7.7.— *Let $(M, \mathcal{G})$ be a manifold-with-corners having empty boundary. If $(M, \mathcal{G})$ is hyperbolic, then (the differentiable manifold) M is hyperbolic.*

Proof.— There exists a geometrically finite group G and an isometric embedding of M into $M(G)$ which realizes M as a convex of $M(G)$ (cf. §1). For sufficiently small δ, $N_\delta(G)$ is contained in M. Then the retraction r_δ (cf. §1) allows to construct an homeomorphism between M and the hyperbolic manifold $N_\delta(G)$. It follows from the Cerf theorem and from the uniqueness of the rounding that (the differentiable manifold) M is diffeomorphic to $N_\delta(G)$. □

Definition.— Let H be a finite group of diffeomorphisms of a manifold M. We say that M *is H-equivariantly hyperbolic* if M carries an *H-equivariant hyperbolic metric*, i.e. a hyperbolic metric for which H acts by isometries.

Let $(M, \mathcal{G}, \partial^0 M)$ be a manifold-with-corners and let $\mathcal{M}$ be its set of mirrors. Let $H = (\mathbb{Z}/2\mathbb{Z})^{\mathcal{M}}$ be the mirror group of $\widehat{M}$.

Fact 7.8.— *The manifold-with-corners $(M, \mathcal{G}, \partial^0 M)$ is hyperbolic if and only if $\widehat{M}$ is H-equivariantly hyperbolic.*

Proof.— Suppose that $(M, \mathcal{G}, \partial^0 M)$ is hyperbolic. Since the mirrors of $(M, \mathcal{G})$ are totally geodesic and meet orthogonally, the "product" hyperbolic metric on $M \times (\mathbb{Z}/2\mathbb{Z})^{\mathcal{M}}$ is invariant under the equivalence relation $\mathcal{R}$. It defines therefore a hyperbolic metric on $\widehat{M}$. Therefore $\widehat{M}$ is hyperbolic. By construction H acts by isometries. Therefore $\widehat{M}$ is H-equivariantly hyperbolic.

Conversely, suppose that $\widehat{M}$ has a H-equivariant hyperbolic metric. The fixed point set of a reversing orientation diffeomorphism of a hyperbolic 3-manifold is a totally geodesic submanifold of codimension 1 which meets the boundary orthogonally. Furthermore, if two distinct orientation reversing isometries of an hyperbolic manifold commute, the components of their fixed point sets are disjoint or intersect orthogonally. It follows then from the way the manifold-with-corners $(M, \mathcal{G}, \partial^0 M)$ can be reconstructed from the action of H on $\widehat{M}$, that $(M, \mathcal{G}, \partial^0 M)$ is hyperbolic. □

Let $(M, \mathcal{G})$ be a manifold-with-corners having empty boundary. Let S be a splitting surface for $(M, \mathcal{G})$. Let $(M_S, \mathcal{G}'_S)$ be a manifold-with-corners having empty boundary such that $S' \subset \partial M_S$ is a union of mirrors of $(M_S, \mathcal{G}'_S)$. Suppose that $(M_S, \mathcal{G}_S, S')$ is obtained from $(M_S, \mathcal{G}'_S)$ by erasing the mirrors that are contained in S'. With these notations, we have:

Fact 7.9.— *If $(M_S, \mathcal{G}'_S)$ is hyperbolic, then*

(i) *$(M_S, \mathcal{G}_S, S')$ is hyperbolic, and*

(ii) *$\widehat{M}_S$ is H-equivariantly hyperbolic.*

Proof.— We begin by proving (ii). Recall that $\widehat{M}_S$ is H-equivariantly diffeomorphic to the manifold obtained by rounding the corners of $\widehat{M}'_S$. When M_S carries a hyperbolic metric arising from a hyperbolic metric on $(M_S, \mathcal{G}'_S)$, the "product" hyperbolic metric on $M_S \times (\mathbb{Z}/2\mathbb{Z})^{\mathcal{M}}$ is invariant under the equivalence relation $\mathcal{R}_S$. It defines therefore a H-equivariant hyperbolic metric on the differentiable manifold with corners $\widehat{M}'_S$. There is a geometrically finite group G such that $N_\delta(G)$ is naturally embedded in $\widehat{M}'_S$ (cf. the proof of Fact 7.7). The group H preserves $N_\delta(G)$ and commutes with the retraction r_δ. It follows that $N_\delta(G)$ is H-equivariantly homeomorphic (and hence diffeomorphic, by uniqueness of the rounding) to the manifold obtained from $\widehat{M}'_S$ by rounding the corners. Therefore, $\widehat{M}_S$ is H-equivariantly hyperbolic.

Since $\partial N_\delta(G)$ is invariant under H, the fixed point set of any non-trivial isometry in H is either disjoint from $\partial N_\delta(G)$ or intersects $N_\delta(G)$ orthogonally. It follows then from the way that $(M_S, \mathcal{G}_S, S')$ can be reconstructed from the action of H on $\widehat{M}_S$ that $(M_S, \mathcal{G}_S, S')$ is also hyperbolic (cf. Fact 7.8). □

CHAPTER 8

Proof of Thurston's hyperbolization theorem

Hyperbolization theorem for manifolds-with-corners.— *Let $(M, \mathcal{G})$ be an irreducible and atoroidal manifold-with-corners having empty boundary. If M is Haken, then $(M, \mathcal{G})$ is hyperbolic.*

In this chapter, we prove this theorem when M is not fibered. The major step of the proof is the following result.

Gluing theorem.— *Let $(M, \mathcal{G})$ be an irreducible and atoroidal manifold-with-corners having empty boundary such that M is not fibered. Let S be a good splitting surface for $(M, \mathcal{G})$ and let $(M_S, \mathcal{G}_S, S')$ be the manifold-with-corners obtained by splitting $(M, \mathcal{G})$ along S. Suppose that $(M_S, \mathcal{G}_S, S')$ is hyperbolic. Then $(M, \mathcal{G})$ is hyperbolic.*

Proof.— We may suppose that M is connected. Let $\mathcal{M}$ be the set of mirrors of $(M, \mathcal{G})$ and set $H = (\mathbb{Z}/2\mathbb{Z})^{\mathcal{M}}$. Let $\widehat{M}$ be the mirrored manifold of $(M, \mathcal{G})$ and let $\widehat{M}_S$ be the partially mirrored manifold of $(M_S, \mathcal{G}_S, S')$. The group H acts by diffeomorphisms on $\widehat{M}$ and on $\widehat{M}_S$. There is an orientation reversing involution τ' of S' which permutes the components and such that $(M, \mathcal{G})$ is diffeomorphic to the quotient space $(M_S, \mathcal{G}_S, S')/\tau'$. This involution τ' lifts to an orientation reversing involution τ of $\partial\widehat{M}_S$ which permutes the components by pairs and which commutes with the action of H. Also $\widehat{M}$ is H-equivariantly diffeomorphic to the quotient space $\widehat{M}_S/\tau$.

By hypothesis and by Fact 7.9, $\widehat{M}_S$ is hyperbolic. Since S is a good splitting surface, Proposition 7.6 says that $\widehat{M}_S$ has incompressible boundary and that $\widehat{M}_S$ is not an interval bundle. Since $(M, \mathcal{G})$ is irreducible and atoroidal, $\widehat{M}$ is atoroidal (Proposition 7.5). Let

$$\tau^* : \mathcal{T}(\overline{\partial\widehat{M}_S}) \to \mathcal{T}(\partial\widehat{M}_S)$$

be the map induced by τ. Let $\sigma : \mathcal{T}(\partial\widehat{M}_S) \to \overline{\mathcal{T}(\partial\widehat{M}_S)}$ be the skinning map associated to $\widehat{M}_S$. By Thurston's fixed point theorem (§6), $\tau^* \circ \sigma$ has a fixed point. This implies that $\widehat{M} = \widehat{M}_S/\tau$ is hyperbolic (cf. §2).

In order to prove that $(M, \mathcal{G})$ is hyperbolic, we need to prove that $\widehat{M}$ is H-equivariantly hyperbolic (Fact 7.8). We introduce first some notations. Since $\widehat{M}_S$ is H-equivariantly hyperbolic, there exists a geometrically finite group G (resp. geometrically finite groups $G_1, \cdots, G_p$, when S is separating) with an isometric action of H on $M(G)$ (resp. on the disjoint union of $M(G_1), \cdots, M(G_p)$) such that $\widehat{M}_S$ is H-equivariantly diffeomorphic to $\overline{M}(G)$ (resp. the disjoint union of $\overline{M}(G_1), \cdots, \overline{M}(G_p)$). To each $s \in \mathcal{T}(\partial\widehat{M}_S)$, it corresponds via the Ahlfors-Bers map a quasi-conformal deformation $(\rho, \widetilde{\varphi})$ of G (resp. quasi-conformal deformations $(\rho_i, \widetilde{\varphi}_i)$ of G_i for $i = 1, \cdots, p$) such that $\partial(\rho, \widetilde{\varphi}) = s$ (resp. $(\partial(\rho_i, \widetilde{\varphi}_i) = s)$. We denote by $\widehat{M}_S^s$ the manifold $\overline{M}(\rho(G))$ (resp. the disjoint union of $\overline{M}(\rho_1(G_1)), \cdots, \overline{M}(\rho_p(G_p))$).

The group H acts on $\partial\widehat{M}_S$ by diffeomorphisms but, since the elements of H don't all preserve the orientation, H does not act in a direct way on $\mathcal{T}(\partial\widehat{M}_S)$. An element $h \in H$ induces a map h^* from $\mathcal{T}(\partial\widehat{M}_S)$ to itself or from $\mathcal{T}(\partial\widehat{M}_S)$ to $\overline{\mathcal{T}(\partial\widehat{M}_S)}$ according as h preserves or not the orientation (cf. §1). We define an action of H on $\mathcal{T}(\partial\widehat{M}_S)$ as follows. Let $s \in \mathcal{T}(\partial\widehat{M}_S)$. For an element $h \in H$ which preserves the orientation, we set $h'(s) = h^*(s)$. If $h \in H$ reverses the orientation, we set $h'(s) = \overline{h^*(s)}$, where $\overline{h^*(s)}$ is the complex conjugated of $h^*(s)$ (cf. §1). Since h^* commutes with the complex conjugation (cf. §1), $h \to h'$ defines an action of H on $\mathcal{T}(\partial\widehat{M}_S)$. Denote by $\mathcal{T}(\partial\widehat{M}_S)_H$ the fixed point set of this action. Since H acts by isometries on $\partial\widehat{M}_S$, the point $\partial\widehat{M}_S \in \mathcal{T}(\partial\widehat{M}_S)$ is fixed by H. Hence $\mathcal{T}(\partial\widehat{M}_S)_H \neq \emptyset$.

Fact 8.1.— *The map $\tau^* \circ \sigma$ leaves $\mathcal{T}(\partial\widehat{M}_S)_H$ invariant.*

Proof.— Consider an element $h \in H$ which preserves the orientation. From the definition of σ, it follows that σ and h^* commute. Therefore, since τ and h commute, $\tau^* \circ \sigma$ and $h' = h^*$ commute also.

For an element $h \in H$ which reverses the orientation, the same conclusion holds. One needs only to observe that the skinning map $\bar{\sigma}$ for the hyperbolic manifold $\widehat{M}_S$ with the reversed orientation satisfies $\bar{\sigma}(\bar{s}) = \sigma(s)$. This implies Fact 8.1. □

Since $\mathcal{T}(\partial\widehat{M}_S)_H$ is a closed non-empty subset of $\mathcal{T}(\partial\widehat{M}_S)$ and since $\tau^* \circ \sigma$ is contracting, its fixed point belongs to $\mathcal{T}(\partial\widehat{M}_S)_H$. Let $s_0 = \partial(\rho, \widetilde{\varphi})$ be this fixed point. Then H acts on $\partial\widehat{M}_S^{s_0}$ by isometries. Since the action of H on $\partial\widehat{M}_S^{s_0}$ extends to an action on $\widehat{M}_S^{s_0}$ by diffeomorphisms, it follows from the injectivity of the Ahlfors-Bers map that it extends also to an action by isometries. Also, $\widetilde{\varphi}$ induces a quasi-conformal homeomorphism $\varphi : \partial\widehat{M}_S \to \partial\widehat{M}_S^{s_0}$ which conjugates the actions of H. Denote by Φ the natural extension of φ (cf. §1). Then $\Phi : \widehat{M}_S \to \widehat{M}_S^{s_0}$ is a homeomorphism. By naturality, since φ conjugates the isometric actions of H on $\partial\widehat{M}_S$ and $\partial\widehat{M}_S^{s_0}$, Φ conjugates the isometric actions of H on $\widehat{M}_S$ and on $\widehat{M}_S^{s_0}$. The uniqueness of the differentiable structure on the manifold-with-corners $\widehat{M}_S/H$

implies that Φ is isotopic to a diffeomorphism which conjugates the actions of H. Therefore, $\widehat{M}_S$ is H-equivariantly diffeomorphic to $\widehat{M}_S^{s_0}$.

In order to show that $\widehat{M}_S/\tau$ is H-equivariantly hyperbolic, we need to recall the proof of Maskit's combination theorem given in §2. Using harmonic functions, we constructed a codimension 0 submanifold N contained in the interior of $\widehat{M}_S^{s_0}$. By definition, N is invariant under the action of H. Since s_0 is a fixed point of $\tau^* \circ \sigma$, τ induces an orientation reversing isometry $\mathfrak{I}$ of ∂N (cf. §2). Since $s_0 \in \mathcal{T}(\partial \widehat{M}_S)_H$, $\mathfrak{I}$ commutes with the action of H. Therefore the quotient space $M' = N/\mathfrak{I}$ is a hyperbolic manifold on which H acts by isometries. We need to show that M' is H-equivariantly diffeomorphic to $\widehat{M}_S$. Like in §2, we distinguish two cases.

1) ∂N *is incompressible in* M'.

To conclude then, we need an equivariant version of Stallings theorem, which in our case goes as follows. By construction, ∂N divides $\widehat{M}_S^{s_0}$ into two manifolds. Let V be the one which contains $\partial \widehat{M}_S^{s_0}$. Consider the product action of H on $\partial \widehat{M}_S^{s_0} \times I$ where H acts on the standard way on $\partial \widehat{M}_S$ and acts by the identity on the second factor. The quotient $(\partial \widehat{M}_S^{s_0} \times I)/H$ can be viewed then as a manifold-with-corners, naturally isomorphic to the product of a surface-with-corners (S', P) by I (its boundary is $S' \times \{0\} \cup S' \times \{1\}$ and its mirrors are squares, product of mirrors of (S', P) by I). From the proof of Maskit's combination theorem, V embeds H-equivariantly in $\widehat{S}' \times I$ in such a way that $\partial \widehat{M}_S^{s_0}$ maps to $\widehat{S}' \times \{1\}$. Then the surface-with-corners $\partial N/H$ embeds in $(S', P) \times I$. Its boundary $\partial(\partial N/H)$ is contained in the reunion of the mirrors and $\partial N/H$ divides $(S', P) \times I$ into two manifolds, one of which being V/H. Since ∂N is incompressible, $(\partial N/H, \partial(\partial N/H))$ is an incompressible surface in the pair $(S', \partial S') \times I$. Therefore by Stallings theorem, $\partial N/H$ is isotopic to $S' \times \{1/2\}$. In particular, $\partial N/H$ is homeomorphic to S' and each component of $\partial(\partial N/H)$ intersects as many mirrors as the corresponding component of $\partial S'$. Therefore the Euler characteristic of every component of ∂N is smaller than the Euler characteristic of $\widehat{S}'$ with equality if and only if each component of $\partial(\partial N/H)$ intersects each mirror in a single arc. Since each component of ∂N is homeomorphic to $\widehat{S}'$, this must be the case. Since the mirrors are squares, it follows that we can suppose that the isotopy in Stallings theorem respects the mirrors. Thus N is H-equivariantly diffeomorphic to $\widehat{M}_S$ and therefore to $\widehat{M}_S^{s_0}$ also. Under this identification τ and $\mathfrak{I}$ are homotopic diffeomorphisms and they commute with the action of H. By the equivariant version of Nielsen theorem (which can also be deduced in this special case from the classical version), τ and $\mathfrak{I}$ are H-equivariantly isotopic. This implies that M' and $\widehat{M}_S/\tau$ are H-equivariantly diffeomorphic.

2) ∂N **is not incompressible in** M'.

Then there is a compression disc for ∂N in M'. By the equivariant Dehn's Lemma, there is such a disc D which is either disjoint or equal to any of its translates by H. By doing equivariant surgery to ∂N along the H-orbit of D, we define a H-invariant manifold N' contained in the interior of $\widehat{M}_S^{s_0}$ with an involution $\mathfrak{I}'$ of $\partial N'$, such that M' is H-equivariantly diffeomorphic to the quotient space $N'/\mathfrak{I}'$ (cf. §2).

If $\partial N'$ is incompressible in M', then the reasoning of case 1) can be applied. If not, we do equivariant surgery again. This process stops after a finite number of steps. □

Proof of the Hyperbolization theorem for manifolds-with-corners. —

The proof is by induction on $\ell(M)$.

If $\ell(M) = 0$, M is a handlebody. Let S be a good splitting surface for $(M, \mathcal{G})$ (S is a system of meridians). Let $(M_S, \mathcal{G}_S, S')$ be the manifold-with-corners obtained by splitting $(M, \mathcal{G})$ along S. Then M_S is diffeomorphic to the 3-ball. Let $(M_S, \mathcal{G}'_S)$ be the manifold-with-corners with empty boundary provided by Proposition 7.4. Since $(M_S, \mathcal{G}'_S)$ is irreducible and atoroidal, the intersection of two distinct mirrors of $(M_S, \mathcal{G}'_S)$ is either empty or a single edge. Therefore $(M_S, \mathcal{G}'_S)$ can be viewed as a polyhedron whose faces are in correspondance with the mirrors of $(M_S, \mathcal{G}'_S)$). Saying that $(M_S, \mathcal{G}'_S)$ is hyperbolic means that this polyhedron can be realized in $\mathbb{H}^3$ with all dihedral angles equal to $\pi/2$. This turns out to be a special case of a theorem of Andreev which provides sufficient conditions on a polyhedron with prescribed dihedral angles to be realizable in $\mathbb{H}^3$. Here is a formulation of Andreev's theorem, when the prescribed dihedral angles all equal $\pi/2$ and in our particular case (recall that any vertex of $\mathcal{G}'_S$ has valence 3).

Theorem ([An], [Th1]). — *Let $\mathcal{P}$ be a polyhedron distinct from the tetrahedron or from the triangular prism. Then $\mathcal{P}$ can be realized in $\mathbb{H}^3$ with all dihedral angles equal to $\pi/2$ if and only if*

(i) *every cycle of faces of length 3 surrounds a vertex of $\mathcal{P}$, and*

(ii) *every cycle of faces of length 4 surrounds an edge of $\mathcal{P}$.*

Definition. — A *cycle of faces of length k of $\mathcal{P}$* is a finite sequence of faces $(f_i)_{i=0,\cdots,k}$ with $f_0 = f_k$ and such that any two successive faces share exactly an edge in common.

By the construction of $\mathcal{G}'_S$, the polyhedron $(M_S, \mathcal{G}'_S)$ has too many faces to be isomorphic to a tetrahedron or to a triangular prism. Also the hypothesis of Andreev's theorem are satisfied since $(M_S, \mathcal{G}'_S)$ is irreducible and atoroidal. Hence $(M_S, \mathcal{G}'_S)$ is hyperbolic. By Fact 7.9, $(M_S, \mathcal{G}_S, S')$ is hyperbolic. Since $(M, \mathcal{G})$ is irreducible and atoroidal and since $\partial M \neq \emptyset$ (cf. Proposition 7.6), the Gluing theorem implies that $(M, \mathcal{G})$ is hyperbolic. This proves the initial step of the induction.

Let now $(M, \mathcal{G})$ be an irreducible and atoroidal manifold-with-corners such that M is a Haken and suppose the theorem true for all manifolds-with-corners $(M', \mathcal{G}')$ with $\ell(M') < \ell(M)$. Let S be a good splitting surface for $(M, \mathcal{G})$. Let $(M_S, \mathcal{G}_S, S')$ be the manifold obtained by splitting $(M, \mathcal{G})$ along S. Since the length of any component of M_S is smaller than $\ell(M)$ (cf. §7) the manifold-with-corners $(M_S, \mathcal{G}'_S)$ provided by Proposition 7.4 is hyperbolic. By Fact 7.9, $(M_S, \mathcal{G}_S, S')$ is hyperbolic. Since M is not fibered, the Gluing theorem implies that $(M, \mathcal{G})$ is hyperbolic also.

This proves the Hyperbolization theorem for manifold-with-corners. □

Thurston's hyperbolization theorem. — *Let M be an irreducible and atoroidal manifold. If M is Haken, then M is hyperbolic.*

Proof. — By Proposition 7.2, there is a graph $\mathcal{G} \subset \partial M$ such that $(M, \mathcal{G})$ is an irreducible and atoroidal manifold-with-corners. If M is Haken, $(M, \mathcal{G})$ is hyperbolic by Thurston's hyperbolization theorem for manifold-with-corners. By Fact 7.7, M is hyperbolic. □

BIBLIOGRAPHY

[Ah1] L. Ahlfors, Finitely generated Kleinian groups, Amer. J. Math., 86 (1964), 413–429; Amer. J. Math., 87 (1965), 759.

[Ah2] L. Ahlfors, "Lectures on Quasiconformal Mappings", Van Nostrand, Princeton, 1966.

[Ah3] L. Ahlfors, "Complex Analysis", McGraw-Hill, New-York, 1953.

[An] E. Andreev, On convex polyhedra in Lobacevskii space, Math. Sb., 81 (1970), 445–478; English translation, Math. USSR, Sb 10 (1970), 413–440.

[AB] L. Ahlfors and L. Bers, Riemann's mapping theorem for variable metrics, Ann. of Math., 72 (1960), 385–404.

[Bea] A. Beardon, "The Geometry of Discrete Groups", Springer-Verlag, New-York, 1983.

[Ber] L. Bers, On boundaries of Teichmüller spaces and on Kleinian groups, Ann. of Math., 91 (1970), 570–600.

[Bo] A. Borel, Compact Clifford-Klein forms of symmetric spaces, Topology, 2 (1963), 111–122.

[BD] D. Barrett and J. Diller, Contraction properties of the Poincaré series operator, preprint (1994).

[BP] R. Benedetti, C. Petronio, "Lectures on Hyperbolic Geometry", Springer-Verlag, Berlin, 1992.

[BS] F. Bonahon and L. Siebenmann, The characteristic toric splitting of irreducible compact 3-orbifolds, Math. Annalen, 278 (1987), 441–479.

[C] J. Cerf, "Sur les Difféomorphismes de la Sphère de Dimension Trois ($\Gamma_4 = 0$)", LNM 53, Springer-Verlag, Berlin, 1968.

[CEG] R. Canary, D. Epstein and P. Green, Notes on notes of Thurston, in "Analytical and Geometric Aspects of Hyperbolic Spaces,", London Math. Soc. lecture notes series, 111, Cambridge Univ. Press, 1985, 3–92.

[Di] J. Diller, A canonical $\bar{\partial}$-problem for Riemann surfaces, Indiana Univ. Math. J., (to appear).

[Do] A. Douady, Arrondissement des arêtes, Séminaire Henri Cartan (Topologie différentielle) (1961/62).

[Du] M. J. Dunwoody, An equivariant sphere theorem, Bull. London Math. Soc., 17 (1985), 437–448.

[EM] D. Epstein and A. Marden, Convex hulls in hyperbolic space, a theorem of Sullivan and measured pleated surfaces, in "Analytical and Geometric Aspects of Hyperbolic Spaces,", London Math. Soc. lecture notes series 111, Cambridge Univ. Press, 1985, 113–253.

[F] C. Feustel, On the torus theorem for closed 3-manifolds, Trans. Amer. Math. Soc., 217 (1976), 45–57.

[Ga] F. Gardiner, "Teichmüller Theory and Quadratic Differentials", John Wiley, New-York, 1987.

[Gr] F. Greenleaf, "Invariant Means on Topological Groups and their Applications", Van Nostrand, New-York, 1969.

[Hak] W. Haken, Über das Homöomorphieproblem der 3-Mannifaltigkeiten, I, Math. Z., 80 (1962), 89–120.

[Hat] A. Hatcher, On the boundary curves of incompressible surfaces, Pac. J. Math., 99 (1982), 373–377.

[He1] J. Hempel, "3-Manifolds", Ann. of Math. Studies 86, Princeton University Press, 1976.

[He2] J. Hempel, Orientation reversing involutions and the first Betti number for finite coverings of 3-manifolds, Invent. Math., 67 (1982), 133–142.

[HR] C. Hodgson and I. Rivin, A characterization of compact convex polyhedra in hyperbolic 3-space, Invent. Math., 111 (1993), 77–111.

[HT] A. Hatcher and W. Thurston, Incompressible surfaces in 2-bridge knots complements, Inv. Math., 79 (1985), 225–246.

[Ja] W. Jaco, "Lectures on Three-Manifold Topology", CBMS, 53, 1980.

[Jo1] K. Johannson, Equivalences d'homotopie des variétés de dimension 3, CRAS, 181 (1975), 1009–1010.

[Jo2] K. Johannson, "Homotopy Equivalence of 3-Manifolds with Boundaries", Lecture Notes in Mathematics 761, Springer-Verlag, Berlin, 1979.

[JR] W. Jaco and H. Rubinstein, PL minimal surfaces in 3-manifolds, Adv. in Math., 73 (1989), 149–191.

[JS1] W. Jaco and P. Shalen, A new decomposition theorem for irreducible sufficiently large 3-manifolds, Proc. Sympos. in Pure Mathematics, A. M. S., vol. 32 (1977), 71–84.

[JS2] W. Jaco and P. Shalen, "Seifert Fibered Spaces in 3-Manifolds", Memoirs of the Amer. Math. Soc., vol. 220, 1979.

[Ka] M. Kapovitch, "Hyperbolic Manifolds and Discrete Groups: Notes on Thurston's Hyperbolization Theorem", University of Utah lecture notes, 1995.

[Kn] H. Kneser, Geschlossene Flächen in dreidimensionale Mannigfaltigkeiten, Jahresbericht der Deut. Math. Verein., 38 (1929), 248–260.

[Kr] I. Kra, Canonical mappings between Teichmüller spaces, Bull. Amer. Math. Soc., 4 (1981), 144–178.

[L] F. Löbell, Beispiele geschlossener dreidimensionaler Clifford-Kleinscher Räume negativer Krümmung, Ber. d . Sächs. Akad. d. Wiss., 83 (1931), 167–174.

[LV] O. Lehto and K. Virtanen, "Quasiconformal Mappings in the Plane", Springer-Verlag, Berlin, 1973.

[Mard] A. Marden, The geometry of finitely generated Kleinian groups, Ann. of Math., 99 (1974), 383–462.

[Marg] G. A. Margulis, Discrete groups of motions of manifolds of non-positive curvature, Proc. Int. Congr. Math., Vancouver, Vol. 2, 1974, 21–34.

[Mas1] B. Maskit, On Klein's combination theorem, III, in "Advances in the theory of Riemann Surfaces", L. V. Ahlfors et al. editors (Proc. Conf. Stony Brook, 1969) Ann. of Math. Studies, Princeton University Press.

[Mas2] B. Maskit, "Kleinian Groups", Springer Verlag, Berlin, 1988.

[McM1] C. McMullen, Amenability, Poincaré series and quasiconformal maps, Invent. Math., 97 (1989), 95–127.

[McM2] C. McMullen, Iteration on Teichmüller space, Invent. Math., 99 (1990), 425–454.

[McM3] C. McMullen, "Renormalization and 3-Manifolds which Fiber over the Circle", Annals of Math. Studies, Princeton University Press, 1996.

[Mi1] J. Milnor, A unique factorization theorem for 3-manifolds, Amer. J. of Math., 84 (1962), 1–7.

[Mi2] J. Milnor, Hyperbolic geometry: the first 150 years, Bull. Amer. Math. Soc., 6 (1982), 9–24.

[Mor] J.W. Morgan, On Thurston's uniformization theorem for three-dimensional manifolds, in "The Smith Conjecture", (H. Bass and J. Morgan ed.) Academic Press, Orlando, Fl, 1984, 37–125.

[Mos] G.D. Mostow, "Strong Rigidity of Locally Symmetric Spaces", Ann. of Math. Studies 86, Princeton University Press, 1973.

[Mu] D. Mumford, A remark on Mahler's compactness theorem, Proc. Amer. Math. Soc., 28 (1971), 289–294.

[My] R. Myers, Contractible open 3-manifolds which are not covering spaces, Topology, 27 (1988), 27–35.

[MS] W. Meeks and P. Scott, Finite group actions on 3-manifolds, Invent. Math., 86 (1988), 287–346.

[MY] W. Meeks and S. Yau, The equivariant Dehn's lemma and loop theorem, Comment. Math. Helv., 56 (1981), 225–239.

[Nic] P. Nichols, "The Ergodic Theory of Discrete Groups", London Math. Soc. lecture notes series 143, Cambridge Univ. Press, 1989.

[Nie] J. Nielsen, Untersuchungen zur Topologie der geschlossenen Flächen, Acta Math., 53 (1929), 1–76.

[O'B] B. O'Byrne, On Finsler geometry and applications to Teichmüller spaces, in "Advances in the Theory of Riemann Surfaces, Ann. of Math. Stud. 66", Princeton Univ. Press, 317–328, 1971.

[O] J.-P. Otal, "Le Théorème d'Hyperbolisation pour les Variétés Fibrées de Dimension 3", Astérisque 235, 1996.

[OP] J.-P. Otal and F. Paulin, "Géométrie Hyperbolique et Groupes Kleiniens", in preparation.

[Pa] C. Papakyriokopoulos, On Dehn's lemma and the asphericity of knots, Ann. of Maths, 66 (1957), 1–26.

[Ra] J. G. Ratcliffe, "Foundations of Hyperbolic Geometry", Springer Verlag, New-York, 1994.

[Re] H. Reimann, Invariant extensions of quasiconformal deformations, Ann. Aca. Scien. Fenn., 10 (1985), 281–288.

[Ri] R. Riley, A quadratic parabolic group, Math. Proc. Cambridge Philos. Soc., 77 (1975), 281–288.

[Ro] C. Rourke, Convex ruled surfaces, in "Analytical and Geometric Aspects of Hyperbolic Spaces, London Math. Soc. lecture notes series 111", Cambridge Univ. Press, 1985, 255–272.

[Sc] P. Scott, The geometries of 3-manifolds, Bull. London Math. Soc., 15 (1983), 401–487.

[Sei] H. Seifert, Topologie dreidimensionalen gefaserten Räume, Acta Math. (1933), 147–238.

[Sel] A. Selberg, On discontinuous groups in higher-dimensional symmetric spaces, in "Contrib. Function Theory", Int. Colloq. Bombay, 1960, 147–164.

[Sh] P. Shalen, Infinitely divisible elements in 3-manifold groups, in "Knots, Groups, and 3-Manifolds,", L. Neuwirth editor, Ann. of Math. Studies 84, Princeton Univ. Press, 1975, 293–335.

[Sta1] J. Stallings, On fibering certain 3-manifolds, in "Topology of 3-manifolds and related topics", Englewwod Cliffs, Prentice Hall, 1962, 95–100.

[Sta2] J. Stallings, "Group Theory and Three-dimensional manifolds", Yale Math. Monographs, Yale University Press, 4, 1971.

[Str] K. Strebel, "Quadratic Differentials", Springer-Verlag, Berlin, 1980.

[Su] D. Sullivan, Travaux de Thurston sur les groupes quasi-fuchsiens et sur les variétés hyperboliques de dimension 3 fibrées sur S^1, in "Séminaire Bourbaki 554", Lecture notes in mathematics 842, Springer-Verlag, Berlin, 1981.

[Th1] W.P. Thurston, "The geometry and topology of 3-manifolds", Lecture notes, Princeton Univ., 1977.

[Th2] W.P. Thurston, On the geometry and dynamics of diffeomorphisms of surfaces, Bull. Amer. Math. Soc., 19 (1988), 417–432.

[Th3] W.P. Thurston, Three-dimensional manifolds, Kleinian groups and hyperbolic geometry, Bull. Amer. Math. Soc., 6 (1982), 357–381.

[Th4] W.P. Thurston, Hyperbolic structures on 3-manifolds I. Deformation of acylindrical manifolds, Ann. of Math., 124 (1986), 203–246.

[Th5] W.P. Thurston, Hyperbolic structures on 3-manifolds II. Surface groups and manifolds which fiber over the circle, Princeton Univ. preprint (1986).

[To] J. Tollefson, Involutions on sufficiently large 3-manifolds, Topology, 20 (1981), 323–352.

[Ts] M. Tsuji, "Potential Theory in Modern Function Theory", Chelsea, New-York, 1959.

[Vi] E. B. Vinberg, "Geometry II", Encyclopedia of Mathematical Sciences, vol. 29, Springer Verlag, Berlin, 1993.

[Wa1] F. Waldhausen, Eine Klasse von 3-dimensionalen Mannigfaltigkeiten I, II, Invent. Math., 3 (1967), 308–333; Invent. Math., 4 (1967), 87–117.

[Wa2] F. Waldhausen, On irreducible 3– manifolds which are sufficiently large, Ann. of Math., 87 (1968), 56–88.

[Wa3] F. Waldhausen, On the determination of some bounded 3-manifolds by their fundamental groups alone, Proc. Sympos. on Topology and its Applications, (Hercez-Novi, 1968) (1969), 332–333.

[Wh] J. H. C. Whitehead, A certain open manifold whose group is unity, Q. J. Math., 6 (1989), 268–279.

[WS] C. Weber and H. Seifert, Die beiden Dodekaederräume, Math. Z., 37 (1933), 237–253.

INDEX OF NOTATIONS

SUBJECT INDEX

QUASI-MINIMAL SEMI-EUCLIDEAN LAMINATIONS IN 3-MANIFOLDS

David Gabai

§0

Introduction

Thurston's hyperbolization conjecture [Th] asserts that a closed, atoroidal, irreducible 3-manifold with infinite fundamental group has a metric of constant negative curvature. A more modest form of the conjecture asks whether the fundamental group of such a manifold is negatively curved in the sense of Gromov [Gr]. (E.g. see [Bu], [Sch], [Mo1], [MO] or [Ka].) A main result of this paper, the Ubiquity Theorem, provides a technique for addressing the group negative curvature conjecture. We use it in [GK] to show that an atoroidal 3-manifold with a genuine essential lamination has group negative curvature. (These manifolds appear to form a vast subset among the irreducible atoroidal 3-manifolds with infinite π_1. See the survey [G4].)

*Partially supported by NSF Grant DMS-9505253 and the MSRI

In this paper the ubiquity theorem is used to derive new information about quasi least area semi-Euclidean laminations in 3-manifolds.

Gromov stated 6.8.S [Gr] that if M is a compact manifold or finite simplicial complex whose fundamental group is nonnegatively curved, then there is a non constant least area conformal map $f : \mathbb{R}^2 \to M$. Mosher and Oertel [MO] provided the first detailed proof (of a sharper version) in the context of finite 2-complexes. Recently Kleiner [Kl] showed that if M is a closed Riemannian n-manifold such that $\pi_1(M)$ is non negatively curved, then there exists a branched Lipschitz conformal least area immersion $f : \mathbb{R}^2 \to M$. If $n = 3$, then f is an immersion, though the induced metric on $\mathbb{R}^2$ may be incomplete.

In §1 we state basic definitions regarding immersed laminations and branched surfaces in 3-manifolds. Then we outline and discuss the proof of the main technical result of [MO] in the 3-manifold context. Their proof makes essential use of a theorem of Ghys and the Plante argument for immersed surfaces. In §2 we show how to extend [MO] to obtain our first main result.

Theorem 2.1. *If M is a closed non negatively curved 3-manifold, then there exists an immersed strongly least area, Euler characteristic 0, measured semi-Euclidean lamination λ. The induced metric on each leaf is complete.*

Here M has a triangulation τ, and the various measurements of length and area are computed simplically. Strongly least area [MO] means that if L is a leaf and D is an embedded disc in the universal covering of L, then the induced map $i : D \to M$ is a least area map. *Semi-Euclidean* means that the set of leaves, conformally equivalent to the Euclidean plane, are of full measure and dense in λ.

By replacing immersed by branched immersed in the above statement, we obtain the exact translation of the main result of [MO] to 3-manifolds. The content of §2 is how to eliminate the branched points.

Our second main result states that if k is a smooth simple closed curve in the irreducible group nonnegatively curved 3-manifold M, then either k lies in a 3-cell, or M is toroidal or among least area discs D_α of small isoperimetric ratio, area(D_α)

is approximately the geometric intersection number of D and k. (By definition M is toroidal if and only if $Z \oplus Z \subset \pi_1(M)$. By [CJ], [G2] the topology of closed irreducible toroidal manifolds is completely understood.) More precisely we have the following equivalent results.

Theorem 5.1. *Let k be a smooth simple closed curve in the closed irreducible Riemannian 3-manifold M. Then either M is toroidal or k is contained in a 3-cell or there exists a constant $C > 0$ such that if D is a least area disc with $\partial D \cap k = \emptyset$, then $area(D) \leq C(wr(\partial D, k) + length(\partial D))$.*

Ubiquity Theorem 5.2. *Let $k \not\subset B^3$ be a smooth simple closed curve in the closed, atoroidal, irreducible 3-manifold M. There exists constants K and L such that if D is a least area disc with $\partial D \cap k = \emptyset$ and $length(\partial D)/area(D) < L$, then $wr(\partial D, k)/area(D) > K$.*

If k is a simple closed curve in M, and α is a homotopically trivial curve in M disjoint from k, then we define the *wrapping number* $\mathrm{wr}(\alpha, k)$ to be the minimal geometric intersection number between k and all immersed discs D_α spanning α.

Corollary 5.3. *Let λ be a quasi least area semi-Euclidean lamination in the closed atoroidal irreducible 3-manifold M. If $k \subset M - \lambda$ is a smooth simple closed curve, then k lies in a 3-cell.*

Chapter §3 explains how to use cellulations of hyperbolic 3-manifolds by ideal polyhedra to obtain PL versions of various results known in the Riemannian world. In §4 we prove the Ubiquity Theorem in the PL category. Chapter §5 is devoted to proving the Ubiquity Theorem in the smooth category.

In §6 we give a very brief survey of progress on the Thurston geometrization conjecture.

Since the unresolved parts of Thurston's conjecture only concern orientable 3-manifolds, we will assume that **all 3-manifolds in this paper are orientable.**

Readers Advisory. If you are interested in Ubiquity theorem, then skip directly to §4 or §5 and refer back as needed. Except for minor references to earlier stated definitions or figures, §3-6 are completely independent from the earlier chapters. To get right to the heart of the matter in the PL category, read 4.1-4.4 and then read the proof of the case, $M - k$ is hyperbolic. To read §5 one need only refer back to three self contained arguments of §4.

I would like to thank Igor Rivin for several key conversations about hyperbolic cellulations.

§1

Immersed Branched Surfaces and Laminations

Definition 1.1. A *generic immersed branched surface* $B = (B^*, \phi)$ in a 3-manifold M is a finite 2-complex B^* which is mapped via ϕ into M as follows. Each $x \in B^*$ has a neighborhood U whose image appears as in Figures 1.1 a-d. Furthermore if $\phi(U)$ appears as in Figures 1.1 a-c, then $\phi|U$ is an embedding. (Figures 1.1 a-c are the standard local models of a generic embedded branched surface [FO].) If $\phi(U)$ appears as in Figure 1.1 d) then U is homeomorphic to the 2-complex shown in Figure 1.1e) and $\phi|U$ fails to be an immersion only at x. Furthermore $\phi(U)$ has an arc of double points which starts at $\phi(x)$. The model shown in Figure 1.1 d) is the *crossed maw* discovered by Joe Christy [Ch]. The singularity is similar to the *undrawable singularity* of Poenaru [Po]. We assume that the various self intersections of ϕ are transverse and generic. Define the *branch locus* $\gamma(B)$ of B to be the set of nonmanifold points of B^*, define a *sector* to be a component of $B^* - \gamma(B)$, and define ∂B to be the union of free edges of B^*. We will often abuse notation by referring to the branch locus and boundary of B as objects in M. We will also refer to generic immersed branched surfaces simply as *branched surfaces*.

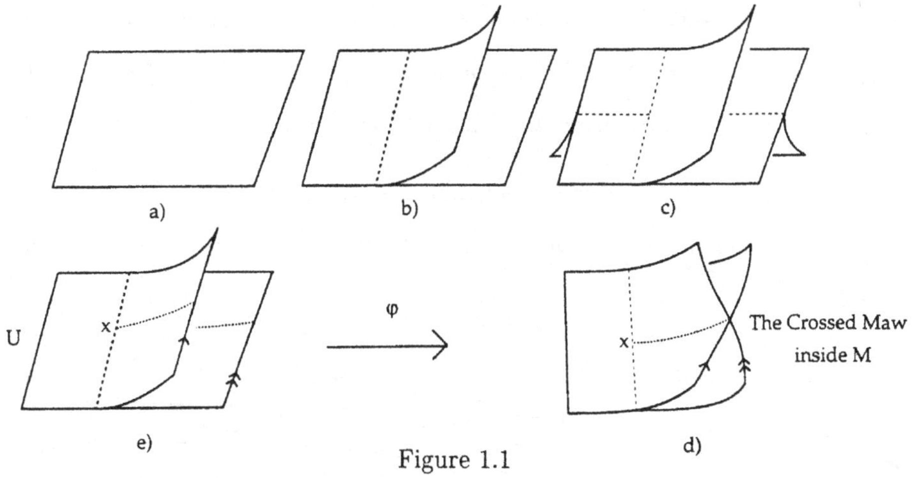

Figure 1.1

Associated to the branched surface B we define the *normal neighborhood* $\mathcal{N}(B) = (\mathcal{N}(B^*), \phi)$ as follows. If B^* is a locally finite union of embedded discs D_i, then define $\mathcal{N}(B^*) = \cup D_i \times [-1,1]$ modulo the equivalence relation, where $x \times [-1,1]$ is identified with $y \times [-1,$disjoint from ∂E1$]$, if $x = y \in B^*$. The identification is either the identity or $t \to -t$ depending on whether or not the local orientations on D_i and D_j agree when viewed inside of M. The singular immersion $\phi : B^* \to M$ induces a singular immersion $\phi : \mathcal{N}(B^*) \to M$. The mapping will be non immersive exactly along $\Gamma(B)$ the *normal branched locus* which we define to be $\pi^{-1}(\gamma(B))$ where $\pi : \mathcal{N}(B^*) \to B^*$ denotes the natural projection. See Figure 1.2. We identify B^* with the 0-section of $\mathcal{N}(B^*)$. The notation $\mathcal{N}(B)$ should not be confused with the similar notation $N(B)$ for *fibred neighborhood* of an embedded branched surface.

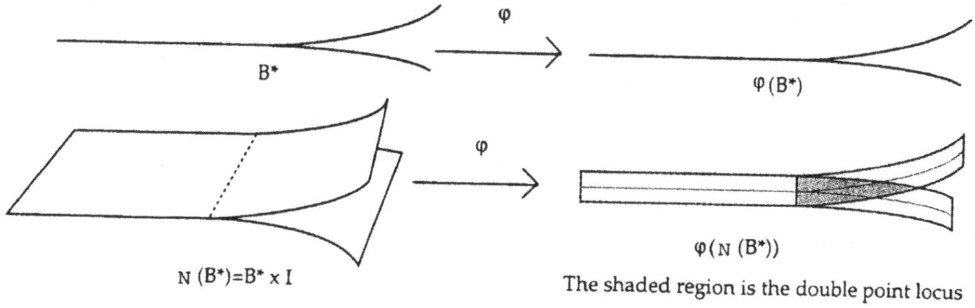

2-dimensional version of the *normal neighborhood*

Figure 1.2

We say that the branched surface B_1 is obtained from B by *splitting* if there is a lift of $\phi_1 : B_1 \to M$ to an immersion B_1^* to $\mathcal{N}(B^*)$ such that B_1^* is transverse to the I-fibres, B_1^* intersects each I-factor nontrivally, and finally $\partial B_1^* \subset \pi^{-1}(\partial B^*)$. Conversely we say that B is obtained from B_1 by *squeezing.*

Remark 1.2. Figure 1.3a shows a splitting one dimension lower. Figure 1.3b describes a splitting in a 3-manifolds. Note that one crossed maw is created in Figure 1.3b.

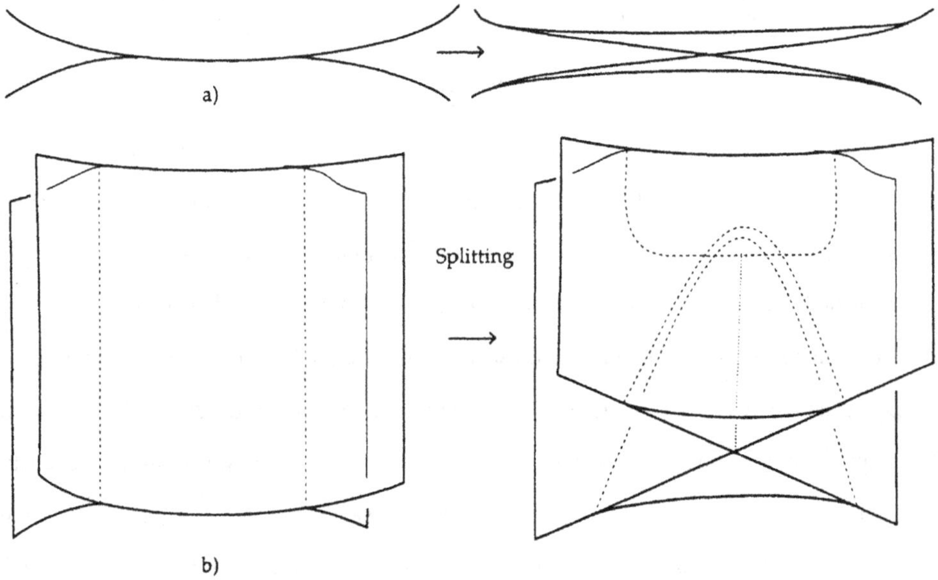

Figure 1.3

Lemma 1.3. *If B_1 is obtained from B by splitting, then B_1 has a normal neighborhood $\mathcal{N}(B_1)$ which lifts to a singular immersion $\mathcal{N}(B_1^*) \subset \mathcal{N}(B^*)$ such that each I-fibre of $\mathcal{N}(B_1^*)$ is properly contained in the associated I-fibre of $\mathcal{N}(B^*)$.* □

Definition 1.4. Suppose that we are given a Riemannian metric on B. If $C > 0$, then we define a *C-splitting* to be a splitting $B \to B_1$ such that if $i : \mathcal{N}(B_1^*) \to \mathcal{N}(B^*)$ is the associated singular immersion then there does not exist paths $\alpha, \beta : [0, t+\epsilon] \to \mathcal{N}(B_1^*)$ such that $\alpha(0), \beta(0)$ lie in the same I-fibre of $\mathcal{N}(B_1^*)$, for all $s \in [0,t], i(\alpha(s)), i(\beta(s))$ lie in the same I-fibre of $\mathcal{N}(B^*)$, and for $s > t; i(\alpha(s)), i(\beta(s))$

lie in distinct I-fibres of $\mathcal{N}(B^*)$. Finally length $\pi(i(\alpha([0,t]))) < C$. Intuitively this means that the branched locus of B has been blasted open at least distance C.

Definition 1.5. By a 2-dimensional ***abstract lamination*** μ we mean a topological space covered by charts of the form $U_i = T_i \times \mathbb{R}^2$, where T_i is a closed subset of $[0,1]$. Furthermore if $U_i \cap U_j \neq \emptyset$, then the coordinate transformations are of the form $(x,y) \to (\psi(x), \phi(x,y))$ where $x \in T_i, y \in \mathbb{R}^2$. This is a specialization of a more general definition given by [Ca]. There Alberto Candel initiates a deep investigation of the differential geometry of 2-dimensional abstract laminations.

Throughout this paper we will refer to abstract laminations simply as laminations. We say that the lamination μ is immersed in M^3 if there is a continuous map $J : \mu \to M^3$ whose restriction to each leaf of μ is a smooth or PL immersion. We say that μ is *carried* by the branched surface B if the mapping J factors through an immersion into $\mathcal{N}(B^*)$ such that the leaves of μ are transverse to I-fibres of $\mathcal{N}(B^*)$. If μ intersects each I-fibre of $\mathcal{N}(B^*)$, then μ is *fully carried* by B.

Definition 1.6. If λ is carried by B, then B_1 is said to be a ***λ-splitting*** of B if there exists a lift of λ to $\mathcal{N}(B^*)$ which after normal homotopy lifts to $\mathcal{N}(B_1^*)$.

Definition 1.7. Generalizing the similar notion of [GO] we say that an infinite sequence of branched surfaces $B_1, B_2, \cdots$ is a ***full splitting*** if there exists $C > 0$ such that for each i there exists a $j(i)$ such that the splitting $B_i \to B_{j(i)}$ is a C-splitting. If B is the first term of a full splitting, then we say that B is *fully splittable.*

The following result is the direct analogue of Lemma 4.3 [GO] to immersed laminations. See [MO] for a version for laminations in finite 2-complexes.

Theorem 1.8. *B is fully splittable if and only if B fully carries a lamination.*

Proof. Suppose that B is fully splittable. After making the lengths of the I-fibres of $\mathcal{N}(B_i)$ go to 0 as $i \to \infty$, then the inverse limit of the B_i's is a lamination fully carried by B. Conversely if B fully carries λ, then λ provides the clue to constructing a full splitting. The resulting lamination λ' arising from the inverse limit construction will generally be distinct from λ, however it will be normally homotopic to a sublamination of λ in $\mathcal{N}(B^*)$. $\square$

Definition 1.9. Let M be a triangulated 3-manifold, with triangulation τ. Let τ^i denote its i-skeleton. Following Haken [Ha], if J is a 1-manifold, then we say that $f : J \to \tau^2$ is *normal* if f is a PL-immersion transverse to τ^1 and for each edge e of the induced triangulation on J, $f|e$ is an embedding which sends the vertices of e to distinct 1-simplices of τ^1. A normal 2-disc is a properly embedded 2-cell D in a 3-simplex σ such that ∂D is a normal curve crossing a given 1-simplex of σ at most once. If S is a surface, then we say $f : S \to M$ is *normal* or S is a *normal surface* if f is a branched immersion transverse to τ^2 and each 2-cell η in the induced cellulation κ on S gets mapped to a normal disc. Thus, all branch points occur in τ^1. For $v \in \kappa^0$ define $s(v)$, the *spinning* of v, to be the number of 2-cells of κ which meet v. If e is a 1-simplex of τ, define $V(e)$, the *valence* of e, to be the number of 2-cells of τ which meet e. Also for $v \in \kappa^0$, define $V(f(v))$ to be $V(e)$ where $f(v) \in e$. Thus $s(v)/V(f(v))$ measures the local branching of f at v. We say that f has *fake boundary branching* if there exists $v \in \kappa^0 \cap \partial S$, such that f is an immersion at v, but $s(v)/V(f(v)) \geq 1$.

Define $S(D) = \sum_{a \in \kappa^0 \cap \partial D} s(a)$ to be the *total spinning* of D. Define $A(D) = S(D)/(|\kappa^0 \cap \partial D|)$ to be the *average spinning* of D.

Proposition 1.10. *Let τ be a triangulation of the 3-manifold M. There exists a branched surface $\mathcal{B}$, called the canonical normal branched surface, with the following property. If λ is any immersed lamination in M whose leaves are normal with respect to τ, then after normal homotopy λ is carried by $\mathcal{B}$. In particular the conclusion holds if λ is a compact normal surface, possibly with boundary.*

Proof. Figure 1.4 shows how to construct a canonical branched surface B on a 3-simplex σ. It is symmetric under the symmetries of σ and carries the isotopy class of each normal disc in a unique way. Thus by putting this branched surface in each 3-simplex of τ one obtains the desired $\mathcal{B}$. To start with, the three distinct normal quadralaterals and the four distinct normal triangles in σ can be normally isotoped to appear in σ as follows. The quadralaterals pairwise intersect in arcs whose endpoints lie on edges of σ. Also if η is a triangle then $\eta \cap (\cup \text{ quads }) = \partial\eta$. Thus the union of the quadralaterals and triangles is isomorphic to the cone on the 1-skeleton of an octahedron together with four faces of the octahedron which pairwise meet in points. (I.e. one color of a two coloring of the octahedron.) See Figure 1.4a. By squeezing the faces in the appropriate manner, we obtain the desired branched surface B. See Figure 1.4b. □

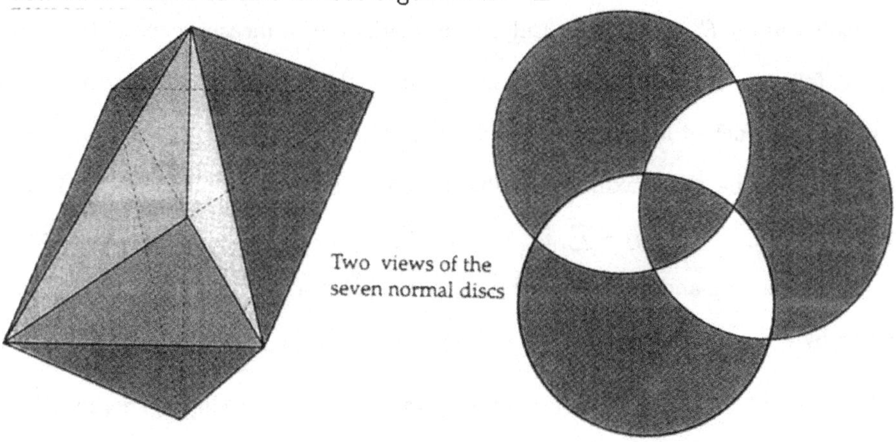

a)

The shaded regions are the triangles. Attach hemispheres to each circle to obtain the quadralerals.

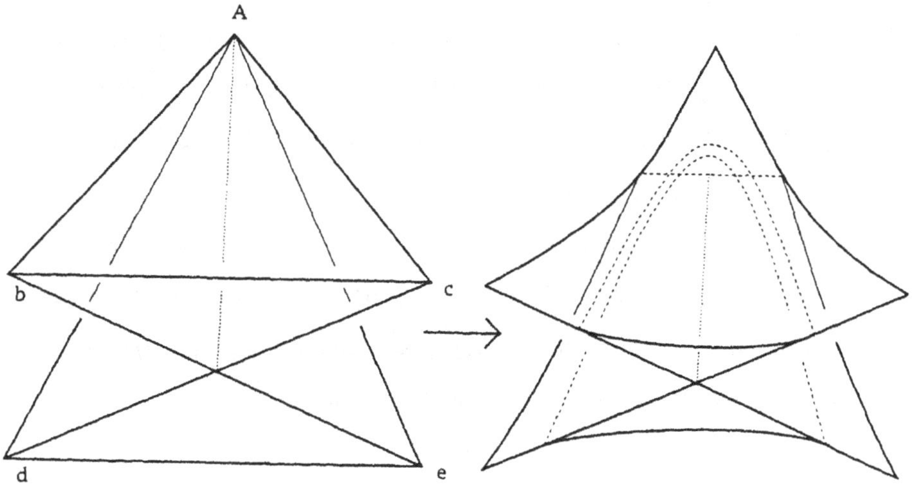

Attach discs to each of Abc, Ade, Abe, Adc.

Smoothing very near corner A of the octahedron creates the branched surface B

b)

Figure 1.4

Definition 1.11. A *measured branched surface* is a branched surface B together with a non negative valued function m : {Sectors of B} $\to \mathbb{R}$ such that if sectors s_1, s_2 merge into sector s_3, then $m(s_1) + m(s_2) = m(s_3)$. A *measured lamination* is a lamination λ together with a transverse measure m. I.e. for each lamination chart $T_i \times \mathbb{R}^2$ there is a Borel measure supported on T_i. Furthermore the transversal overlap functions are measure preserving. If the measured lamination (λ, m) is carried by the branched surface B, then m induces a measure also called m on B. Define the Euler characteristic of a measured branched surface as follows. If (B, m) is a branched surface, then put a cell structure on B so that the branched locus is contained in the 1-skeleton. If t is a vertex or edge, define $m(t)$ to be the maximal m value of a sector which contains t and define $\chi(B, m) = \sum_{vertices} m(v_i) - \sum_{edges} m(e_j) + \sum_{faces} m(f_k)$. This formula is invariant under splitting and choice of cell structure. If (λ, m) is a measured lamination, carried by a branched surface (B, m), then define $\chi(\lambda, m) = \chi(B, m)$.

Definition 1.12. We need the notion of *diagram* [LS] which is a slight generalization of a disc. (The first time reader is strongly advised to pretend that all diagrams are discs.) A diagram D is a simply connected finite 2-complex that is embedded in the plane. Define $\overset{\circ}{D}$ to be the interior of D as a subset of $\mathbb{R}^2$. Define $\partial(D) = D - \overset{\circ}{D}$ and Bd(D) to be a simplicial map of the circle to the curve which traverses around D. See Figure 1.5. Call a diagram map $f : D \to M$ *admissible* if ∂D has a triangulation ϕ such that f takes each 1-cell of ϕ to a normal arc and f is transverse to τ^2, except at isolated points where f appears as in Figure 1.6. We say that $g : S^1 \to \tau^2$ *extends* to the diagram $G : D \to M$ or G *spans* g, if $G \circ \mathrm{Bd}(D)$ is normally homotopic to g. If $f : D \to M$ is an admissible diagram map, then define length(∂f) to be the number of 1-cells in the induced triangulation on Bd(D).

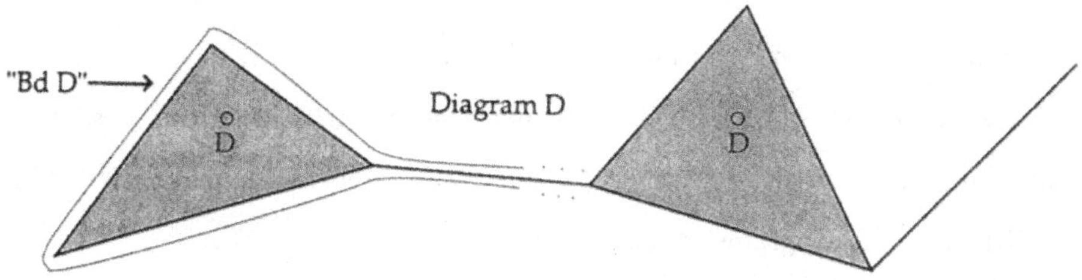

Figure 1.5

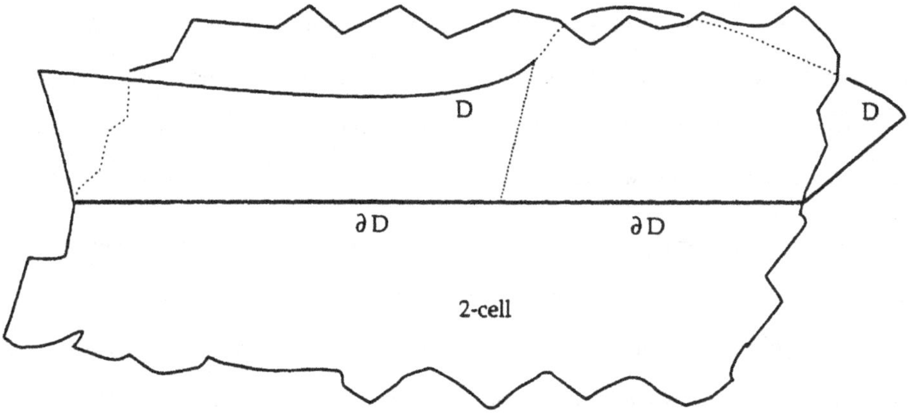

Figure 1.6

Definition 1.13. If $f : S \to M$ is a normal surface, then define "area (f)" $= \sum_{i=1}^{n} s(v_i)/V(f(v_i))$, where $\{v_1, \cdots, v_n\}$ are the vertices in the induced cellulation κ on S. Thus if $f : S \to M$ is a normal immersion of a closed surface, then area(f) is the number of vertices in κ, i.e. the Haken weight. This definition of area takes into account local branching and is additive under union. In a similar manner we can define area(f) where $f : D \to M$ is an admissible diagram map, or if f is a map of a compact surface such that $f|\partial D$ is normal and f is transverse to τ^2 except as in Figure 1.6. In these cases one uses the induced stratification on D to compute $s(v_i)$ and sum over points of $f^{-1}(\tau^1)$.

A diagram $f : D \to M$, and in particular a normal map of a disc, is said to be *least area* (resp. *k-quasi least area*) if area(D) $\leq$ area(E) (resp. area(D) $\leq k$ area(E)) among all diagrams E spanning ∂D.

A normal surface $f : L \to M$ is *least area* if the restriction of f to embedded discs is a least area diagram. L is *strongly least area* [MO] if the induced immersion of the universal covering of L into M is least area. In a similar manner define *strongly k-quasi least area for a normal map* $f : L \to M$. The leaves of a lamination are *quasi strongly least area* if there exists a k such that for each leaf L, the induced immersion of L into M is strongly k-quasi least area.

The following Proposition is a restatement of the main technical result of Mosher and Oertel [MO]. We translate their work about PL maps of discs into finite 2-complexes into a Proposition about maps of discs into 3-manifolds. We generalize slightly to allow for *k-quasi least area* maps.

Proposition 1.14 [MO]. *Let $D_1, D_2, \cdots$ be a sequence of immersed normal discs, in the triangulated compact 3-manifold M, such that $\lim_{i\to\infty}$ length(∂D_i)/area(D_i) $\to 0$. If no D_i has fake boundary branching, then after passing to a subsequence, D_i converges to a normal Euler characteristic-0 measured lamination λ. If each of the discs is k-quasi least area, then each leaf of λ is strongly k-quasi least area. If each D_i is least area, then λ has no 2-sphere leaves. Finally, the induced metric on each leaf is complete.* □

Remarks. The proof of the existence of λ in [MO] is greatly complicated by the fact that it is happening in the setting of branched immersed discs into 2-complexes. In the setting of immersed normal discs without fake boundary branching in 3-manifolds the argument is considerably simpler. Mosher and Oertel do a competent job of explaining their proof of strong area minimization. Indeed, their argument shows that if λ had a 2-sphere leaf, then for i sufficiently large, some component of D_i would be a 2-sphere, which gives a contradiction. By construction each leaf of λ is complete. One should be aware that the discussion of diagrams is implicit, though suppressed in [MO].

Idea of the proof of existence of λ. Each D_i is carried by $\mathcal{B}$, the canonical normal branched surface. Fix a small $C > 0$. By the usual diagonal argument one passes to a subsequence of the D_i's and finds a sequence $\mathcal{B} = B_0, B_1, \cdots$ such that B_i is obtained from B_{i-1} via C-splitting where each B_j carries every D_i for $i \geq j$. (This step would fail if there was uncontrolled fake boundary branching.) By counting the number of times a given D_i crosses a sector of B_j, we obtain a function m_{ij} on the sectors of B_j which fails to be a transverse measure only because of the boundary of D_i. Let a_{ij} be the maximal value of m_{ij}. By considering m_{ij}/a_{ij} and using the length area hypothesis, then by passing to a subsequence of the D_i's the measures m_{i1}/a_{i1} converge to a fixed transverse measure m_1 on B_1. By passing to another subsequence we can assume that the D_i's induce a measure m_2 on B_2 so that as measured laminations (B_2, m_2) is obtained by splitting (B_1, m_1). Thus by applying Theorem 1.8 to the sequence $\{(B_i, m_i)\}$ we obtain a measured lamination λ'. Finally take λ to be the sublamination which is the support of the measure m. This type of argument, in tamer form, goes back to Plante [Pl].

Since $\chi(D_i) = 1$, the Euler characteristic of the approximate measure m_{ij}/a_{ij} on B_j is approximately $1/a_{ij}$ and $m_{ij}/a_{ij} \to m_j$, it follows that the Euler characteristic of λ is zero.

Definition 1.15. Let λ be a lamination with a piecewise Riemannian metric on each leaf, which varies continuously in the transverse direction. (E.g. a metric induced from a 3-manifold) We say λ is *conformally Euclidean* if each leaf is conformally equivalent to the Euclidean plane. A measured Riemannian lamination λ is said to be semi-Euclidean if the set of leaves conformally equivalent to the Euclidean plane are of full measure and dense in λ.

The following result provides the crucial link between the topology of the measured laminations and their geometry.

Theorem (Ghys [Gh] 1.16). *Let λ be a measured 2-dimensional abstract Riemannian lamination with out spherical leaves. $\chi(\lambda) = 0$ if and only if almost all leaves are conformally Euclidean.* □

Remark 1.17. This was proven originally for 2-dimensional foliations of manifolds, but Candel [Ca] pointed out that the proof holds for 2-dimensional abstract Riemannian laminations.

§2

Immersed Least Area Semi-Euclidean Laminations

Theorem 2.1. *If M is a closed non negatively curved aspherical 3-manifold, then there exists an immersed strongly least area, Euler characteristic 0, measured semi-Euclidean lamination λ. The induced metric on each leaf is complete.*

Remark 2.2. i) Here M has a triangulation τ, and the various measurements of length and area are computed simplically.

ii) Theorem 2.1 should be viewed as a refinement of [MO]. Their work translated to closed 3-manifolds yields the above theorem with immersed replace by branched immersed.

Conjecture 2.3. *If K_0 is a non negatively curved 2-complex, then K_0 is simple homotopy equivalent to a 2-complex K_1 such that K_1 supports an immersed least area Euler characteristic zero semi-Euclidean lamination.*

The following is the main result of this section.

Proposition 2.4. *Let M be a closed 3-manifold with triangulation τ. $\pi_1(M)$ is not negatively curved if and only if there exists a sequence of least area normal immersed discs $f_i : D_i \to M$, such that length$(\partial f_i)/$area$(f_i) \to 0$ and no f_i has fake boundary branching.*

Remark. This result is a serious technical advance beyond the obvious translation of Gromov's theorem, *group negative curvature implies linear isoperimetric inequality,* to normal surfaces in 3-manifolds. The new wrinkle is that "no f_i has fake boundary branching."

Proof of Proposition 2.4.

Step 0. $\pi_1(M)$ is non negatively curved if and only if there exists a sequence of least area normal immersed discs $f_i : D^2 \to M$, such that $\text{length}(\partial f_i)/\text{area}(f_i) \to 0$.

Proof of Step 0. Since $\pi_1(\tau^2) = \pi_1(M)$ is not negatively curved, there exists by Gromov [Gr], [Bo] a sequence of simplicial maps of discs $a_i : A_i \to \tau^2$ such that $\text{length}_\tau(A_i)/\text{area}_\tau(A_i) \to 0$. Where $\text{length}_\tau(\partial A_i)$ (resp. $\text{area}_\tau(A_i)$ is the number of 1-simplices (resp. 2-simplices) in ∂A_i (resp. in A_i.) Furthermore $\text{area}_\tau(A_i)$ is minimal among all such simplicial maps of diagrams that span $a_1|\partial A_i$.

Its routine to translate this statement about 2-complexes to the statement of Step 0. □

From now on we will assume that $\pi_1(M)$ is non negatively curved and that Proposition 2.4 is false. Thus there exists a constant $K_0 \leq 1$ such that if $f : D \to M$ is least area normal immersed disc with no fake boundary branching, then $\text{length}(\partial f)/\text{area}(f) > K_0$.

Here is how we derive a contradiction. As a result of Steps 0 and 1 we find a sequence satisfying the conclusion of Proposition 2.4, except that each disc has fake boundary branching at a single vertex. We then delete a small neighborhood of the branching vertex and take (in Step 2) a weak limit of these trimmed discs to obtain an immersed least area measured lamination λ' of Euler characteristic 0, with boundary $S^1 \times K, K$ a closed subset of $[0,1]$. By capping off the boundary circles with discs, we obtain a measured lamination λ of positive Euler characteristic and derive a contradiction to the least area property of λ.

Step 1. There exists a sequence of least area normal immersed discs $f_i : D^2 \to M$, such that $\text{length}(\partial f_i)/\text{area}(f_i) \to 0$. Furthermore all the fake boundary branching of f_i occurs at a single vertex w_i of ∂D_i.

Proof of Step 1. Let f_i be as in Step 0 and let Δ_i denote the cellulation induced on D_i by τ. If J is a subcomplex of a cell complex, then let $\mathrm{St}(J)$ denote the union of all open cells whose closures intersect J and $\bar{S}t(J)$ to denote the closure of $\mathrm{St}(J)$. Define $D_i^{trim} = D_i - St(\partial D_i)$.

Claim 0. We can assume that D_i^{trim} is connected, that Δ_i has no separating 1-cells, and $\bar{S}t(v) \cap D_i^{trim}$ is nonempty and connected for each vertex $v \in \partial D_i$.

Proof of Claim 0. Consider a maximal collection of properly embedded arcs δ_j in D_i such that $\partial\delta_j \subset \Delta^0$, interiors of the δ_j's are pairwise disjoint and the δ_j's satisfy the following additional property. Either δ_j is a separating 1-cell of Δ_i, or $\overset{\circ}{\delta}_j \subset \Delta_i - \Delta_i^2$ and each component of $D_i - \delta_j$ contains at least one open 2-cell of Δ_i. The number of such paths between two given vertices is uniformly bounded by a constant K_1, since M is compact and D_i is least area. Thus for Euler characteristic reasons there are at most $K_1\mathrm{length}(\partial D_i)$, such δ_j's. If C_n' is a component of $D_i - \cup\delta_j$, then let C_n be the smallest subcomplex of Δ_i which contains C_n'. Since $\sum_n \mathrm{length}(\partial C_n) \leq 4K_1(\mathrm{length}(\partial D_i))$, some C_n will have very small isoperimetric ratio, if D_i does. Finally C_n satisfies the conclusions of Claim 0. Thus by cutting down the D_i's we can assume that they satisfy the conclusions of both Step 0 and Claim 0. $\square$

We will supress cumbersome language by writing as if each D_i^{trim}, as well as other to be defined objects are discs rather than diagrams.

If $v \in \Delta^0 \cap \partial D_i$, then define $D_i^v = D_i^{trim} \cup \bar{S}t(v)$. D_i^v is an immersed disc with possible fake boundary branching only at v_i. Since D_i^{trim} is an immersed least area disc with no fake boundary branching, it follows that for all i, $\mathrm{length}(\partial D_i^{trim})/\mathrm{area}(D_i^{trim}) > K_0$.

Claim 1. The average spinning $A(D_i) \to \infty$.

Proof. Otherwise $A(D_i) \leq A$ all i implies $\mathrm{length}(\partial D_i^{trim})/\mathrm{area}(D_i^{trim}) \leq 2S(D_i)/(\mathrm{area}(D_i) - S(D_i)) \leq 2A\mathrm{length}(\partial D_i)/(\mathrm{area}(D_i) - A\mathrm{length}(\partial D_i)) \to 0$ as

$i \to \infty$. □

We now define $\text{Fan}_j(v)$, which more or less is the smallest simply connected subset of $D_i^v(v)$ which contains all simplices distance j from v. More precisely, if v is a vertex of ∂D_i then define inductively define $St_n^{trim}(v) = \bar{S}t(St_{n-1}^{trim}(v))$ where $St_1^{trim}(v) = \bar{S}t(v)$, all these complexes being computed in D_i^v. Let $\text{Fan}_j(v)$ be the minimal simply connected subcomplex of D_i^v which contains $St_j^{trim}(v)$.

Claim 2. Given p, there exists $B(p)$ independent of i, such that for any $v_r, v_s \in \partial D_i$, $\text{area}(St_p^{trim}(v_r) \cap St_p^{trim}(v_s)) < B(p)$. (Recall $St_p^{trim}(v_r) \subset D_i^{v_r}$ and $St_p^{trim}(v_s) \subset D_i^{v_s}$.)

Proof. If x is a vertex in $St_p^{trim}(v_r) \cap St_p^{trim}(v_s)$, then there exists a simplicial path α_x of length $\leq 4p$ from v_r to v_s which passes through x. The union of two such paths α_x, α_y gives rise to a least area diagram $D_{xy} \subset D_i$, with $\partial D_{xy} \subset \alpha_x \cup \alpha_y \subset D_{xy}$ and hence length $\partial D_{xy} \leq 8p$. Since M is compact, the number of such diagrams is bounded and hence if $x, y \in St_p^{trim}(v_r) \cap St_p^{trim}(v_s)$, then the distance between x and y in D_i^{trim} is uniformly bounded, independent of i. Since the valence of vertices of D_i^{trim} are uniformly bounded, independent of i, Claim 2 follows. □

Claim 3. Given p, there exists a constant K_3 independent of i such that at most $K_3(\text{length}(\partial D_i))$ distinct pairs $(\text{Fan}_p(v_r), \text{Fan}_p(v_s))$ intersect.

Proof. If $\text{Fan}_p(v_r) \cap \text{Fan}_p(v_s) \neq \emptyset$, then define α_{rs} to be a shortest simplicial path in 1-cells of $D_i^{v_r} \cup D_i^{v_s}$ from v_r to v_s. Thus α_{rs} is embedded of length $\leq 4p$. By rechoosing α_{rs} we can assume that $D_i^{trim} \cap \alpha_{rs}$ is connected, and using Claim 0 that either $\alpha_{rs} \cap D_i^{trim} \neq \emptyset$ or $\alpha_{rs} \subset \partial D_i$. Furthermore, if $\alpha_{rs} \cap D_i^{trim} \neq \emptyset$, then after possibly deleting the first and/or last 1-cells of α_{rs}, the remainder is properly embedded in D_i. Also if $\alpha_{rs} \subset \partial D_i$, then $\text{length}(\alpha_{rs}) \leq 4$. Thus there are at most $4\,\text{length}(\partial D_i)$ peripheral α_{rs}'s and at most $5N$ non peripheral α_{rs}'s, where N is the number of properly embedded α_{rs}'s. Since at most finitely many paths α_{rs} can pass through a given vertex of D_i^{trim}, it follows that there exists a uniform bound K_2 on the number of α_{rs}'s which can intersect a given α_{uv} inside of D_i^{trim}. The collection of properly embedded α_{rs}'s can be partitioned into K_2+1 subsets, with the property that if α_{rs}, α_{uv} are distinct elements of a given subset, then $\mathring{\alpha}_{rs} \cap \mathring{\alpha}_{uv} = \emptyset$. For Euler characteristic reasons each subset has at most length$(\partial D_i) - 1$ elements, so $N \leq (K_2+1)(\text{length}(\partial D_i) - 1)$ and Claim 3 follows with $K_3 = 5(K_2+1)+4$. $\square$

Claim 4. For every $k > 0$, there is a $N > 0$ such that $i > N$ implies that there exists a $v \in \Delta^0 \cap D_i$ and a j such that $\text{length}(\partial \text{Fan}_j(v))/\text{area}(\text{Fan}_j(v)) < 1/k$.

Proof. Fix k and i. If v is a vertex in ∂D_i, then let $\partial_{ext}\text{Fan}_j(v)$ denote $\partial \text{ Fan}_j(v) \cap \mathring{D}_i^v$. If there exists j such that $\text{length}(\partial_{ext}\text{Fan}_j(v)) \leq s(v)/2k$, then $\text{length}(\partial\text{Fan}_j(v))/\text{area}(\text{Fan}_j(v)) \leq (s(v)/2k+ \text{length}(\text{Fan}_j(v) \cap \partial(D_i^v)))/s(v)$ and hence either $\text{length}(\partial\text{Fan}_j(v))/\text{area}(\text{Fan}_j(v)) < 1/k$ would hold or $\text{length}(St_j^{trim}(v) \cap \partial D_i^v) = \text{length}(\text{Fan}_j(v) \cap \partial D_i^v) \geq s(v)/2k$. In the latter case, for all $p \geq j$, $\text{length}(St_p^{trim}(v) \cap \partial D_i^v) \geq s(v)/2k$.

Now choose p so that $p/(6k) \geq 8/K_0$. Therefore for all vertices v of ∂D_i

i) $\text{length}(\partial\text{Fan}_j(v))/\text{area}(\text{Fan}_j(v)) < 1/k$ for some $j \leq p$ or

ii) $\text{length}(St_p^{trim}(v) \cap \partial D_i^v) \geq s(v)/2k$ or

iii) $\text{area}(St_p^{trim}(v)) \geq 8s(v)/K_0$.

Indeed if neither i) nor ii) held for v, then for all $j \leq p$, $\text{length}(\partial_{ext}\text{Fan}_j(v)) \geq s(v)/2k$ and hence $\text{area}(St_p^{trim}(v)) \geq (1/3)\sum_{j=1}^{p}\text{length}(\partial_{ext}\text{Fan}_j(v)) \geq ps(v)/6k \geq 8s(v)/K_0$. This argument was inspired by [Pa].

Let $v_1, \cdots, v_m$ denote the vertices of ∂D_i. It follows that either

a) $\text{length}(\partial\text{Fan}_j(v_s))/\text{area}(\text{Fan}_j(v_s)) < 1/k$ for some vertex v_s of ∂D_i and some $j \leq p$ or

b) $\sum_{n=1}^{m}\text{length}(St_p^{trim}(v_n) \cap \partial D_i^v) \geq S(D_i)/4k$ or

c) $\sum_{n=1}^{m}\text{area}(St_p^{trim}(v_n)) \geq 4S(D_i)/K_0$.

We show that b) does not hold if i is sufficiently large. Claims 2,3 imply that the sum of the pairwise area overlaps of the St_p^{trim}'s is bounded above by $B(p)K_3\text{length}(\partial D_i)$. Possibility b) implies that the sum of the total pairwise area overlaps of the St_p^{trim}'s is $\geq S(D_i)/12k = A(D_i)\text{length}(\partial D_i)/12k$. Now apply Claim 1.

We show that c) does not hold if i is sufficiently large. If c) holds then we obtain area$(D_i^{trim}) \geq$ area$((\cup_{n=1}^m St_p^{trim}(v_n))-$ St$(\partial D_i)) \geq 4S(D_i)/K_0 - B(p)K_3$ length(∂D_i) $-$ $S(D_i)$ $\geq$ $2S(D_i)/K_0$ $+$ $S(D_i)$ $-$ $B(p)K_3$length(∂D_i) $\geq$ length$(\partial D_i^{trim})/K_0 + (A(D_i) - B(p)K_3)$length$(\partial D_i) >$ area(D_i^{trim}). By Claim 1, the latter inequality holds if i is sufficiently large.

Thus for fixed k, a) holds for i sufficienlty large. □

Thus we can assume that each D_i has fake branching except at the vertex w_i. Furthermore by the compactness of M we can assume that for all i, $f(w_i) = w \in \tau^1$. Let $E_i = D_i - \overset{\circ}{St}(v_i)$ and $h_i = f_i|(D_i - \overset{\circ}{St}(w_i))$. Let γ be the 1-simplex of τ which contains w and let $\alpha_i = \partial$ St$(w_i) - \partial D_i$. Since h_i has no branching or fake boundary branching we can assume that for all i, length(∂h_i)/area$(h_i) > K_0$. Thus there exists $K_7 > 0$ so that for i sufficiently large length (α_i)/area$(E_i) > K_7$ which in particular implies that length$(\partial E_i - \alpha_i)$/length$(\alpha_i) \to 0$. Since the restriction of f_i to $St(w_i)$ is an immersion, $h_i(\alpha_i)$ is an embedded arc which spirals around $\partial\bar{S}t(\gamma)$.

Step 2. Either Proposition 2.4 is true or there exists an immersed normal measured lamination (λ, m) by strongly least area leaves such that $\chi(\lambda) = 0$. Furthermore $\partial\lambda$ is a lamination of the form $S^1 \times K \subset \partial\bar{S}t(\gamma)$ where K is compact and $m(K) = 1$.

Proof of Step 2. Let n_i be the number of times $h(\alpha_i)$ spirals around St(γ). Here we abuse notation by letting St(γ) denote the union of all 3-simplices of τ which nontrivially intersect $\overset{\circ}{\gamma}$. As in the proof of Proposition 1.14 normalize the various weights associated to the discs E_i by dividing by n_i. Since area$(D_i)/n_i$ is uniformly bounded, thus we can apply the proof of Proposition 1.14 to let the normalized discs E_i/n_i weakly converges to the measured lamination (λ, m). Since $\partial\lambda$ is the weak limit of the sequence $\partial E_i/n_i$ and length$(\partial E_i - \alpha_i)/n_i = 0$, any weak limit of $\partial E_i/n_i$ is a weak limit of $h_i(\alpha_i)/n_i$. Because each D_i is immersed in M, α_i is an embedded spiral in ∂St(γ), so it follows that any weak limit is of the form $S^1 \times K$ where K is compact and $m(K) = 1$. □

Step 3. Obtain the contradiction.

Proof of Step 3. Let (λ', m') be the immersed measured lamination obtained from (λ, m) by capping off each component of $S^1 \times K$ by a disc. $\chi(\lambda') = 1$ and so by the Connes sphere theorem [Co] (or [MO1] for an elementary proof) there exists a measured sublamination $\mathcal{S}$ of λ' consisting only of 2-spheres and $\chi(\mathcal{S}) \geq 1$.

Since λ is strongly area minimizing it has no 2-sphere leaves. The rest of the argument follows implicitly from the proof of "strong area minimization of leaves theorem" [MO]. See the next paragraph for a hint. If some leaf S of $\mathcal{S}$ intersects $D^2 \times K$ in a single disc F, then for i sufficiently large, there exists a subdisc H_i of D_i which is normally parallel to $S - \mathring{F}$, and $\partial H_i \subset \alpha_i$. This implies that D_i is a sphere. Now suppose that some leaf S of $\mathcal{S}$ intersects $D^2 \times K$ in exactly 2-discs F_1, F_2. Given $n \in \mathbb{N}$, and an essential arc β in the annulus $A - \mathring{F}_1 \cup \mathring{F}_2$, then for i sufficiently large, there exists a subdics H_i of D_i of the form $[0,1] \times [0,n]$ where $(\partial[0,1]) \times [0,n] \subset \alpha$, and H_i naturally projects to A as part of an n-fold cyclic cover. Finally each component of $[0,1] \times \{0,n\}$ naturally projects isomorphically onto β. Thus by coning off $H_i \cap \alpha_i$ one obtains a subdisc of D_i whose boundary length is constant, independent of i, but whose area $\to \infty$ as $i \to \infty$. This thereby contradicts the least area property of D_i. A similar argument works when $S \cap D^2 \times K$ has $r > 2$ components. However the cases $r \leq 1$, together with $m(K) = 1$, implies $\chi(\mathcal{S}) = 1$ and that the generic leaf of $\mathcal{S} - D^2 \times K$ is an annulus thereby contradicting the $r = 2$ argument.

Here is the idea behind what was used in the previous paragraph. Recall that we had, by Proposition 1.14, a sequence of measured branched surfaces (B_i, m_i) which converged to a measured lamination with measure m, and that λ was the minimal sublamination which supported m. Also $j > i$ implies that (B_j, m_j) is obtained by splitting the measured branched surface (B_i, m_i). Therefore if T is a compact leaf of λ, then for i sufficiently large T embeds in B_i. If $r = 1$, then T is a disc so for C sufficiently large, any C-splitting either isolates or destroys T. Since the latter does not occur, T must appear as an isolated sector in B_j, for j sufficiently large. If

$r = 2$, then T is an annulus and a holonomy argument shows that for i sufficiently large near $T \subset B_i$, part of each $D_j, j \geq i$, spirals around T. The spiraling must get arbitrarily large to obtain the annulus T in the limit. Our $[0,1] \times [0,n]$ is a piece of that spiral. This completes the proof of Proposition 2.4 □

Proof of Theorem 2.1. Combine Propositions 2.4 and 1.14 with Theorem 1.16. □

§3

HYPERBOLIC CELLULATIONS

Definition 3.1. Let ψ be a piecewise linear cellulation of the 3-manifold M. This means that M is obtained by taking a disjoint union of 3-dimensional polyhedra $\sigma_1, \cdots, \sigma_n$ and pairwise identifying 2-dimensional faces in a PL fashion. Thus to each cell C_i of ψ, there is a piecewise linear mapping $f_i : C_i \to M$ which restricts to an embedding on $\overset{\circ}{C}_i$. We will usually suppress mentioning the f_i's and will denote the i-skeleton of ψ by ψ^i.

As in the setting of triangulations define a *normal curve* (resp. *local ψ-embedding*) to be an immersion $f : \alpha \to \psi^2$ transverse to ψ^1, where α is a compact 1-manifold and for each 1-simplex η in the induced triangulation on $\alpha, f|\alpha = f_i \circ t$, where $t : \eta \to C_i$ is an embedding into the 2-cell C_i with $\partial\eta$ going to distinct edges (resp. η not necessarily going to distinct edges.).

Similary define a *normal disc* to be a map $f : D^2 \to M$, which factors through an embedding into a 3-cell σ of ψ, so that $f|\partial D \to M$ is normal and (the lift into σ) crosses a given edge of σ at most once. If S is a surface, define a map $f : (S, \partial S) \to (M, \psi^2)$ to be *normal* if f is transverse to ψ^2, $f|\partial S$ is a normal curve, and for each 2-cell η in the induced cellulation on S, $f|S$ is a normal disc. If S is a surface, define a map $f : (S, \partial S) \to (M, \psi^2)$ to be *a-transverse* to ψ^2 if f is transverse to ψ^2 except at isolated points of ∂S where f appears as in Figure 1.6 and furthermore $f|\partial S$ is a local ψ-embedding. We say that $f : S \to N$ is *normal*, or S is a *normal surface*, if f is a branched immersion a-transverse to ψ^2, inducing a cellulation κ on S such that each 2-cell of κ gets mapped to a normal disc in ψ.

In particular $f|\partial S$ is a normal curve.

In order to suppress notation, we will often view S as lying in M. E.g. if Y is a submanifold of M we may refer to $S \cap Y$ rather than $f^{-1}(Y)$.

If α is a curve in ψ^2 transverse to ψ^1, then define *length*(α) to be the number of 1-cells crossed by α. If $f : S \to N$ is a-transverse to ψ^2, then f induces a stratification κ on S, where $f^{-1}(\psi^i) = \kappa^i$ the i-th strata. We'll abuse notation by calling a component of $\kappa^i - \kappa^{i-1}$ an i-cell. Finally define *area*(S) $= \sum_{i=1}^{n} s(v_i)/\text{valence}(f(v_i))$. Here $v_1, \cdots, v_n$ are the vertices of κ, $s(v_i)$ is the local number of 2-cells of κ touching v_i, and *valence*($f(v_i)$) is the local number of 2-cells of ψ which come into $f(v_i)$. For example if $f : S \to M$ is an immersion of a closed surface, then area(S) is simply the intersection number of S with ψ^1. This definition of area keeps track of the local branching at vertices and is additive under unions, e.g. if $S = S_1 \cup S_2, S_1 \cap S_2 \subset \partial S_1 \cap \partial S_2$, and $f|S$ is normal, then area(S_1)+ area(S_2) = area(S).

View the mass of M as being concentrated near ψ^1. Think of a 1-cell ϕ as a $D^2 \times I$. If k 3-cells of ψ touch ϕ, then pretend the $D^2 \times I$ is subdivided into k wedges, one for each 3-cell. Finally imagine that each $D^2 \times t, t \in I$ is subdivided into k pie slices, the area of each slice being $1/k$.

We may use the notation length$_\psi(\alpha)$ or area$_\psi(D)$ to make clear that we are measuring with respect the cellulation ψ.

Lemma 3.2. *If $f : S \to M$ is a mapping of a closed surface into the aspherical manifold M with cellulation ψ, then f can be homotoped to a normal surface, provided that for each essential simple closed curve in S, $1 \neq [f_*(\alpha)] \in \pi_1(M)$ (i.e. f injects on simple loops).* □

Proof. First homotope f to a generic least area immersion transverse to ψ^2. After a further homotopy we can assume that the induced statification κ is a cellulation. Among all such maps choose one that minimizes double points of $f|f^{-1}(\psi^2)$. If η is a 2-cell of κ which maps into the 3-cell σ, then $f|\partial\eta$ is an embedding (when lifted) into σ. Otherwise a homotopy of f supported near η reduces the number of double points. Thus we can assume that the restriction of f to each 2-cell is an embedding (when lifted) into its associated 3-cell. Such an f is normal. □

Lemma 3.3. *(Epstein - Penner [EP]). If N is a complete noncompact hyperbolic 3-manifold of finite volume, then N has a cellulation by ideal polyhedra.* □

Question 3.4. (Igor Rivin) Does every complete noncompact hyperbolic 3-manifold of finite volume have a cellulation by ideal triangles?

Definition 3.5. If the complete noncompact finite volume orientable hyperbolic 3-manifold N has a cellulation by ideal polyhedra, then removing a neighborhood of the ends of N one obtains a compact manifold M whose boundary is a non empty union of tori. The induced cellulation on M is a cellulation by truncated ideal polyhedra. We call such a cellulation a *relative hyperbolic cellulation.* Call the newly created 2-cells (on ∂M) the *facets.*

Definition 3.6. Define a *combinatorial geometry* (κ, angle) on the compact surface S to be a cellulation κ together with a function angle:{vertices of 2-cells of ψ} $\to [0, \pi]$ such that if $v \in \kappa^0$, $w_1, \cdots, w_n$ are the vertices of 2-cells of κ which are identified with the vertex v of κ and Angle$(v) \overset{def}{=} \sum_{i=1}^n$ angle(w_i), then Angle$(v) = 2\pi$ for $v \in \overset{\circ}{S}$.

If C is a 2-cell with vertices $v_1, \cdots, v_n$, then define $\int_C K = \sum_{i=1}^n \text{angle}(v_i) - (n-2)\pi = 2\pi - \sum_{i=1}^n(\pi - \text{angle}(v_i))$ and $\int_S K = \sum_{i=1}^m \int_{C_i} K$, where $C_1, \cdots, C_m$ are the 2-cells of κ. If v is a vertex of $\kappa, v \in \partial S$, then define $\gamma_g(v) = \pi - \text{Angle}(v)$ and $\int_{\partial S} \gamma_g = \sum_{v_i \in \kappa^0 \cap \partial S} \gamma_g(v_i)$. Call a 2-cell C respectively *negatively curved, flat* or *positively curved* if $\int_C K$ is respectively $< 0, 0$, or > 0.

PL Gauss Bonnet theorem [We] **3.7.** *If S has a combinatorial geometry (ψ,angle), then $\int_S K + \int_{\partial S} \gamma_g = 2\pi\chi(S)$.* □

Remark 3.8. The proof is an elementary combinatorial exercise. See [We] for the case of a closed surface.

Remark 3.9. The following result is the obvious generalization of an observation of Andrew Casson which was stated for normal closed surfaces with respect to ideal triangulations on hyperbolic 3-manifolds.

Lemma 3.10. *Let ψ be a relative hyperbolic cellulation on Y. If S is an immersed surface in Y which is normal with respect to ψ. Then f induces on S a combinatorial geometry (κ, angle) where κ is the induced cellulation on S and if $v \in \overset{\circ}{Y}$, then angle(v) is the dihedral angle at the edge of the associated 3-cell of ψ. Otherwise angle(v) $= \pi/2$.* □

Corollary 3.11. *(Casson [Ca]) Let ψ be a relative hyperbolic cellulation on Y. If S is an immersed closed surface in Y which is is normal with respect to ψ, then $\chi(S) \leq 0$. Equality holds if and only if S is a normally parallel to a peripheral torus.*

Proof. Combine the the following result with Lemma 3.10 and Theorem 3.7. □

Rivin's Lemma 3.12 [Ri]. *If α is a normal closed curve on the boundary of an ideal hyperbolic polygon which crosses (not necessarily distinct) edges $e_1, \cdots, e_n$, then $2\pi \leq \sum_{i=1}^{n}(\pi - d(e_i))$, where $d(e_i)$ is the dihedral angle at e_i. Equality holds if and only if α is a simple closed curve which separates off a single facet from the rest.*

Proof. Use the fact that any ideal polygon is the union of ideal tetrahedra and use the fact that for an ideal tetrahedron, dihedral angles of opposite edges are equal. □

Remark 3.13. Its a famous theorem of Igor Rivin [Ri], that the converse is also true. That is any combinatorial geometry on a polygon satisfying the conclusion of Lemma 3.12 arises from a unique ideal hyperbolic polygon.

Remark 3.14. A closed surface of genus g which is (Riemannian) least area in a hyperbolic 3-manifold has area bounded above by $(2g-2)\pi$. The rest of §3 is devoted to establishing a similar result for least area surfaces with respect to relative hyperbolic cellulations. Unlike the Riemannian setting, it may happen that $\int_C K > 0$, where C is a normal disc. However, we show that if S is a normal surface in the relative hyperbolic cellulation, then after a small homotopy and redefinition of the combinatorial geometry, all such "positively curved" discs can be eliminated. It will then follow that the total integral over the negatively curved normal discs is bounded below by $2\pi\chi(S) - \pi$ length(∂S). Therefore by the finiteness of normal disc types the negative curvature is concentrated in a finite area subsurface. After shoving, the flat part of S out of Y, we are left with a surface S_Y whose area is bounded above by $C_0($ length$(\partial S) - \chi(S))$ where C_0 is a constant depending only on ψ.

Definition 3.15. A compact immersed surface S in the 3-manifold Y with relative hyperbolic cellulation ψ is ***h-normal*** if S is a normal surface with respect to ψ and satisfies the following additional properties.

i) $\partial S \cap \partial Y = \emptyset$,

ii) S has no fake boundary branching, and

iii) for each 1-cell β in κ (the induced cellulation on S) which lies in the 2-cell η of ψ, if β separates in η, a single vertex from the rest, then either $\beta \subset \partial Y$, or $\beta \cap \partial S, \neq \emptyset$. Call an arc of the latter type ***bad***. See Figure 3.1.

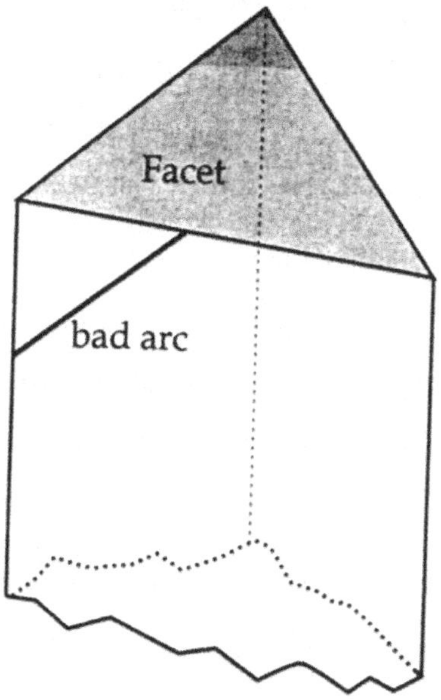

Figure 3.1

The next lemma is a version of Lemma 3.2 for surfaces in relative hyperbolic cellulations. It says that at the cost of pushing part of S out of Y, S can be homotoped to be h-normal, without decreasing $\chi(S)$.

Lemma 3.16. *Let Y be a codimension-0 submanifold of the compact aspherical 3-manifold M with $Y \cap \partial M = \emptyset$. Let ψ be a relative hyperbolic cellulation on Y and $f : S \to M$ satisfy $\partial S \subset \mathring{Y}, f|\partial S$ is normal, f injects on simple loops, f is transverse to ψ^2 near ∂Y (and in particular transverse to ∂Y) and f is boundary incompressible in the following sense. There is no properly embedded arc $\alpha \subset S \cap Y$ with endpoints in distinct 1-cells of the induced triangulation on ∂S, such that $f|\alpha$ can be homotoped rel $\partial\alpha$ into a 2-cell of ψ. Assume also that no component of $f^{-1}(Y)$ is a 2-disc disjoint from ∂S. Then S can be homotoped rel ∂S to S_1 such that $S_1 \cap Y$ is h-normal respect to ψ and $\chi(S_1 \cap Y) \geq \chi(S \cap Y)$. Furthermore no component of $S_1 \cap Y$ is a disc disjoint from ∂S_1.*

Proof. Using the boundary incompressibility of f, first homotope f rel $\partial S \cup (S \cap \partial Y)$ to an immersion a-transverse to ψ^2. Homotope f near ∂S to eliminate fake boundary branching at the expense of introducing intersections as in Figure 1.6. If some component of $f^{-1}(Y) - f^{-1}(\psi^2)$ is not a disc, then f can be homotoped rel ∂S to reduce $|S \cap \psi^1|$. If for some 2-cell C of $\kappa, f(\partial C)$ crossed a 1-cell e of ψ more than once, then a homotopy of f reduces $|S \cap \psi^1|$. (But $\chi(S \cap Y)$ may rise if $e \subset \partial Y$.) If S had a bad arc β, then a small homotopy eliminates this arc thereby reducing $|S \cap \mathring{Y} \cap \psi^1|$ and possibly increasing $|S \cap \psi^1|$. Argue as in the proof of Lemma 3.2 to homotope f so that the restriction of f to each 2-cell of κ is an embedding into its associated 3-cell. All of these homotopies can be accomplished without decreasing $\chi(S \cap Y)$ or introducing disc components of $S \cap Y$. The result follows by induction on $(|S \cap \mathring{Y} \cap \psi^1|, |S \cap \psi^1|)$ ordered lexicographically. □

Proposition 3.17. *Let ψ be a relative hyperbolic triangulation on the compact 3-manifold Y which is contained in the interior of the compact 3-manifold M. Let $f : S \to M$ be a map of a compact surface transverse to ∂Y such that $f|\partial S \subset \mathring{Y}$. There exists a constant C_0 such that if $S \cap Y$ is an h-normal surface, then S is homotopic* *rel ∂S to a surface S_1, such that area($S_1 \cap Y$) $\leq C_0$(length(∂S) $- \chi(S \cap Y)$).*

Corollary 3.18. *Under the hypothesis of Proposition 3.17, there exists a C_1, such that length($\partial(S_1 \cap Y)$) $\leq C_1$(length(∂S) $- \chi(S \cap Y)$).*

Proof. Apply the following Lemma 3.19 taking $C_1 = vC_0$. □

Remark. If S is a Riemannian least area surface in Y which goes far out the cusps, then the region out in the cusp accounts for little Riemannian area, but large PL area. The passage from S to S_1 amounts to shoving the cusp stuff out of Y.

Lemma 3.19. *Let v be the maximal valence of an edge of ψ. If T is a-transverse to ψ, then area(T) $\geq$ length(∂T)$/v$.* □

Proof of Proposition 3.17. Let κ be the induced cellulation on $S \cap Y$. Define a combinatorial geometry (κ,angle) on $S \cap Y$ as follows. Let v be a vertex of the 2-cell η of κ.

If $f(v) \in \partial Y$, define angle(v) $= \pi/2$.

If $v \in \partial S \cap \psi^1$, then define angle($v$) $= 0$.

If $v \in \partial S - \psi^1$, then define angle($v$) $= \pi/2$ (Such vertices arise as in Figure 1.6.)

Otherwise angle(v) is the associated dihedral angle of the 3-cell σ of ψ which contains η.

Claim. If C is a 2-cell of κ, then C is nonpositively curved. If C is flat, then either

a) $f(C)$ is parallel to a facet and C is disjoint from ∂S, or

b) $f(\partial C)$ encircles a vertex of ψ, or

c) $f(C)$ encircles a single edge e on the boundary of the 3-cell which contains $f(C)$, $e \not\subset \partial Y$ and $C \cap \partial S = \emptyset$. See Figure 3.2.

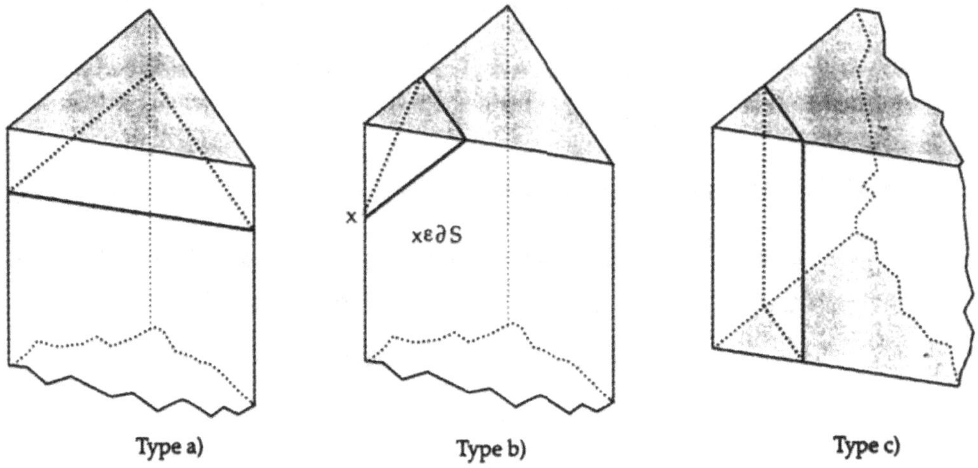

Flat 2-cells
Figure 3.2

Proof of Claim. If σ is the 3-cell which contains $f(C)$, then $f|C$ lifts to an embedding (also called f) of C into σ. First suppose that $f(\partial C)$ is disjoint from the facets. Let $\hat{C}$ denote C with the following combinatorial geometry. The 1-skeleton of $\hat{C}$ equals $\partial\hat{C}$, $f^{-1}(\psi^1)\cap C$ are the vertices of $\partial\hat{C}$, and each point x of $f^{-1}(\psi^1)$ is assigned the dihedral angle of σ at $f(x)$. Thus by Rivin's lemma, $\hat{C}$ is either negatively curved or is flat and parallel to a facet. If $C\cap\partial S=\emptyset$, then C and $\hat{C}$ will have the same combinatorial geometry. If $C\cap\partial S\neq\emptyset$, then by comparing the combinatorial geometry of $\hat{C}$ with that of C it follows that $\int_C K<\int_{\hat{C}} K\leq 0$.

If ∂C crosses exactly one facet and has no bad subarcs, then by pushing ∂C off of the facet, in at least one of the two possible ways, one obtains a new normal curve bounding a disc C' which is not more curved than C. Any two facets of a truncated polyhedron are connected by at most one edge. Thus if C' is flat, then C is negatively curved. If ∂C has a bad subarc β, then by decree angle$(v)=0$ for the non facet vertex v of $\partial\beta$. Another application of Rivin's Lemma implies that such a C is flat if and only if ∂C crosses exactly 3 edges of ψ.

If ∂C crosses exactly two facets, and $C\subset\sigma$ is a 3-cell of ψ, then a similar argument to the one above shows that either C is negatively curved or ∂C separates off a single edge of σ disjoint from ∂S.

If ∂C crosses more than two facets, then C is negatively curved. □

Even accounting for the possibility that part of a 2-cell lies in ∂S, there are only a bounded number of combinatorial possibilities for negatively curved h-normal 2-cells in ψ. Thus there exists $c_1 > 0$ which depends only on ψ such that $-c_1$ is the maximal possible value of $\int_C K$ where C is a negatively curved h-normal 2-cell.

By definition for $v \in \kappa^0 \cap \partial S$ either $f(v) \in \psi^1$ in which case $\gamma_g(v) = \pi$ or $f(v) \notin \psi^1$ in which case $\gamma_g(v) = 0$. Also if α is a component of $f^{-1}(\partial Y)$, then $\int_\alpha \gamma_g = 0$. Therefore $\int_S K = 2\pi\chi(S) - \pi\text{length}(\partial S)$ and hence there are at most $\frac{1}{c_1}(\pi\text{length}(\partial S) - 2\pi\chi(S))$ negatively curved 2-cells of κ.

Let c_2 be the maximal valence of an edge of ψ. The Claim and the fact that there is no fake boundary branching implies that there are at most c_2 length(∂S) flat 2-cells of type b). Since a flat 2-cell of type c) can only share an edge with either another type c) flat 2-cell or a negatively curved 2-cell, and that a string of c_2 type c) flat 2-cells leads to an area reducing homotopy of S, it follows that after homotopy there can be at most $\frac{c_3 c_2}{c_1}(\pi\text{length}(\partial S) - 2\pi\chi(S))$ type c) flat 2-cells. Here c_3 is the maximal length of the boundary of an h-normal 2-cell. Thus for some constant c_4, there are at most $c_4(\text{length}(\partial S) - \chi(S))$ 2-cells of κ which are not flat of type a). Now let $W \subset S$ be the union of type a) flat 2-cells. Homotope S to S_1 by a homotopy supported in $\mathring{W}$ to push most of W out of Y. So if T denotes W with a small collar of ∂W removed, then $S_1 \cap Y = ((S \cap Y) - \mathring{T}) \cup A$, where A is a union of annuli connecting ∂T straight to ∂Y. Therefore $\text{area}(S_1 \cap Y) < \text{area}((S - \mathring{W}) \cap Y) + 2c_2\text{length}(\partial W) < c_5\text{area}((S - \mathring{W}) \cap Y) < c_5 c_3 c_4(\text{length}(\partial S) - \chi(S)) = C_0(\text{length}(\partial S) - \chi(S))$. □

Proposition 3.20. *Let ψ be a relative hyperbolic triangulation on the compact 3-manifold $Y \subset \mathring{M}$. Suppose that $f : S \to M$ is a map of a compact surface such that for each component δ of ∂S, either $\delta \cap Y = \emptyset$ or $f|\delta$ is an immersed curve in $\psi^2 \cap \mathring{Y}$ which is transverse to ψ^1, f injects on simple loops, f is transverse to ∂Y, f is a-transverse to ψ^2 and no component of $f^{-1}(Y)$ is a disc disjoint from ∂S. Then S can be homotoped rel ∂S to S_1 such that area$(S_1 \cap Y) \leq$ Max$(0, C_0($length$((\partial S) \cap Y) - \chi(S \cap Y)))$.*

Proof of Proposition 3.20. The main improvement of this Proposition over the last is that $f : S \to M$ is not h-normal. We argue by induction on $c(S) = \text{length}((\partial S) \cap Y) - \chi(S \cap Y)$ and can assume that S is connected. If $c(S) = -2$, then S is a 2-sphere in Y and can be homotoped out of Y. Now assume that the Proposition is true for $c(S) < k$. Using the induction hypothesis, its routine to verify the Proposition if $(\partial S) \cap Y$ is not a union of normal curves.

Suppose that there is an arc $\alpha \subset f^{-1}(Y)$ with endpoints in distinct 1-cells of κ such that $f|\alpha$ is homotopic into a 2-cell of ψ, via a homotopy fixing $\partial\alpha$ and supported in Y. Boundary compress along α to get a new map $f_1 : T \to M$ which satisfies the hypothesis of the Proposition, and $c(f_1) < c(f)$. By induction, f_1 is homotopic rel ∂T to g_1 which satisfies the conclusions of the Proposition. If T_1 is the resulting surface, then attach a little 1-handle to T_1 (missing ψ^1) to obtain a surface S_1 which is homotopic to S rel ∂S and satisfies the conclusion.

Therefore we can assume that f satisfies the hypothesis of Lemma 3.16. Now apply Lemma 3.16 and Proposition 3.17. □

Corollary 3.21. *If S_1 is as in Proposition 3.20, then length$(\partial(S_1 \cap Y))$ $\leq$ Max$(0, C_1(\text{length}((\partial S) \cap Y) - \chi(S \cap Y)))$.* □

Remark 3.22. The main results of §1-2 can all be stated in terms of cellulations rather than triangulations. In particular Proposition 1.14, Theorem 2.1 and Proposition 2.4 hold by substituting triangulation with cellulation. The proofs parallel those of §1-2. Of course a version of Proposition 1.10 also holds for cellulations, but there is not an obvious canonical branched surface which carries all normal laminations. For the sake of reference we single out

Proposition 3.23. *Let M be a closed 3-manifold with cellulation ψ. $\pi_1(M)$ is not negatively curved if and only if there exists a sequence of least area normal immersed discs $f_i : D_i \to M$, such that length$(\partial f_i)/$area$(f_i) \to 0$ and no f_i has fake boundary branching.* □

§4

The Ubiquity Theorem

Definition 4.1. If α is a homotopically trivial curve in the 3-manifold M and $k \subset M - \alpha$ is a smooth simple closed curve, then define the *wrapping number* $\mathrm{wr}(\alpha, k) = \min\{|E \cap k|$ where $E \subset M$ is an immersed 2-disc transverse to k such that $\partial E = \alpha\}$.

Theorem 4.2. *Let k be a smooth simple closed curve in the closed, irreducible 3-manifold M with cellulation ψ such that k is transverse to ψ^2, disjoint from ψ^1 and for each 2-cell $\eta \subset \psi, |k \cap \eta| \leq 1$. Then either M is toroidal, or k is contained in a 3-cell or there exists a constant C such that for any least area disc D a-transverse to ψ^2 with $\partial D \cap k = \emptyset$ we have area$(D) \leq C($wr$(\partial D, k)+$length$(\partial D))$.*

Remark. Here length and area are measured as in Definition 3.1. Recall that normal immersed discs are examples of a-transverse discs.

The following result is an immediate consequence of Theorem 4.2.

Ubiquity Theorem 4.3. *Let $k \not\subset B^3$ be a smooth simple closed curve in the closed, atoroidal, irreducible 3-manifold M with cellulation ψ such that k is transverse to ψ^2, disjoint from ψ^1 and for each 2-cell $\eta \subset \psi, |k \cap \eta| \leq 1$. There exists constants K and L such that if D is a least area disc a-transverse to ψ^2 with $\partial D \cap k = \emptyset$ and length$(\partial D)/$area$(D) < L$, then wr$(\partial D, k)/$area$(D) > K$.* $\square$

Corollary 4.4. *Let λ be a quasi-least area semi-Euclidean lamination in the closed atoroidal irreducible 3-manifold M. If $k \subset M - \lambda$ is a smooth simple closed curve, then k lies in a 3-cell.*

Proof of Corollary 4.4. Suppose that λ is normal with respect to the cellulation ψ and all measurements are taken with respect to ψ. After passing to a subdivision of ψ and isotopy of k we can assume that for each 2-cell η of ψ that $|k \cap \eta| \leq 1$. Note that the property of a leaf being conformally Euclidean or quasi-least area is preserved after passing to subdivision. If L is a conformally Euclidean leaf of λ, then by the Ahlfors Lemma, [Ca] p. 499, there exists a sequence of embedded discs $\tilde{E}_1 \subset \tilde{E}_2 \subset \cdots \subset \tilde{L}$ the universal covering of L, such that length$(\partial\tilde{E}_i)/$area$(\tilde{E}_i) <$

$1/i$. Since λ is quasi least area there exists a constant K_1 such that for each i, $\text{area}(\tilde{E}_i) \leq K_1 \text{area}(D_i)$ where D_i is a least area disc in $\tilde{M}$ with $\partial \tilde{E}_i = \partial D_i$. If E_i is a slightly perturbed projection of $\tilde{E}_i$ to M such that ∂E is embedded, then $\text{wr}(\partial E_i, k) = 0$ all i. Also if $k \not\subset B^3$, and i is sufficiently large then we obtain a contradiction to Theorem 4.3. □

Remarks 4.5. i) In words the Ubiquity Theorem says that the area of a least area disc D of small isoperimetric ratio is proportional to the wrapping number of ∂D and k.

ii) The proof we give relies on Thurston's hyperbolization theorem for Haken 3-manifolds. One can give a much less elegant proof which uses only [Gr] and standard 3-manifold topology.

Proof of Theorem 4.2. To obtain a clue, read the proof of case 1. For a complete and more detailed argument read the proof of case 2.

Case 1. $M - k$ is hyperbolic.

Proof of Case 1. It suffices to consider the case that ψ is obtained by attaching a 2-cell and a 3-cell to a relative hyperbolic cellulation on $M - \overset{\circ}{N}(k)$. And it suffices to prove Theorem 4.2 for ψ-least area discs D such that $\partial D \subset M - N(k)$. Given such a disc D, let $E \subset M$ be an immersed disc such that $\partial E = \partial D$ and $\text{wr}(\partial D, k) = |E \cap k| = |E \cap \partial N(k)|$. By Proposition 3.20 and Corollary 3.21 E can be homotoped to a disc E_1 rel ∂E, such that $\text{area}(D) \leq \text{area}(E_1) = \text{area}(E_1 \cap (M - \overset{\circ}{N}(k))) + \text{area}(E_1 \cap N(k)) \leq \text{area}(E_1 \cap (M - \overset{\circ}{N}(k))) + \text{length}(\partial E_1 \cap N(k)) \leq (C_0 + C_1)(\text{length}(\partial E) - \chi(E \cap (M - \overset{\circ}{N}(k)))) = (C_0 + C_1)(\text{length}(\partial D) + \text{wr}(\partial D, k) - 1) < (C_0 + C_1)(\text{length}(\partial D) + \text{wr}(\partial D, k))$. □

Case 2. General case

Proof of Case 2.

If k does not lie in a 3-cell, then $M - \overset{\circ}{N}(k)$ is irreducible and hence there exists by the characteristic manifold theory of [JS], [J] a collection of π_1-injective pair wise disjoint embedded tori $T_0, T_1, \cdots, T_n \subset M$ such that each component of

$(M-\overset{\circ}{N}(k))-\cup_{i=0}^{n}T_i$ is either a finite volume hyperbolic 3-manifold or the interior of a graph manifold. (Recall that all 3-manifolds in this paper are orientable and that a graph manifold is the union of finitely many Seifert fibred spaces glued together along some of their boundary components.) Here $T_0 = \partial N(k)$. By Thurston [Th] $M-k$ is hyperbolic if and only if $\{T_0, T_1, \cdots, T_n\} = \{T_0\} = \{\partial N(k)\}$ and $M-\overset{\circ}{N}(k)$ is a not a graph manifold.

If T is an embedded π_1−injective non boundary parallel torus in $M-\overset{\circ}{N}(k)$, then either T is π_1-injective in M and hence Theorem 4.2 is proved or T is compressible. By standard 3-manifold topology, if T is compressible, then either T bounds a unique solid torus W in M or T bounds a *cube with knotted hole* W. In the former case $k \subset W$. In the latter case there exists an embedded $B^3 \subset M$ and a knotted properly embedded arc $\beta \subset B^3$ such that $W = B^3 - \overset{\circ}{N}(\beta)$ and thus $k \cap W = \emptyset$. Note that $k \not\subset B^3$ implies that T cannot bound both a solid torus and a cube with knotted hole. For each i, let W_i be the associated solid torus or cube with knotted hole bounded by T_i. Observe that if $T_i \subset \overset{\circ}{W}_j$, then $W_i \subset \overset{\circ}{W}_j$.

The W_i's are partially ordered by inclusion, so reorder the T_i's so that W_0 is the maximal solid torus region (which is necessarily nonempty since for some $i, W_i = N(k)$) and $W_1, \cdots W_r$ are the maximal cube with knotted hole regions. Let $Y = M - \overset{\circ}{W}_0 \cup \cdots \cup \overset{\circ}{W}_r$.

Claim 1. Either Theorem 4.2 is true or Y is atoroidal and hence by Thurston [Th] has a complete hyperbolic metric of finite volume.

Proof of Claim 1. By construction Y is either atoroidal or is a graph manifold. We will assume the latter and derive a contradiction. If $r > 0$, then Figure 4.1 shows how to reembed $Y \cup W_0 \cup \cdots \cup W_{r-1}$ into M in such a way that $\partial W_r'$ bounds a solid torus V_r in $M - Y' \cup W_0' \cup \cdots \cup W_{r-1}'$, where W_i' (resp. $\partial W_r', Y'$) denotes the reembedded W_i (resp. $\partial W_r, Y$). There are an infinite number of such reembeddings, by "twisting the neck". Now fix a graph structure on Y. This structure induces a Seifert fibering on ∂W_r. At most one reembedding corresponds to spanning Seifert fibres of $\partial W_r'$ by meridinal discs of V_r. After any other reembedding, $M - W_0' \cup$

$\cdots \cup W'_{r-1}$ is a graph manifold. By repeating this operation $r-1$ more times, we conclude that M is obtained by attaching a solid torus to the boundary of a graph manifold. Since such a manifold is either reducible, has finite π_1 or has a π_1-injective immersed torus, Claim 1 is established. Note that if $\pi_1(M)$ is finite, then M has a linear isoperimetric inequality. $\square$

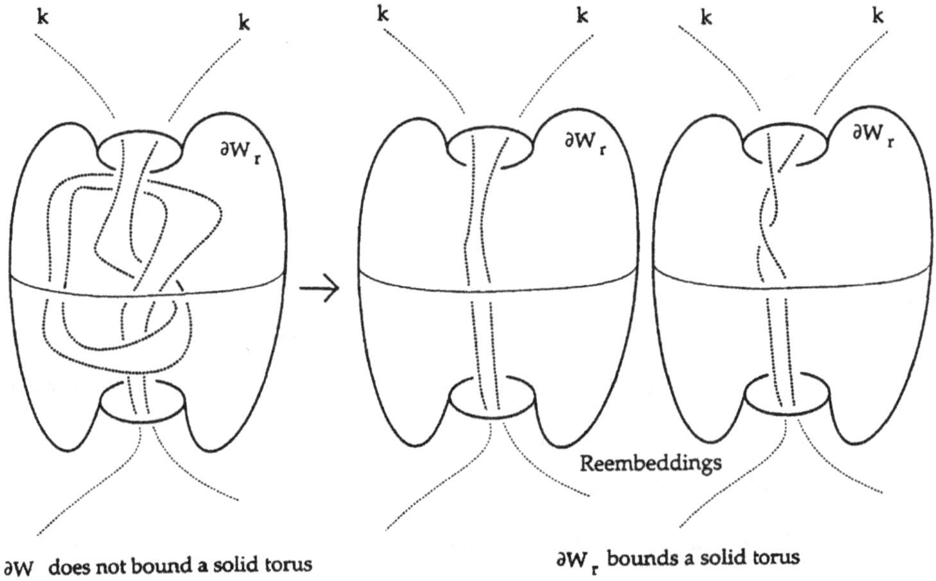

Figure 4.1

Given a ψ-least area disc D, define

$$W_\psi(D,k) = \frac{\mathrm{wr}(\partial D, k) + \mathrm{length}_\psi(\partial D)}{\mathrm{area}_\psi(D)}.$$

We may suppress the subscript ψ when the cellulation is understood.

Claim 2. We can assume that ψ satisfies

a) $\psi|Y$ is a relative hyperbolic cellulation and in particular for each $i \le r, \psi$ restricts to a cellulation on W_i.

b) There exists a $D^2 \times S^1 \subset \mathring{W}_0(k)$ called $N_1(k)$ such that $k \subset \mathring{N}_1(k)$ is a core of $N_1(k)$, $\psi^2 \cap N_1(k)$ consists of $\partial N_1(k)$ and a finite number of pairwise disjoint meridinal discs. Furthermore if σ is a 3-cell of $\psi|N_1(k)$, then $k \cap \sigma$ is a properly embedded unknotted arc. Note that $\psi^1 \cap \mathring{N}_1(k) = \emptyset$. See Figure 4.2

A schematic view of $N_1(k)$

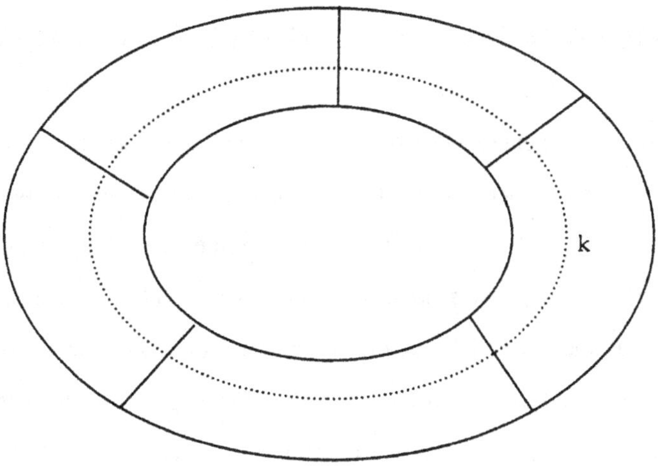

Figure 4.2

Proof of Claim 2. If ϕ is a subdivision of ψ and Theorem 4.2 holds for ϕ, then Theorem 4.2 holds for ψ. Use the fact that there exists a $K_1 > 0$ such that if D is a ψ-least area disc, then there is a ψ least area disc, also called D, with $W_\psi(D,k)$ unchanged such that D is a K_1-quasi ϕ-least area disc. Furthermore $\text{length}_\psi(\partial D) \leq \text{length}_\phi(\partial D) \leq K_1\text{length}_\psi(\partial D)$. The last inequality uses the fact that an embedded arc α in a 2-cell of ψ is isotopic rel $\partial\alpha$ to an arc crosing ϕ^1 a uniformly bounded number of times, via an isotopy disjoint from k. (This is where we use the hypothesis that for each 2-cell η of ψ, $|k \cap \eta| \leq 1$.) Thus a sequence $\{D_i\}$ of ψ-least area discs with $W_\psi(D_i) \to 0$ gives rise to a sequence $\{E_i\}$ of ϕ-least area discs with $W_\phi(E_i) \to 0$.

Conversely suppose that ϕ is a subdivision of ψ such that if σ is a 3-cell of ψ and $\sigma \cap k \neq \emptyset$, then $\phi|\sigma = \psi|\sigma$. Then standard arguments show that there exists a $K_2 \geq 1$ such that a ϕ-least area disc D give rise to a ψ-least area disc E where $\frac{\text{length}_\psi(\partial E)}{\text{area}_\psi(E)} \leq \frac{K_2\text{length}_\phi(\partial D)}{\text{area}_\phi(D)}$ and ∂E is homotopic to ∂D via a homotopy disjoint from k. Thus if Theorem 4.2 holds for ψ, then it holds for ϕ. Similarly this result follows in the following case. Here if σ is a 3-cell of ψ, then either $k \cap \sigma = \emptyset$ or $k \cap \sigma$

is an unknotted arc and ϕ is a subdivision of ψ such that $\phi^2 \cap k = \psi^2 \cap k$.

Given ψ, we first construct a subdivision ψ_1 with the property that for each 3-cell σ of $\psi_1, k \cap \sigma = \emptyset$ or $k \cap \sigma$ is an unknotted arc. The subdivision is created on each 3-cell B of ψ as follows. Let $t = B \cap k$. Think of B as $D^2 \times I$ with $\partial t \subset D^2 \times 0$ and t in generic position with respect to the height function coming from the I-factor. Now attach a finite number of pairwise disjoint properly embedded 2-cells $C_1, \cdots, C_n \subset B$, one for each local minimum of $t \subset \mathring{B}$, with the property that for each i, C_i separates off a 3-cell B_i with $B_i \cap t$ is an unknotted arc and $t \cap (B - \cup_{i=1}^n \mathring{B}_i)$ is a trivial B^3-link on s components. Now subdivide each C_i along a properly embedded arc such that each "half" of of C_i hits t exactly once. See Figure 4.3. After subdividing along another $s-1$ 2-cells we obtain a cell division on B such that each 3-cell intersects t in an unknotted arc. ψ_1 is the result of these $n+s-1$ subdivisions of ψ. One readily checks in $n+s-1$ steps that Theorem 4.2 holds for ψ if and only if it holds for ψ_1.

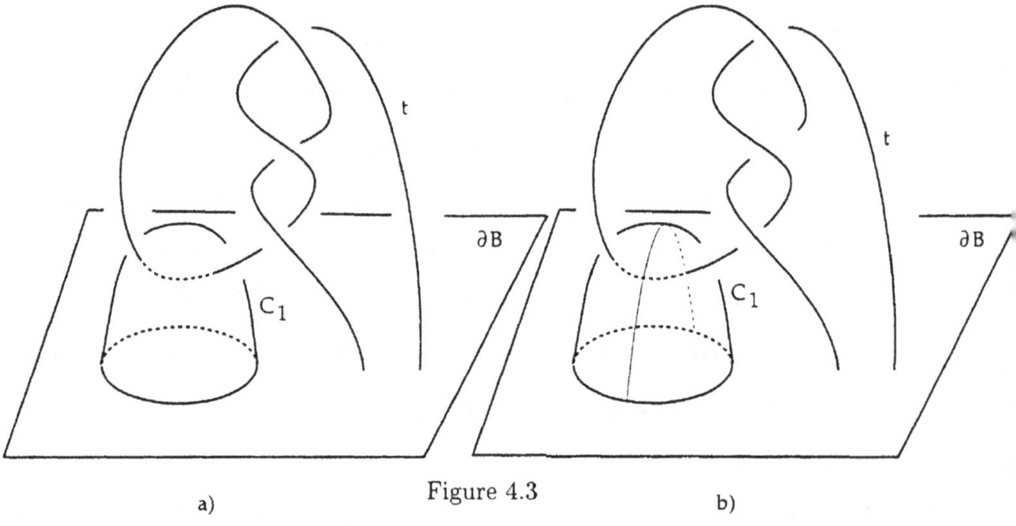

Figure 4.3

Let $N_1(k)$ be a $D^2 \times S^1$ in $\mathring{W}_0$ whose core is k such that $N_1(k) \cap \psi_1^1 = \emptyset$ and $N_1(k) \cap \psi_1^2$ is a disjoint union of meridinal discs. Subdivide ψ_1 to ψ_2 so that $N_1(k) \cap \psi_2^2 = N_1(k) \cap \psi_1^2 \cup \partial N_1(k)$. By the second paragraph of this proof, Theorem 4.2 holds for ψ_2 if and only if it holds for ψ_1.

By [EP] there exists a hyperbolic cellulation on Y and hence there exists a cellulation ψ_3 on M satisfying a) and b) and which agrees with ψ_2 near $N_1(k)$. By the uniqueness of PL structures for 3-manifolds [Mo], there exists a cellulation ψ_4 on M isomorphic to ψ_3 via an isomorphism which restricts to the identity on $N_1(k)$ such that ψ_4 and ψ_2 have a common subdivision ψ_5 and $\psi_5|N_1(k) = \psi_4|N_1(k) = \psi_2|N_1(k)$. By the first two paragraphs Theorem 4.2 holds for ψ_4 if and only if it holds for ψ_5 if and only if it holds for ψ_2. Therefore Theorem 4.2 holds for ψ_4 if and only if it holds for ψ. Since ψ_4 satisfies the conclusions of Claim 2, that result follows. □

Claim 3. It suffices to prove Theorem 4.2 for discs D such that $\partial D \subset \mathring{Y}$.

Proof of Claim 3. It suffices to prove Theorem 4.2 for discs D such that $\partial D \cap N_1(k) = \emptyset$. Indeed by construction of $N_1(k)$, a small homotopy disjoint from k, pushes ∂D off of $N_1(k)$ and so D gives rise to a least area disc E a-transverse to ψ with $E \cap N_1(k) = \emptyset, \mathrm{wr}(\partial E, k) = \mathrm{wr}(\partial D, k)$ and D, E have approximate isoperimetric inequalities.

Since balls and solid tori have linear isoperimetric inequalities it follows that there exists a K_3 such that if γ is a locally ψ-embedded homotopically trivial curve in $M - N_1(k)$ then there exists a homotopy $H : S^1 \times I \to M$ of γ to γ_1 such that γ_1 is locally ψ-embedded, $\gamma_1 \subset \mathring{Y}$ and $\mathrm{area}(H) \leq K_3\mathrm{length}(\gamma)$. This implies that if D (resp. E) is a least area disc spanning γ, (resp. γ_1), then

$$\mathrm{area}(E) \geq \mathrm{area}(D) - \mathrm{area}(H)$$

$$\mathrm{length}(\gamma_1) \leq \mathrm{length}(\gamma) + v(\mathrm{area}(H)) \leq (1 + vK_3)\mathrm{length}(\gamma)$$

$$\mathrm{wr}(\gamma_1, k) \leq \mathrm{wr}(\gamma, k) + 3\mathrm{area}(H) \leq \mathrm{wr}(\gamma, k) + 3K_3\mathrm{length}(\gamma)$$

The first and third inequalities are immediate. The second follows from Lemma 3.19. To obtain the fourth, note that H can be chosen so that if η is a 2-cell in the induced cellulation on $S^1 \times I$, and $H(\eta) \cap \mathring{N}_1(k) \neq \emptyset$, then $H|\eta$ is an embedding and $H(\eta) \cap \psi^1 \neq \emptyset$. (This uses the fact $\gamma \cap N_1(k) = \emptyset$.) Since H can be chosen so that $|\eta \cap k| \leq 1$ and $\mathrm{area}(H|\eta) \geq \frac{1}{3}$ it follows that $|H^{-1}(k)| \leq 3\mathrm{area}(H)$.

Thus a sequence of discs $\{D_i\}$ such that $W(D_i, k) \to 0$ gives rise to a sequence $\{E_i\}$ with $W(E_i, k) \to 0$ and $\partial E_i \subset \overset{\circ}{Y}$. $\square$

Let E be an immersed disc in M a-transverse to ψ^2 such that $\partial E = \partial D$ and $|E \cap k| = wr(\partial D, k)$. Among all such discs with these properties choose one that minimizes $|\partial Y \cap E|$. Let $E_Y = E \cap Y$. Assume that length$(\partial E) > 0$.

Claim 4. $-\chi(E_Y) < \text{wr}(\partial E, k)$

Proof of Claim 4. Since Y has no essential annuli, and $|\partial Y \cap E|$ is minimal, each component of E_Y disjoint from ∂E is a disc with at least two open discs removed. Define a partial ordering on the components $F_1, \cdots, F_m$ of E_Y, by $F_i < F_j$ if F_i is separated from ∂E by F_j. Let $\tilde{F}_i = \cup_{F_j \leq F_i} F_j$ and $\hat{F}_i$ be the subdisc of E bounded by the outermost component of ∂F_i. Since for some $i, E_Y = \tilde{F}_i$ and $E = \hat{F}_i$ it suffices to show that for each i, $-\chi(\tilde{F}_i) \leq |\hat{F}_i \cap k| - 1$. To prove this let F_i be a minimal component of E_Y. All but one component of ∂F_i bounds a disc in E whose interior is disjoint from F_i. Let G be one such disc. The incompressibility of ∂W_i in $W_i, i > 0$ implies that $G \subset W_0$. The incompressibility of ∂W_0 in $W_0 - k$ implies that $G \cap k \neq \emptyset$. Hence $-\chi(F_i) < |\hat{F}_i \cap k|$. Now inductively assume that for all $F_i < F_j, -\chi(F_i) \leq |\hat{F}_i \cap k| - 1$. Again let $G \subset E$ be the disc spanning an innermost component of ∂F_j. By either the induction hypothesis, if $\overset{\circ}{G} \cap E_Y \neq \emptyset$ or the previous argument, if $\overset{\circ}{G} \cap E_Y = \emptyset$ it disjoint from ∂E follows that $-\chi(G \cap E_Y) \leq |G \cap k| - 1$ and therefore $-\chi(\tilde{F}_j) \leq |\hat{F}_j \cap k| - 1$. $\square$

Claim 5. E can be homotoped to E' rel ∂E such that if F is the component of E'_Y which contains ∂E, then

1) area$(F) < C_0(\text{wr}(\partial D, k) + \text{length}(\partial D))$ and
2) length$(\partial F) < C_1(\text{wr}(\partial D, k) + \text{length}(\partial D))$. $\square$

Proof. Let S be the compact codimension-0 submanifold of E such that $S \cap Y$ is exactly the component of $E \cap Y$ which contains ∂E. By Proposition 3.20 and Claim 4, S can be homotoped rel ∂S to S_1 such that area$(S_1 \cap Y) \leq C_0(\text{length}((\partial S) \cap Y) - \chi(S \cap Y)) < C_0(\text{length}(\partial E) + \text{wr}(\partial E, k))$. Since the homotopy of S to S_1

extends trivially to a homotopy of E and $\partial E = \partial D$ conclusion 1) follows. (Here E' is the homotoped E and F is the component of $S_1 \cap Y$ containing ∂E.) Conclusion 2) follows from Proposition 3.21. □

Claim 6. There exists a constant k_2 such that if α is a closed curve in W_i, which is homotopically trivial in M, then α bounds a disc A such that area(A) $< k_2(\text{length}(\alpha))$. □

Proof. Since balls and solid tori have linear isoperimetric inequalities and each $W_i, i \geq 1$, lies in a 3-cell, Claim 6 holds (using the same constant k_3) for all $\alpha \subset W_i, i \geq 1$, and for all $\alpha \subset W_0$ which are homotopically trivial in W_0.

Let $\beta \subset W_0$ be a normal curve which generates $\pi_1(W_0)$. There exists a k_4 such that if $\alpha \subset W_0$ is not null homotopic in W_0, then α is homotopic in W_0 to $\beta^n, n \neq 0$ via a homotopy $L : S^1 \times I \to M$ such that area$(L) + |n| \leq k_4$length(α). Since α is homotopically trivial in M, β^n and hence β (since $\pi_1(M)$ is torsion free) are homotopically trivial in M. If B (resp. A) is a least area disc bounded by β (resp. α), then

$$\text{area}(A) \leq |n|\text{area}(B) + \text{area}(L) \leq (1 + \text{area}(B))(|n| + \text{area}(L))$$

$$\leq 2(1 + \text{area}(B))k_4\text{length}(\alpha) = k_5\text{length}(\alpha),$$

where $k_5 = (1 + \text{area}(B))k_4$. Finally, take $k_2 = \max(k_3, k_5)$. □

To complete the proof of Theorem 4.2 observe that ∂D bounds a disc J which is the union of F (which was defined in Claim 5) and least area discs bounded by $\partial F - \partial D$. Thus by Claims 5 and 6 we obtain area$(D) \leq$ area$(J) \leq$ area(F) + $k_2$length$(\partial F) \leq (C_0 + C_1k_2)(\text{wr}(\partial D, k) + \text{length}(\partial D))$. □

§5

Ubiquity In The Smooth Category

Theorem 5.1. *Let k be a smooth simple closed curve in the closed irreducible Riemannian 3-manifold M. Then either M is toroidal or k is contained in a 3-cell* *or there exists a constant* $C > 0$ *such that if* D *is a least area disc with* $\partial D \cap k = \emptyset$, *then* $area(D) \leq C(wr(\partial D, k) + length(\partial D))$.

Proof: Step 1. Either $k \subset B^3$ or M is toroidal or $M = Y \cup W_0 \cup \cdots \cup W_r$ where W_0 is a $D^2 \times S^1, k \subset \overset{\circ}{W}_0, \partial W_0$ is π_1-injective in $W_0 - k, W_i \cap W_j = \emptyset$ if $i \neq j$, and if $i \geq 1$, then W_i lies in a 3-cell and ∂W_i is a torus which is π_1-injective in $\overset{\circ}{W}_i$. Furthermore $Y \neq \emptyset$ and $\overset{\circ}{Y}$ has a hyperbolic structure of finite volume. Finally $Y \cap (W_0 \cup \cdots \cup W_r) = \partial Y$.

Proof of Step 1. For the proof of step 1, read (in §4) from the beginning of the proof of case 2, through the end of the proof of Claim 1. □

Step 2. It suffices to establish Theorem 5.1 for any Riemannian metric on M.

Proof of Step 2. Use the fact that changes in the Riemannian metric changes arc length as well as area by uniformly bounded multiplicative amounts. □

Let X be a space diffeomorphic to $\overset{\circ}{Y}$ and give X a complete hyperbolic metric. Let $T \times [0, \infty)$ cut off the ends of X and be parametrized so that each $T \times i$ is a union of $r + 1$ horospherical tori and for $z \in T, z \times [0, \infty)$ is a geodesic parametrized by arc length.

Let $N(\partial Y) \subset (W_0 \cup \cdots \cup W_r) - k$ be a closed regular neighborhood of ∂Y. Let g be a Riemannian metric on M such that $g|Y$ is isometric to $X - T \times (0, \infty)$ and $g|N(\partial Y)$ is isometric to $T \times [0, 1]$. I.e. $g|Y \cup N(Y)$ is the pullback metric induced from a diffeomorphism $f : Y \cup N(Y) \to X - T \times (1, \infty)$ such that $f(Y) = X - T \times (0, \infty)$ and $f(N(\partial Y)) = T \times [0, 1]$.

Step 3. a) There exists $\gamma > 0$ with the following property. It suffices to prove Theorem 5.1 for discs D where the geodesic curvature at each point of $\partial D \cap Y$ is bounded above by γ.

b) We can assume that $\partial D \cap N(\partial Y)$ is a union of curves of the form $z \times [0, 1]$.

c) We can assume the ∂D is embedded in M. □

Let D be a least area disc in M such that $\partial D \cap k = \emptyset$. Let E be a smooth disc which spans ∂D, is transverse to $k \cup N(\partial Y)$, $|E \cap k| = \text{wr}(\partial D, k)$ and satisfies the following two conditions.

i) No component of $E \cap Y$ disjoint from ∂E is either a disc or an annulus

ii) If $\alpha \subset E \cap Y$ is a simple closed curve, then either α bounds a disc in $E \cap Y$ or α is homotopically nontrivial in Y.

Remark. The reader may have noticed that we have abused notation by identifying the disc E with its image in M thereby suppressing the mapping $f : E \to M$ which defines that image. Thus it would have been more precise to have first defined $f : E \to M$ and then rephrased i) by "no component of $f^{-1}(Y)$ is a disc or annulus disjoint from ∂E" and ii) by "If α is a simple closed curve in $f^{-1}(Y)$, then either α bounds a disc in $f^{-1}(Y)$ or $f|\alpha$ is homotopically nontrivial as a map into Y. In the following statement $E \cap Y$ denotes a subset of the disc E.

Step 4. $-\chi(E \cap Y) < \mathrm{wr}(\partial E, k)$

Proof of Step 4. By hypothesis $\mathrm{wr}(\partial E, k) = |E \cap k|$. Define a partial ordering on the components $F_1, \cdots, F_m$ of $E \cap Y$, by $F_i < F_j$ if F_i is separated from ∂E by F_j. Let $\tilde{F}_i = \cup_{F_j \leq F_i} F_j$ and $\hat{F}_i$ be the subdisc of E bounded by the outermost component of ∂F_i. Since $E \cap Y$ is the union of $\tilde{F}_{i_1}, \cdots, \tilde{F}_{i_n}$ where $F_{i_1}, \cdots, F_{i_n}$ are the maximal components, it suffices to show that for each j, $-\chi(\tilde{F}_{i_j}) \leq |\hat{F}_{i_j} \cap k| - 1$. The proof of this fact is given after Claim 4 §4. Ignore the assertion, true only in §4, that $\partial E = \partial F_i$ for some i. In that proof E_Y denotes $E \cap Y$. [Note that for each i, no component of $\hat{F}_i \cap Y$ is a disc or annulus disjoint from ∂E, each disc component of $\hat{F}_i \cap W_j$ disjoint from ∂E must lie in W_0 and nontrivially intersect k, and that $\mathrm{wr}(\partial \hat{F}_i, k) = |\hat{F}_i \cap k|$.)] $\square$

Step 5. There exists constants C_4, C_5 such that after homotopy of E rel ∂E, there exists a codimension-0 submanifold H of E such that $E \cap Y \subset H \subset Y \cup N(\partial Y)$ and

a) $\mathrm{area}(H) \leq C_4(\mathrm{length}(\partial E) + \mathrm{wr}(\partial E, k))$

b) $\mathrm{length}(\partial H) \leq C_5(\mathrm{length}(\partial E) + \mathrm{wr}(\partial E, k))$

Proof of Step 5. Let $H_1 = E \cap Y$ and view H_1 as lying in $X - T \times (0, \infty)$. After a homotopy of E rel ∂E, supported near ∂Y we can assume that $H_1 \cap T \times 0$ is straight in the Euclidean tori $T \times 0$. Let $H_2 \subset X$ be $H_1 \cup ((H_1 \cap T \times 0) \times [0, s])$. [More precisely, if $f : H_1 \to Y$, then define $H_2 = H_1 \cup (f^{-1}(\partial Y) \times [0, s]) / \sim$ where $x \in f^{-1}(\partial Y)$ is identified with $x \times 0$. Finally extend f to H_2 by $f(x, t) = (f(x), t) \in T \times t$.] Choose s sufficiently large so that, length$(H_2 \cap (T \times s)) < \frac{1}{2}$. After an extremely tiny perturbation of ∂H_2 near $T \times s$ we can assume that ∂H_2 is embedded in X. By Theorem 3 of Meeks - Yau [MY] (generalizing [SY]) H_2 is homotopic rel ∂H_2 (in X) to a least area surface H_3. It follows that for s sufficiently large we obtain

$$\begin{aligned}
\operatorname{area}(H_3) &= \int_{H_3} dA \\
&\le \int_{H_3} -K dA \\
&= -2\pi\chi(H_3) + \int_{\partial H} \gamma_g \\
&\le -2\pi\chi(H_3) + \gamma \operatorname{length}(\partial H_3 \cap Y) + \pi |\partial E \cap (M - \mathring{Y})| + \frac{1}{2} \\
&< 2\pi \operatorname{wr}(\partial E, k) + \gamma \operatorname{length}((\partial E \cap Y)) + \frac{\pi}{2} \operatorname{length}(\partial E \cap N(\partial Y)) \\
&\le C_4(\operatorname{length}(\partial E) + \operatorname{wr}(\partial E, k))
\end{aligned}$$

where $C_4 = \max(2\pi, \gamma)$. The second equation follows from the fact that H_3 is a minimal surface and hence its sectional curvature is ≤ -1 everywhere. The third equation is Gauss - Bonnet. To establish the fourth equation use the fact that $\partial H_3 \cap Y$ has geodesic curvature bounded above by γ, $\partial H_3 \cap T \times [0, s)$ is geodesic and hence has zero geodesic curvature and ∂H_3 near $T \times s$ is essentially horospherical, hence has geodesic curvature equal extremely close to 1. Finally note that there are two right angled corners of ∂H_3 for each arc component of $(\partial H_1) \cap \partial Y$. The number of such components is equal to $|\partial E \cap (M - \mathring{Y})|$. To obtain the fifth equation observe that $-\chi(H_3) = -\chi(H_2) = -\chi(H_1) = -\chi(E \cap Y) < \operatorname{wr}(\partial E, k) - \frac{1}{2}$, the last inequality following by Step 4. Also each component of $\partial E \cap (M - \mathring{Y})$ corresponds to two components of $\partial E \cap N(\partial Y)$ each of which has length 1.

After a small perturbation of H_3 we can assume that the conclusion of the above series of equations still hold and except for finitely many levels, H_3 is transverse to each $T \times i$. Since $\int_0^1 \text{length}(H_3 \cap (T \times t))dt \leq \text{area}(H_3 \cap (T \times [0,1])) \leq \text{area}(H_3)$, it follows that for some $t \in [0,1), H_3$ is transverse to $T \times t$ and $\text{length}(H_3 \cap (T \times t)) \leq \text{area}(H_3) \leq C_4(\text{length}(\partial E) + \text{wr}(\partial E, k))$. Let $H = H_3 \cap X - T \times (t, \infty)$. Thus $\text{area}(H) \leq \text{area}(H_3) \leq C_4(\text{length}(\partial E) + \text{wr}(\partial E, k))$ and $\text{length}(\partial H) \leq \text{length}(\partial E) + \text{length}(H_3 \cap T \times t) \leq 2C_4(\text{length}(\partial E) + \text{wr}(\partial E, k))$. Let $C_5 = 2C_4 + 1$. To complete the proof of Step 5, view H as lying in $Y \cup N(\partial Y)$ and observe that there is a homotopy of E rel ∂E supported in $Y \cup N(\partial Y)$ such that after the homotopy $E \cap (Y \cup T \times [0,t]) = H$. □

Proof of Theorem 5.1. Let G be the closure of $E - H$. By the proof of Step 5 $\text{length}(\partial G) \leq \text{length}(\partial E) + \text{length}(H_3 \cap T \times t) \leq C_5(\text{length}(\partial E) + \text{wr}(\partial E, k))$. Each component of $\partial G \subset W_i$ for some i and is homotopically trivial in M. The argument of Claim 6 (§4) shows that there exists a constant C_6 such that if α is a closed curve in some W_i and α is homotopically trivial in M, then α bounds a disc $D_\alpha \subset M$ such that $\text{area}(D_\alpha) \leq C_6 \text{length}(\alpha)$. Therefore we can span ∂G by a collection of discs $\mathcal{D}$ such that $\text{area}(\mathcal{D}) \leq C_6 \text{length}(\partial G)$.

One can construct a disc F spanning ∂D by piecing together a subset of the components of H together with a subset of the least area discs bounded by components of ∂G. In fact, use those components of H which, as subsets of E, nontrivially intersect ∂E and those components of ∂G which are outermost in E. It follows that $\text{area}(D) \leq \text{area}(F) \leq \text{area}(\mathcal{D} \cup H) \leq \text{area}(H) + C_6 \text{length}(\partial G) \leq (C_4 + C_6 C_5)(\text{length}(\partial E) + \text{wr}(\partial E, k))$. □

Ubiquity Theorem 5.2. *Let $k \not\subset B^3$ be a smooth simple closed curve in the closed, atoroidal, irreducible 3-manifold M. There exists constants K and L such that if D is a least area disc with $\partial D \cap k = \emptyset$ and $\text{length}(\partial D)/\text{area}(D) < L$, then $\text{wr}(\partial D, k)/\text{area}(D) > K$.* □

Corollary 5.3. *Let λ be a quasi-least area semi-Euclidean lamination in the closed atoroidal irreducible 3-manifold M. If $k \subset M - \lambda$ is a smooth simple closed curve, then k lies in a 3-cell.* □

Remark 5.4. By defining $\text{wr}(\alpha, k)$ for all reasonable homotopically trivial curves α

(rather than just α with $\alpha \cap k = \emptyset$) in the appropriate manner we can then obtain 5.1, 5.2, 5.3 with the hypothesis $\partial D \cap k = \emptyset$ eliminated.

§6 A Very Brief Survey On The Geometrization Of 3-Manifolds

Thurston Geometization Conjecture 6.1 [Th]. *Let M be a closed connected irreducible 3-manifold, then either*

a) $M = S^3/\Gamma$ where $\Gamma \subset Isom(S^3)$ or

b) $\mathbb{Z} \oplus \mathbb{Z} \subset \pi_1(M)$ or

c) $M = \mathbb{H}^3/\Gamma$ where $\Gamma \subset Isom(\mathbb{H}^3)$

Remarks 6.2. i) These conclusions exactly parallel the conclusions of the corresponding geometrization theorem for closed surfaces.

ii) The three conclusions are mutually exclusive. Conclusion a) is the geometrization conjecture for 3-manifolds with finite fundamental group. That conjecture is equivalent to positive solutions to each of the following longstanding open problems.

Poincare Conjecture 6.3. *If M is a closed connected simply connected 3-manifold, then M is homeomorphic to S^3.*

Generalized Space Form Conjecture 6.4. *If the finite group Γ acts freely on S^3, then Γ is conjugate to a linear action.*

Remark 6.5. See [Liv], [My], [Ru], and [T] for partial results on this conjecture. In particular by [T], Conjecture 6.4 for cyclic groups implies a positive solution to the following classical question.

Space Form Conjecture 6.6. *If M is covered by S^3, then $\pi_1(M)$ is isomorphic to a subgroup of $SO(4)$*

As of this writing there is one class of groups that still needs to be eliminated [Mi], [Lee].

Remark 6.7. Conclusions b) and c) of the Geometrization conjecture treat the case that $\pi_1(M)$ is infinite. The topology of closed irreducible 3-manifolds with $\mathbb{Z} \oplus \mathbb{Z} \subset \pi_1(M)$ is now completely understood, thanks to the work of [CJ], [G2], [M], and

[Sc1-2]. These works in turn built on ideas of many others e.g. Tukia, Ghering, Martin, Jaco, Shalen, Johannson, Waldhausen, Burde, Zieschang, Neuwirth, and Murasugi. (See [G2] for a more detailed history.) Such manifolds contain either embedded π_1-injective surfaces or are Seifert fibred spaces. The [JS][J]-theory, discussed in §4, details the structure of 3-manifolds with π_1-injective embedded tori. The nature of 3-manifolds M with $\mathbb{Z} \oplus \mathbb{Z} \subset \pi_1(M)$ was not known when the Geometrization conjecture was originally formulated and conclusion b) was spelled out with all its possibilities.

Conclusion c) is also known as the

Thurston Hyperbolization Conjecture 6.8. *Let M be a closed, connected, irreducible 3-manifold with $\pi_1(M)$ infinite and $\mathbb{Z} \oplus \mathbb{Z} \not\subset \pi_1(M)$, then $M = \mathbb{H}^3/\Gamma$ where $\Gamma \subset Isom(\mathbb{H}^3)$.*

Remark 6.9. In 1978 Thurston proved this conjecture for Haken 3-manifolds [Th]. A result known in those days as the Monster Theorem.

At the JDG 96 conference, the author announced the following result [GMT] which was joint with Robert Meyerhoff and Nathaniel Thurston.

Theorem 6.10. *Let N be a closed hyperbolic 3-manifold. Then*

a) If $f : M \to N$ is a homotopy equivalence where M is an irreducible 3-manifold, then f is homotopic to a homeomorphism.

b) If $f, g : M \to N$ are homotopic homeomorphisms, then f is isotopic to g.

c) The space of hyperbolic metrics on N is path connected.

Remark 6.11. i) The analogue of a) for spherical manifolds is false. Its a classical theorem of Reidemeister and Whitehead that the lens space $L(7,1)$ is homotopy equivalent but not homeomorphic to $L(7,2)$. Its a much more recent observation of the author that there exists an orientation preserving self homotopy equivalence of a lens space which is not homotopic to a homeomorphism [G3]. For example, consider the homotopy equivalence of $L(8,1)$ to itself whose action on π_1 is multiplication by 3.

ii) Theorem 6.10 reduces the Hyperbolization Conjecture to

Conjecture 6.12. *If M is a closed, connected, aspherical 3-manifold with $\mathbb{Z} \oplus \mathbb{Z} \not\subset \pi_1(M)$, then M is homotopy equivalent to a hyperbolic 3-manifold. Equivalently [since M is a $K(\pi, 1)$], $\pi_1(M)$ is isomorphic to $\pi_1(N)$ where N is a hyperbolic 3-manifold.*

Conjecture 6.12 is equivalent to the following two well known open problems.

Group Negative Curvature Conjecture 6.13. *If M is a closed, connected, aspherical 3-manifold and $\mathbb{Z} \oplus \mathbb{Z} \not\subset \pi_1(M)$, then $\pi_1(M)$ is negatively curved.*

See [Gr], [Bu], [Sch], [Mo1], [MO], and [Ka] for some contributions towards Conjecture 6.13.

Cannon Conjecture 6.14. *If M^3 is closed, connected, aspherical and $\pi_1(M)$ is negatively curved, then $\pi_1(M)$ is isomorphic to the fundamental group of a closed hyperbolic 3-manifold.*

It was first proved by Casson [Ca] and Poenaru [Po], that if M is closed, aspherical with a negatively curved fundamental group, then the universal covering of M is $\mathbb{R}^3$. Also by [HRS] and [M] a closed irreducible 3-manifold M with $Z \oplus Z \subset \pi_1(M)$ is covered by $\mathbb{R}^3$. Thus a special case of Conjecture 6.13 is the longstanding

Conjecture 6.15. *If M is a closed irreducible connected 3-manifold with $\pi_1(M)$ infinite, then M is covered by $\mathbb{R}^3$.*

Remarks 6.16. i) See p. 149-151 [Ki] for a list of contributions to this problem.

ii) Bestvina - Mess [BM] show that the sphere at infinity of an aspherical 3-manifold with negatively curved fundamental group is a 2-sphere.

Remark 6.17. See [Can], and [CFP] for some very remarkable and beautiful developments towards Jim Cannon's program to establish Conjecture 6.14. In particular the introduction to [CFP] gives a lucid overview of both the general program and the state of their work.

REFERENCES

[BM] M. Bestvina & J. Mess, *The Boundary of Negatively Curved Groups*, J. AMS **4**, 469–481.

[Bo] B. Bowditch, *Notes on Gromov's Hyperbolicity Criterian for Path-Metric Spaces*, Group Theory From a Geometrical Viewpoint (Trieste 1990), World Sci. (1991), 64–167.

[Bu] S. V. Buyalo, *Euclidean Planes in 3-Dimensional Manifolds of Non Positive Curvature*, Math. Notes **43** (1988), 60–66.

[C] A. Casson, U. Montreal Lecture June 1995.

[Can] J. Cannon, *The Combinatorial Riemann Mapping Theorem*, Acta. Math. **173** (1994), 155–234.

[Ca] A. Candel, *Uniformization of Surface Laminations*, Ann. Ecole Norm. Sup. **26** (1993), 489–516.

[CFP] J. Cannon & W. Floyd & W. Parry, *Sufficiently Rich Families of Planar Rings*, preprint.

[Ch] J. Christy, *Immersing Branched Surfaces in Dimension-3*, Proc. AMS **115** (1992), 853-861.

[CJ] A. Casson & D. Jungreis, *Convergence Groups and Seifert Fibered 3-Manifolds*, Invent. Math. **118**, 441-456.

[Co] A. Connes, *A survey on Foliations and Operator Algebras*, Proc. Symp. Pure Math. **38** (1982), 521–628.

[EP] D. Epstein & R. Penner, *Euclidean Decompositions of non Compact Hyperbolic Manifolds*, J. Diff. Geom **27** (1988), 67–80.

[FO] W. Floyd & U. Oertel, *Incompressible Surfaces via Branched Surfaces*, Topology **23**, 117-125.

[G1] D. Gabai, *The Simple Loop Conjecture*, J. Diff. Geom **22** (1985), 143-149.

[G2] ———, *Convergence Groups are Fuchsian Groups*, Ann. Math. **136** (1992), 447–510.

[G3] ———, *On the Geometric and Topologica Rigidity of Hyperbolic 3-Manifolds*, Bull. AMS **31** (1994), 228–232.

[G4] ———, *Problems in the Geometric Theory of Foliations and Laminations on 3-Manifolds*, AMS/IP Stud. Adv. Math, 2.2 (1997), 1–33.

[GK] D. Gabai & W. H. Kazez, *Group Negative Curvature for 3-Manifolds with Genuine Laminations*, Geom. Top., http://www.maths.warwick.ac.uk/gt/, **2** (1998), 65-77.

[GMT] D. Gabai & R. Meyerhoff & N. J. Thurston, *Homotopy Hyperbolic 3-Manifolds are Hyperbolic*, MSRI-preprint.

[GO] D. Gabai & U. Oertel, *Essential Laminations in 3-Manifolds*, Ann. Math. **(2) 130** (1989), 41-73.

[Gh] E. Ghys, *Gauss Bonnet Theorem for 2-Dimensional Foliations*, J. Funct. Anal. **77** (1988), 51–59.

[Gr] M. Gromov, *Hyperbolic Groups*, MSRI Pubs. **8**, 75-264.

[GS] S. Gerston and J. Stallings, *Casson's Idea about 3-Manifolds whose Universal Cover is* $\mathbb{R}^3$, Int. J. Alg. Comp. **1** (1991), 395–406.

[Ha] W. Haken, *Theorie der Normal Flachen*, Acta. Math. **105** (1961), 245–375.

[HRS] J. Hass & H. Rubinstein & G. P. Scott, *Compactifying Coverings of Closed 3-Manifolds*, J. Diff. Geom. **30**, 817–832.

[J] K. Johannson, *Homotopy Equivalences of 3-Manifolds with Boundary*, Springer LNM **761** (1979).

[JS] W. Jaco & P. Shalen, *Seifert Fibered Spaces in 3-Manifolds*, Mem. AMS **21** (1979).

[Ka] M. Kapovich, *Flats in 3-Manifolds*, preprint.

[Kl] B. Kleiner, in preparation.

[Lee] R. Lee, *Semicharacteristic Classes*, Topology **12** (1973), 183-199.

[Liv] G. R. Livesay, *Fixed Point Free Involutions on the 3-Sphere*, Ann. of Math. **59** (1960), 603–611.

[LS] R. Lyndon & P. Schupp, *Combinatorial Group Theory*, Erge. der Math., Springer **89** (1977).

[M] E. Moise, *Affine Structures on 3-Manifolds V*, Ann. Math. **(2) 56** (1952), 96–114.

[Me] J. Mess, *The Seifert Fibred Space Conjecture and Groups Coarse Quasiisometric to Planes*, preprint.

[Mi] J. Milnor, *Groups which Act on* S^n *without Fixed Points*, Am. J. Math. **79** (1957), 623–630.

[Mo] E. Moise, *Affine Structures in 3-Manifolds V*, Ann. Math. **56** (1952), 96–114.

[Mo1] L. Mosher, *Geometry of Cubulated 3-Manifolds*, Topology **34** (1995), 789-814.

[MO] L. Mosher & U. Oertel, *Spaces which are not Negatively Curved*, Comm. Anal. Geom. **6** (1998), 67–140.

[MO1] ———, *Two-dimensional Laminations of Positive Euler-Characteristic*, to appear, Quart. J. Math.

[My] R. Myers, *Free Involutions on Lens Spaces*, Topology **20**, 313–318.

[MY] W. Meeks & S. T. Yau, *The Equivariant Loop Theorem for Three-Dimensional Manifolds and a Review of the Existence Theorems for Minimal Surfaces*, Academic Press, The Smith Conjecture (1984), 153–163.

[Pa] P. Papasoglu, *On the Sub-Quadratic Isoperimetric Inequality*, Ohio St. U. math. Res. Inst. Publ. **3** (1995), 149–157.

[Pl] J. Plante, *Foliations with Measure Preserving Holonomy*, Ann. Math. **(2) 102** (1975), 327-361.

[Po] V. Poenaru, *Almost Convex Groups, Lipshitz Combing and π_1^∞ for various Universal Covering Spaces of Closed 3-Manifolds*, J. Diff. Geom. **35** (1992), 103–130.

[Po1] V. Poenaru, *The Collapsible Pseudo-Spine Representation Theorem*, Topology **31** (1992), 625–656.

[Ri] I. Rivin, *A Characterization of Ideal Polyhedra in Hyperbolic 3-Space*, Ann. Math. **(2) 143** (1996), 51–70.

[Ru] H. Rubinstein, 1996 CBMS Lecture Notes.

[Sc1] G. P. Scott, *A New Proof of the Annulus and Torus Theorems*, Amer. J. Math. **2** (1978), 241–277.

[Sc2] ———, *There are no fake Seifert Fibred Spaces with Infinite π_1*, Ann. Math. **117** (1985), 35–70.

[Sch] V. Schroeder, *Codimension-1 Tori in Manifolds of Non Positive Curvature*, Geom. Ded. **33** (1990), 251–263.

[SY] R. Schoen & S. T. Yau, *Existence of Incompressible Minimal Surfaces and the Topology of Three Dimensional Manifolds with Non-Negative Scalar Curvature*, Ann. Math. **110** (1979), 127–142.

[T] C. B. Thomas, *Elliptic Structures on 3-Manifolds*, LMS lecture notes **104** (1986).

[Th] W. P. Thurston, *Three-Dimensional Manifolds, Kleinian Groups and Hyperbolic Geometry*, Proc. Symp. Pure Math. **39** (1979), 87–111.

[We] J. Weeks, *The Shape of Space*, Pure and Applied Math. 96, Marcel Dekker (1985).

California Institute of Technology, Pasadena, CA 91125-0001 USA
E-mail address: gabai@its.caltech.edu

Embedded surfaces and gauge theory in three and four dimensions

P. B. Kronheimer[1]

Department of Mathematics
Harvard University,
Cambridge MA 02138
kronheim@math.harvard.edu

Introduction

Given a 2-dimensional homology class σ in a smooth 4-manifold X, what is the least possible genus for a smoothly embedded, oriented surface Σ in X whose fundamental class is σ? Gauge theory has been a successful tool in answering a collection of basic questions of this sort. In [22, 25, 20, 24], information extracted from Donaldson's polynomial invariants of 4-manifolds [6] gave some strong lower bounds, which were in many cases sharp. To give just one example, if X is a smooth quintic surface in $\mathbb{CP}^3$ and Σ is a smooth

[1]Partially supported by NSF grant number DMS-9531964

algebraic curve obtained as the intersection of X with any other complex surface $H \subset \mathbb{CP}^3$, then Σ is known to achieve the smallest possible genus in its homology class.

The introduction of the Seiberg-Witten monopole equations and the replacement of Donaldson's polynomial invariants by the apparently equivalent monopole invariants [42] lead to much simpler proofs of essentially the same results, though often in rather greater generality. For example, the statement about the quintic surface is now known to hold for complex algebraic surfaces in general, and the theorems can be extended quite cleanly to the case of symplectic manifolds [36].

At the same time, the simpler understanding that the Seiberg-Witten techniques afforded made clear that the case of complex and symplectic 4-manifolds was rather special. Although gauge theory gives lower bounds on the genus of embedded surfaces in general 4-manifolds, these lower bounds should no longer be expected to be sharp, at least in the form in which they are usually phrased. The situation is clarified by the work of Meng and Taubes [30], in which a 3-dimensional version of the monopole invariants is studied. For 3-manifolds with non-zero first Betti number, the monopole invariants are closely related to Milnor torsion, and are a cousin of the familiar Alexander invariant of a knot. These invariants do contain information about embedded surfaces, but they do not lie very deep: the degree of the Alexander polynomial gives a lower bound for the genus of a knot, but not always a very good one. There seems no reason to expect the outcome in four dimensions to be any better.

With a little more work, however, there is more information to be extracted from the monopole equations, at least on 3-manifolds, where one can use the equations to define a 'Floer homology'. There is much here that is not yet worked out in detail, but it is already clear that one obtains sharp lower bounds for the genus of embedded surfaces in general 3-manifolds.

One of our aims in this article is to give an account of this 3-dimensional story, which is closely tied up with a well-established theory of embedded surfaces, foliations and contact structures, due to Bennequin, Eliashberg, Gabai and Thurston, among others. We include a leisurely summary of those parts of the foliation story that are most relevant to the gauge theory side. We give a basic account of gauge theory on 3-manifolds, defining the Seiberg-Witten monopole equations, touching on Floer homology and explaining the connection with foliations and contact structures.

By and large, the flow of ideas is one-way: from the geometric and topological results concerning foliations of 3-manifolds, we learn something about the gauge theory invariants. But it seems likely that gauge theory has something to offer in return. We mention some potential applications in section 6, among them the question of whether one can make a homotopy 3-sphere by surgery on a knot. For such applications, one needs not only the Seiberg-Witten techniques, but the older Yang-Mills invariants of Donaldson and Floer. We shall also look at 4-manifolds, to explore the limitations of the existing tools.

Acknowledgment. Amongst the material presented here, the theorems in which the author had a hand are the result of joint work with Tom Mrowka, and have all appeared (or are appearing) elsewhere. In particular, the results which collect around Theorem 3.6 are contained in the joint paper [21] or are easily deduced from results presented there. The fact that one can use the Seiberg-Witten monopole invariants to define invariants of 3-manifolds (either the integer invariants that we call $SW(Y)$ in this paper or the more problematic Floer homology) has been pursued by several people, as has the fact that one can bound the genus of embedded surfaces in terms of these invariants. See for example [2, 4, 27, 29, 30, 40].

1 Surfaces in 3-manifolds

The Thurston norm

We begin in three dimensions. For a 3-manifold Y, the fact that any class σ in $H_2(Y;\mathbb{Z})$ is represented by a smoothly embedded surface can be seen as follows. Take a smooth map $f_\sigma : Y \to S^1$ such that the pull-back of the generator of $H^1(S^1)$ is Poincaré dual to σ. Then for any regular value θ of f_σ, the set $f_\sigma^{-1}(\theta)$ is a suitable surface. Any surface representing σ arises in this way for some map f_σ.

It is not always possible to represent σ as the fundamental class of a *connected* surface. Even when a connected representative exists, it is profitable to consider disconnected representatives also and to try and minimize not the genus but the quantity

$$\chi_-(\Sigma) = \sum_{g(\Sigma_i)>0} (2g(\Sigma_i) - 2),$$

over all oriented embedded surfaces Σ whose fundamental class is σ. One might call this the complexity of Σ. The minimum complexity in this sense, as a function of the homology class represented, was considered by Thurston [39], who made the following observation. Define

$$\chi_{\min}(\sigma) = \min\{\, \chi_-(\Sigma) \mid [\Sigma] = \sigma \,\}.$$

Proposition 1.1 (Thurston). *On any closed, oriented 3-manifold Y, the function $\chi_{\min}$ on $H_2(Y;\mathbb{Z})$ satisfies the triangle inequality and is linear on rays, in that $\chi_{\min}(n\sigma) = n\,\chi_{\min}(\sigma)$ for $n \geq 0$. It is the restriction of a semi-norm on $H_2(Y;\mathbb{R})$.*

The triangle inequality and linearity have straightforward geometrical explanations. Suppose that Σ is an oriented embedded surface realizing the minimal complexity in its homology class $\sigma = [\Sigma]$. Let $\tilde{\Sigma}$ be the surface obtained by taking n disjoint, parallel copies of Σ inside a product neighborhood $\Sigma \times [-1,1] \subset Y$. This surface represents the class $n\sigma$ and has complexity

$$\chi_-(\tilde{\Sigma}) = n\chi_-(\Sigma).$$

The linearity assertion is that $\tilde{\Sigma}$ also has minimal complexity in its class. The reason this is true is that *any* surface $\hat{\Sigma}$ representing $n\sigma$ must be a disjoint union of n surfaces, each representing σ, so the complexity of $\hat{\Sigma}$ cannot be less than $n\,\chi_{\min}(\sigma)$. Indeed, we can realize $\hat{\Sigma}$ as a regular inverse image $\hat{f}^{-1}(\hat{\theta})$ for a suitable $\hat{f}\colon Y \to S^1$ as above, and the divisibility of $[\hat{\Sigma}]$ implies that this $\hat{f}$ lifts through the n-fold covering map $\mu_n\colon S^1 \to S^1$:

$$\hat{f} = \mu_n \circ f.$$

Thus $\hat{\Sigma}$ is the disjoint union of the surfaces $f^{-1}(\theta)$, as θ runs through the n preimages of $\hat{\theta}$.

To see that that $\chi_{\min}$ satisfies the triangle inequality, let classes σ and τ be represented by surfaces Σ and T of minimal complexity. If these are moved into general position, they will intersect along a collection of circles. Each of the circles has a neighborhood $D^2 \times S^1$ which meets $\Sigma \cup T$ in a standard $K \times S^1$, where K is a pair of intersecting arcs in D^2. Replacing $K \times S^1$ with $J \times S^1$, where J is a pair of disjoint arcs in D^2 with the same four endpoints (connected differently), we obtain a new surface in Y representing $\sigma + \tau$. Its complexity is the sum of the complexities of Σ and T, thus

$$\chi_{\min}(\sigma + \tau) \leq \chi_{\min}(\sigma) + \chi_{\min}(\tau). \tag{1}$$

We shall write $|\sigma|$ for the norm (or semi-norm) $\chi_{\min}(\sigma)$ in the 3-dimen/-sional case. The dual norm on $H^2(Y;\mathbb{R})$ can be characterized by

$$|\alpha|_* = \sup_{\Sigma} \frac{\langle \alpha, [\Sigma] \rangle}{(2g(\Sigma)-2)}, \qquad \alpha \in H^2(Y;\mathbb{R}), \tag{2}$$

where the supremum need only be taken only over connected embedded surfaces of genus 1 or more on which α has non-zero pairing, with the understanding that the norm is infinite if α has non-zero pairing with an embedded torus. The *Thurston polytope* $B(Y)$ is the unit ball for this dual Thurston norm. It is a convex polytope lying in the subspace of $H^2(Y;\mathbb{R})$ on which the norm is finite. Its vertices are lattice points (that is, they are the reduction of integer classes).

Foliations

If working with a closed 3-manifold is difficult, one can get a feel for the problem of determining the Thurston norm, or equivalently the polytope $B(Y)$, by looking at a version of the question for a manifold with boundary, such as a knot complement. Let $K \subset S^3$ be a knot and Y the 3-manifold obtained by removing an open tubular neighborhood of K. The boundary of Y is a torus T, on which there is a simple closed curve λ having the property that it is homologous to zero in Y: this is the longitude of the knot. Being null-homologous, λ is the boundary of an oriented surface $\Sigma \subset Y$. Any such surface is a *spanning surface* for K. The *genus* of the knot K is the least genus of any spanning surface.

Finding a spanning surface for K is never hard. Harder is to provide a spanning surface Σ with a certificate assuring us that it is of least possible genus. It is another observation of Thurston's [39] that a suitable *foliation* of the 3-manifold may supply such a certificate.

To state this result, we return to the closed case and suppose that the 3-manifold Y has a smooth foliation $\mathcal{F}$ by oriented 2-dimensional leaves L. Such a foliation determines a field of oriented 2-planes $T\mathcal{F} \subset TY$, the tangent directions to the leaves. Let $e(\mathcal{F})$ denote the Euler class of $T\mathcal{F}$ in $H^2(Y;\mathbb{Z})$.

Theorem 1.2 (Thurston). *Suppose the foliation $\mathcal{F}$ of Y has no Reeb components, and suppose that Y is not $S^1 \times S^2$. Then the Euler class $e(\mathcal{F})$ belongs to the polytope $B(Y)$. In other words, if Σ is any embedded surface, its complexity satisfies the lower bound*

$$\chi_-(\Sigma) \geq \langle e(\mathcal{F}), [\Sigma] \rangle.$$

A *Reeb component* is a foliation of a solid torus in which the boundary torus is a leaf and all interior leaves are planes.

Corollary 1.3. *Let Σ be an oriented embedded surface in Y which is a union of compact leaves of an oriented foliation $\mathcal{F}$ without Reeb components. Then Σ has minimal complexity in its homology class.*

The Corollary follows from the Theorem because if Σ is a union of correctly-oriented compact leaves, then

$$\begin{aligned}\langle e(\mathcal{F}), [\Sigma]\rangle &= \langle e(T\Sigma), [\Sigma]\rangle \\ &= -\chi_-(\Sigma),\end{aligned}$$

as long as no component of Σ is a sphere. Changing the orientation of the leaves then gives a foliation $\bar{\mathcal{F}}$ for which the inequality of the Theorem is an equality. Note that the Theorem contains the statement that no compact leaf of $\mathcal{F}$ can be a sphere. The proof of Theorem 1.2 itself is given in [39]; see also Theorem 7.1 of [13], where a detailed proof is given of a more general version of the basic lemma which underlies the result. The original version of the lemma was first proved by Thurston [38] and Roussarie [34]:

Lemma 1.4. *Let $\mathcal{F}$ and Y be as in Theorem 1.2, and let Σ be an incompressible surface in Y. Then there is a surface Σ' isotopic to Σ which is transverse to $\mathcal{F}$ except at a finite number of circle and saddle tangencies.*

In this statement, *incompressible* means as usual that no loop in Σ' bounds a disk in $Y \setminus \Sigma'$, unless it already bounds a disk in Σ'. This condition is certainly necessary if Σ is to be of least complexity, since by cutting along a compressing disk one reduces the complexity of a surface. A *circle* tangency is what one sees along the level rim of a volcano, in the foliation of 3-space by level planes. A *saddle* tangency needs no explanation. In each case, the tangency may be given one of two signs, according as the orientation agrees with that of the leaves or not. One can calculate $\chi_-(\Sigma')$ and the evaluation of $e(\mathcal{F})$ on Σ' in terms of the number of tangencies of each type [39], and the Theorem is an elementary consequence.

Corollary 1.3 gives a criterion with which to ascertain that a given surface is of minimal complexity. But it is of little use in itself unless we also have a handle on constructing foliations without Reeb components. In [14], Gabai gave a practical algorithm for finding the genus of a large class of knots, based on exhibiting a spanning surface as a compact leaf of a foliation of the knot complement. Further, in [13], Gabai proved that every surface of minimal complexity can be certified as such by a suitable foliation. We state the result only in the closed case:

Theorem 1.5 (Gabai). *Let Y be a closed, irreducible, oriented 3-manifold. Let Σ be an embedded surface representing a non-trivial homology class σ. Suppose that $\chi_-(\Sigma)$ is least possible amongst surfaces representing this class. Then there exists a taut, oriented foliation $\mathcal{F}$ of Y of class C^0, having Σ as an oriented union of compact leaves. The foliation can be taken to be smooth except along the components of Σ which are tori.*

Remarks. The *irreducibility* of Y is the condition that every embedded 2-sphere bounds a ball. It excludes $S^1 \times S^2$ as well as non-trivial connected sums. One definition of *taut* is that every leaf L is met by a closed curve γ in Y which is everywhere transverse to the leaves. This is a stronger condition than the absence of Reeb components. In Lemma 1.4 for example, the taut condition allows one to dispose of circle tangencies, leaving only saddles. Generally, in a taut foliation, no correctly-oriented union of compact leaves can bound: if $W \subset Y$ had oriented boundary which was a union of leaves, then no transverse curve which left W to enter $Y \setminus W$ could return. The possibility that the foliation $\mathcal{F}$ in the Theorem may not be smooth when Σ has tori amongst its components presents a difficulty at some points: we will sometimes legislate against it in the statement of our results. In many cases one can construct a foliation that is smooth, despite the presence of tori.

Gabai's result combines with Thurston's to characterize the polytope $B(Y)$, and hence the Thurston norm, in terms of smooth foliations:

Corollary 1.6. *Let Y be a closed, irreducible oriented 3-manifold in which embedded tori do not form a basis for the homology. Then the unit ball $B(Y) \subset H^2(Y;\mathbb{R})$ for the dual Thurston norm is the convex hull of the classes $e(\mathcal{F})$, as $\mathcal{F}$ runs through smooth, taut foliations. In other words, the Thurston norm is given by*

$$|\sigma| = \max_{\mathcal{F}\text{ taut}} \langle e(\mathcal{F}), \sigma \rangle.$$

The extra hypothesis in this Corollary ensures that at least one of the foliations which result for Theorem 1.5 is smooth. When this is the case, we can throw out all the non-smooth foliations corresponding which might arise when Σ contains tori, without changing the convex hull.

In [13], the inequality of Theorem 1.2 is extended to the case that Σ is not an embedded surface, but simply a surface mapped into Y by an arbitrary map, whose image may therefore have singularities. Combining this strengthened inequality with the existence theorem for foliations, one obtains:

Corollary 1.7 (Gabai, [13]). *If a homology class σ in an irreducible 3-manifold Y is represented as $f_*[\Sigma]$ for some map $f : \Sigma \to Y$, then the homology class is also the fundamental class of an embedded surface of the same complexity.*

Non-trivial examples of plane polygons arising as the Thurston polytopes of various 3-manifolds with $b_1 = 2$ are given in Thurston's original paper. Gabai's theorem and its refinements have many applications in 3-manifold topology [13, 15, 16, 17].

2 Gauge theory on 3-manifolds

The Seiberg-Witten *monopole equations* were introduced as a tool in 4-dimensional topology in [42]. One of their first applications was to questions of embedded surfaces [23]. Here we shall explore their relationship to the Thurston norm in dimension 3.

The monopole equations

A Spin^c structure $\mathfrak{c}$ on an oriented Riemannian 3-manifold Y consists of a rank-2 complex bundle $W = W_{\mathfrak{c}}$ with a hermitian metric (the spinor bundle) and an action ρ of 1-forms on spinors, called Clifford multiplication:

$$\rho : T^*Y \to \mathrm{End}(W).$$

If e^1, e^2, e^3 are an oriented orthonormal frame for the cotangent space at a point, then there should be an orthonormal basis for the fiber of W at that point so that the $\rho(e^i)$ are represented by the Pauli matrices. The standard convention on orientations has $\rho(e^1)\rho(e^2)\rho(e^3) = -1$. Clifford multiplication is extended to forms of higher degree, by $\rho(e^1 \wedge e^2) = \rho(e^1)\rho(e^2)$, for example.

If $\mathfrak{c}$ is a Spin^c structure and $e \in H^2(Y;\mathbb{Z})$ a 2-dimensional cohomology class, there is a new Spin^c structure $\mathfrak{c}+e$. Its spin bundle is $W_\mathfrak{c} \otimes L_e$, where L_e is the unique line bundle with first Chern class e. Conversely, if $\mathfrak{c}$ and $\mathfrak{c}'$ are two Spin^c structures, there is a unique difference element $\mathfrak{c}' - \mathfrak{c} \in H^2(Y;\mathbb{Z})$. We write $c_1(\mathfrak{c})$ for the first Chern class of $W_\mathfrak{c}$, and we note that, since

$$c_1(\mathfrak{c} + e) = c_1(\mathfrak{c}) + 2e,$$

the class $c_1(\mathfrak{c})$ determines $\mathfrak{c}$ in the absence of 2-torsion in the cohomology. Because every 3-manifold is parallelizable, there is always at least one Spin^c structure with a topologically trivial spin bundle. It follows that $c_1(\mathfrak{c})$ is always divisible by 2 in $H^2(Y;\mathbb{Z})$. Although our definition involves a Riemannian metric, the set of isomorphism classes of Spin^c structures can be viewed as metric-independent.

Given a Spin^c structure, a unitary connection A on W is a *spin* connection if ρ is parallel; that is, the resulting connection on $\mathrm{End}(W)$ should coincide with the Levi-Civita connection on the image of ρ. This leaves only the central part of A undetermined, so if A and A' are two spin connections, then their difference is scalar:

$$A' = A + a\mathbf{1}_W,$$

for an imaginary-valued 1-form a. A spin connection A is therefore determined by its 'trace', the induced connection $\hat{A}$ in the line bundle $\Lambda^2 W$.

The monopole equations are the following equations for a spin connection A and a section Φ of W on a Riemannian 3-manifold Y equipped with Spin^c structure:

$$\begin{aligned} \rho(F_{\hat{A}}) - \{\Phi \otimes \Phi^*\} &= 0 \\ D_A \Phi &= 0. \end{aligned} \tag{3}$$

In the first equation, $F_{\hat{A}}$ is the curvature of the connection in the line bundle (an imaginary-valued 2-form), and the curly brackets denote the trace-free part of the endomorphism. In the second equation, D_A is the Dirac operator for the spin connection A, which is defined as the composite

$$\Gamma(W) \xrightarrow{\nabla_A} \Gamma(T^*Y \otimes W) \xrightarrow{\rho} \Gamma(W).$$

An application of the Weitzenbock formula

When a Riemannian 3-manifold Y is given and a Spin^c structure $\mathfrak{c}$ is specified, we can ask first whether the monopole equations (3) have any solutions at all. There is a constraint which must be satisfied if a solution is to exist, which comes from the Weitzenbock formula for the Dirac operator D_A:

$$D_A^* D_A \Phi = \nabla_A^* \nabla_A \Phi + \frac{s}{4}\Phi + \frac{1}{2}\rho(F_{\hat{A}})\Phi.$$

Here s is the scalar curvature of the Riemannian metric. If (A, Φ) is a solution of the equations, then the left and right-hand sides are zero. For a solution, then, we calculate

$$\begin{aligned}\Delta|\Phi|^2 &= 2\langle \nabla_A^*\nabla_A\Phi, \Phi\rangle - 2\langle \nabla_A\Phi, \nabla_A\Phi\rangle \\ &\leq -\frac{s}{2}|\Phi|^2 - \langle \rho(F_{\hat{A}})\Phi, \Phi\rangle.\end{aligned}$$

(Our inner products are real.) The first of the equations (3) can be used to rewrite the last term as $|\Phi|^4/2$ (an elementary calculation with vectors in $\mathbb{C}^2$). So we have

$$2\Delta|\Phi|^2 \leq -s|\Phi|^2 - |\Phi|^4, \tag{4}$$

which we may integrate to obtain

$$\int |\Phi|^4 d\mathrm{vol} \leq \int (-s)|\Phi|^2 d\mathrm{vol},$$

and hence

$$\int |\Phi|^4 d\mathrm{vol} \leq \int s^2 d\mathrm{vol} \tag{5}$$

by Cauchy-Schwartz. The first equation of (3) is used again to rewrite this as

$$\int |F_{\hat{A}}|^2 d\mathrm{vol} \leq \int \frac{|s|^2}{4} d\mathrm{vol},$$

which is an inequality between the L^2 norms:

$$\|F_{\hat{A}}\| \leq \|s\|/2. \tag{6}$$

The two form $(i/2\pi)\|F_{\hat{A}}\|$ represents the first Chern class of $\Lambda^2 W_{\mathfrak{c}}$, which is $c_1(\mathfrak{c})$. The de Rham representative of this class with smallest L^2 norm is the *harmonic* representative. Let us define an L^2 norm on $H^2(Y;\mathbb{R})$ by defining $\|\alpha\|$ to be the L^2 norm of the harmonic representative. Then we have, from (6)

$$\|c_1(\mathfrak{c})\| \leq \|s\|/4\pi.$$

(Both sides depend, of course, on the Riemannian metric.) We state the conclusion as a lemma:

Lemma 2.1. *A necessary condition for the existence of a solution to the monopole equations on Y for a given Riemannian metric and a given* Spin^c *structure $\mathfrak{c}$ is that the harmonic L^2 norm of the class $c_1(\mathfrak{c})$ satisfy*

$$\|c_1(\mathfrak{c})\| \leq \|s\|/4\pi. \tag{7}$$

□

Note that we could have taken a little more care over the argument, to arrive at the inequality

$$\|c_1(\mathfrak{c})\| \leq \|s_-\|/4\pi, \tag{8}$$

where s_- is defined pointwise as $\max(0, -s)$.

The calculation above, and the inequality (6), were first described by Witten in [42] (though in a 4-dimensional version). As pointed out there, it follows that, for a fixed Riemannian metric, there can be solutions for only finitely many different Spin^c structures on Y. Indeed, there are only finitely many integer classes with norm less than any given constant, and the cohomology class $c_1(\mathfrak{c})$ (as an integer class) determines the Spin^c structure $\mathfrak{c}$ to within the addition of a 2-torsion element.

One can also obtain a pointwise bound on $|\Phi|$ from the inequality (4). At a point where $|\Phi|$ is maximum, the Laplacian is positive, and as long as $|\Phi|$ is not zero at this point one may divide to obtain the estimate

$$|\Phi|^2 \leq -s$$

at the maximum.

Scalar curvature and the Thurston norm

The relationship between the monopole equations and the genus of embedded surfaces arises in its simplest form from the following simple lemma. Like the inequality (7), it compares the norm of a class in H^2 to the norm of the scalar curvature.

Lemma 2.2. *Let $\alpha \in H^2(Y;\mathbb{R})$ be a two-dimensional cohomology class. Then the dual Thurston norm $|\alpha|_*$ satisfies the inequality*

$$|\alpha|_* \leq 4\pi \sup_h \frac{\|\alpha\|_h}{\|s_h\|_h}, \tag{9}$$

in which the supremum is taken over all Riemannian metrics h, the norms on the right-hand side are the L^2 norms, and s_h denotes the scalar curvature of h.

Proof. Let Σ be an oriented embedded surface of genus 1 or more on which α is non-zero. Let h_1 be a Riemannian metric on Y such that some neighborhood of Σ is isometric to a product $\Sigma \times [0,1]$, with Σ having constant scalar curvature $-4\pi(2g-2)$ and unit area. Let h_r be a metric which contains a product region $\Sigma \times [0,r]$ and is isometric to h_1 outside that region. Then

$$\|s_h\|_h = 4\pi r^{1/2}(2g-2) + O(1)$$

as $r \to \infty$, while any 2-form ω representing a class α must satisfy

$$\|\omega\|_h \geq r^{1/2}\langle \alpha, [\Sigma]\rangle.$$

Thus if g is at least two we have

$$\langle \alpha, [\Sigma]\rangle/(2g-2) \leq \sup_h 4\pi\|\alpha\|_h/\|s_h\|_h,$$

and in the case that Σ is a torus we see that the right-hand side is infinite. This is the desired result, in view of the characterization of the dual Thurston norm at (2). □

Combining this lemma with the previous one, we obtain:

Corollary 2.3. *If $\mathfrak{c}$ is a* Spin^c *structure on Y and the dual Thurston norm of $c_1(\mathfrak{c})$ is bigger than 1, then there exists a Riemannian metric on Y for which the monopole equations have no solution.* □

We shall see later that the inequality in Lemma 2.2 is actually an equality for many 3-manifolds Y (Proposition 3.8).

3 The monopole invariants

Obtaining invariants from the monopole equations

There is a well-understood procedure by which we can extract some metric-independent data from the set of solutions to the monopole equations, to obtain an invariant

$$SW(Y, \mathfrak{c}) \in \mathbb{Z}$$

depending only on Y (a closed, oriented 3-manifold) and the Spin^c structure $\mathfrak{c}$. A careful account of the most general case is given in [4, 27], and there is a model for this construction in [35], where Casson's invariant is given a gauge-theory interpretation. In this subsection, we review this construction, restricting ourselves sometimes to the simpler cases.

The monopole equations on a Riemannian 3-manifold Y are the variational equations for a functional

$$\mathrm{CSD}(A, \Phi) = \frac{1}{2}\int_Y (\hat{A} - \hat{A}_0) \wedge (F_{\hat{A}} + F_{\hat{A}_0}) - \frac{1}{2}\int_Y \langle \Phi, D_A\Phi\rangle \, d\mathrm{vol},$$

the *Chern-Simons-Dirac* functional of A and Φ. (A reference connection A_0 is chosen to define the first term, but a change of reference connection only changes the functional by addition of a constant.) The equations are invariant under the natural symmetry group of the Spin^c bundle $W_{\mathfrak{c}}$, which is the group G of maps $u : Y \to S^1$ acting on $W_{\mathfrak{c}}$ by scalar endomorphisms. This group acts on A and Φ by

$$\begin{aligned} A &\mapsto A - (u^{-1}du)\mathbf{1} \\ \Phi &\mapsto u\Phi. \end{aligned}$$

The functional CSD is invariant only under the identity component of G. The component group of G is the group of homotopy classes of maps $Y \to S^1$, which is isomorphic to $H^1(Y; \mathbb{Z})$: the change in the functional under a general element of G is given by

$$\mathrm{CSD}(A - (u^{-1}du)\mathbf{1}, u\Phi) - \mathrm{CSD}(A, \Phi) = -4\pi^2\big([u] \smile c_1(\mathfrak{c})\big)[Y].$$

So CSD does not descend to a real-valued function on the quotient space $\mathcal{C} = \{(A, \Phi)\}/G$, but does descend to a circle-valued function whose periods are multiples of $4\pi^2$.

The orbit space $\mathcal{C}$ is an infinite-dimensional manifold except at points where the action has non-trivial stablilizer. These are the configurations with $\Phi = 0$ (the *reducible* configurations), whose stabilizer is the circle group of constant maps u. If $c_1(\mathfrak{c})$ is zero or a torsion class, then there are spin connections for which $\hat{A}$ is flat, and these are reducible solutions to the equations.

To define $SW(Y, \mathfrak{c})$, one first chooses a perturbation of the functional CSD so as to make the critical points non-degenerate. To do this, one may add an extra term, setting

$$\mathrm{CSD}_\mu(A, \Phi) = \mathrm{CSD}(A, \Phi) + \int_Y (\hat{A} - \hat{A}_0) \wedge (i\mu),$$

where μ is an exact 2-form. The variational equations are now

$$\begin{aligned} \rho(F_{\hat{A}} + i\mu) - \{\Phi \otimes \Phi^*\} &= 0 \\ D_A \Phi &= 0, \end{aligned} \tag{10}$$

and for a suitable choice of μ (such μ are dense), the Hessian will be non-degenerate at all irreducible solutions in $\mathcal{C}$. A proof is given in [12]. (The condition that μ be exact means, among other things, that the function CSD_μ still descends to a circle-valued function on $\mathcal{C}$ and that its periods are the same as those of CSD.)

Critical points of CSD_μ on $\mathcal{C}$ have infinite index: the Hessian is a self-adjoint operator which, like the Dirac operator, has a discrete spectrum which is infinite in both the positive and negative directions, so one cannot define an index $i(a)$ at a critical point a as the dimension of the sum of the negative eigenspaces. However, there is a way to define a relative index between any two critical points. If a and b are two non-degenerate critical points of CSD_μ and $\gamma(t)$ is a path in $\mathcal{C}$ which joins them, then the Hessians $H_{\gamma t}$ a family of operators for which one can define a spectral flow – the number of eigenvalues which move from negative to positive in the family. One can defines the relative index $i(a, b)$ as the spectral flow. There is a further point to attend to, which is the possible dependence of the spectral flow on the choice of path. The fundamental group of $\mathcal{C}$ is again isomorphic to $H^1(Y; \mathbb{Z})$, and for a *closed* path γ, the spectral flow is given by

$$\mathrm{SF}(\gamma) = \big([u] \smile c_1(\mathfrak{c})\big)[Y],$$

where u is the corresponding element of H^1. As noted above, $c_1(\mathfrak{c})$ is always divisible by 2, so at least the parity of $i(a, b)$ is well-defined. After settling on a convention as to which is to be which, we can divide the critical points into even and odd, using the relative index.

The invariant $SW(Y, \mathfrak{c})$ can now be defined as, roughly speaking, the euler number of the vector field CSD_μ on $\mathcal{C}$. We restrict our attention to the case that $c_1(\mathfrak{c})$ is not a torsion class, so that there are no reducible solutions to the equations, and after choosing μ so that the solutions are non-degenerate we set

$$SW(\mathfrak{c}) = \#(\text{even}) - \#(\text{odd}). \tag{11}$$

The main technical point here is that the set of critical points in $\mathcal{C}$ is compact, and hence finite under our non-degeneracy assumption. The proof of the compactness property of solutions to the monopole equations is a straightforward application of standard techniques, starting from the C^0 bound obtained by applying the maximum principle to (4). The quantity (11) is independent of the choice of Riemannian metric on Y and the choice of μ. Our failure to fix a convention about which is even and which is odd leaves an overall sign ambiguity in the invariant.

If $c_1(\mathfrak{c})$ is torsion, a similar definition leads to an invariant in the case that $b_1(Y)$ is non-zero: one can perturb by a small, non-exact, closed 2-form μ to remove the reducible solutions. The case of $b_1(Y) = 0$ is a little more subtle, but an invariant can be defined: see [4, 27].

Basic classes

We call a Spin^c structure $\mathfrak{c}$ on a closed 3-manifold Y *basic* if the monopole invariant $SW(Y, \mathfrak{c})$ is non-zero. In this case, we also refer to the first Chern class $c_1(\mathfrak{c})$ as a *basic class* if Y. For our present purposes, these definitions are only interesting when $b_1(Y)$ is non-zero and $c_1(\mathfrak{c})$ is not torsion. The definition of the invariant SW means that, if $\mathfrak{c}$ is basic, then the corresponding perturbed monopole equations (10) have at least one solution, for every Riemannian metric and a dense set of exact 2-forms μ. The compactness properties of the equations imply that the non-emptiness of the solution space is an open condition, so it is also true that the original equations (3) have solutions for every Riemannian metric h. This observation, together with Corollary 2.3, yields a relationship between basic classes and the genus of embedded surfaces:

Proposition 3.1. *If Y is a closed, oriented 3-manifold with $b_1 \neq 0$ and α is a basic class on Y, then the dual Thurston norm of α satisfies $|\alpha|_* \leq 1$. In other words, for any oriented embedded surface Σ in Y representing a class σ, we have*

$$\chi_-(\Sigma) \geq \langle \alpha, \sigma \rangle.$$

□

Monopole invariants and the Alexander invariant

Whether the above proposition is useful depends on what else one knows about the invariant SW and the basic classes. Meng and Taubes [30] showed that when $b_1(Y)$ is non-zero, $SW(Y, \mathfrak{c})$ is completely determined by a classical invariant, the Milnor torsion. While it is interesting that torsion is calculated by the gauge theory route, this result does make Proposition 3.1 look less interesting.

The situation is easiest to describe when the 3-manifold Y is the result of zero surgery on a knot K in S^3: that is, Y is obtained by removing a solid torus neighborhood of K and replacing it while interchanging the longitude and meridian curves on its boundary. In this case, the information contained in the Milnor torsion is the *Alexander polynomial* of K. We shall use the symmetrized Alexander polynomial, whose shape is

$$A_K(t) = a_{-r}t^{-r} + \cdots + a_{-1}t^{-1} + a_0 + a_1 t + \cdots + a_r t^r$$

with $a_i = a_{-i}$. We refer to r as the degree of the polynomial. The betti number of $Y = Y(K)$ is 1, and the second cohomology is $\mathbb{Z}$. There is therefore exactly on Spin^c structure $\mathfrak{c}_k$ with $c_1(\mathfrak{c}_k) = 2k$. Its monopole invariant can be expressed in terms of the Alexander polynomial:

Theorem 3.2 (Meng-Taubes [30]). *On the manifold $Y(K)$ obtained by zero-surgery on K, the monopole invariants are given by*

$$SW(Y, \mathfrak{c}_k) = \sum_{j>0} j a_{j+|k|}. \tag{12}$$

In particular, $SW(z, \mathfrak{c}_k) = 0$ for $k > r - 1$. □

(The symmetry $SW(Y, \mathfrak{c}_k) = SW(Y, \mathfrak{c}_{-k})$ is a general property of the monopole invariants, and follows from a symmetry of the equations.)

With this interpretation of the monopole invariants, we can reformulate Proposition 3.1 as

Statement 3.3. *If Σ is a connected, oriented surface in $Y(K)$ representing the generator of H_2, then*

$$g(\Sigma) \geq r,$$

where r is the degree of the Alexander polynomial.

Indeed, the theorem tells us that the extreme basic classes in $H^2 \cong \mathbb{Z}$ arise from the Spin^c structures $\mathfrak{c}_{\pm(r-1)}$, with $c_1(\mathfrak{c})[\Sigma] = \pm(2r-2)$. (The restriction to connected surfaces has no effect on our statement when $b_1 = 1$.)

The statement above, however, is an elementary consequence of the definition of the Alexander polynomial, in the formulation which expresses $A_K(t)$ in terms of the homology of the infinite cyclic cover of $Y(K)$. Furthermore, the inequality between the genus of Σ and the degree of the Alexander polynomial is not a sharp one in general. Here is a relevant result:

Theorem 3.4 (Gabai [15]). *If Σ is an embedded surface in $Y(K)$ representing the generator of H_2, then $g(\Sigma)$ is no smaller than the genus of the knot.*

This theorem is not self-evident. The proof of the result comes from refinement of the existence theorem for taut foliations quoted earlier: a minimal-genus spanning surface for K can be filled out to a taut foliation $\mathcal{F}$ of the complement of a neighborhood $N(K)$ in S^3, and this can be done so that the boundaries of the leaves are a family of longitudinal circles. The solid torus $N(K)$ has a trivial foliation by meridianal disks, and in the surgered manifold $Y(K)$, this foliation joins with $\mathcal{F}$ to give a taut foliation of the closed manifold.

The genus of a knot K and the degree of A_K are in general different. The untwisted Whitehead double of any non-trivial knot has genus 1 and Alexander polynomial 1. Figure 1 shows a Whitehead double of a trefoil knot; a spanning surface is formed from a ribbon which follows the course of the trefoil, together with a small band with one full twist at the clasp. Under the connected sum operation, the genus of a knot is additive, while the Alexander polynomial is multiplicative, so one can easily obtain knots of large genus whose polynomial has small degree.

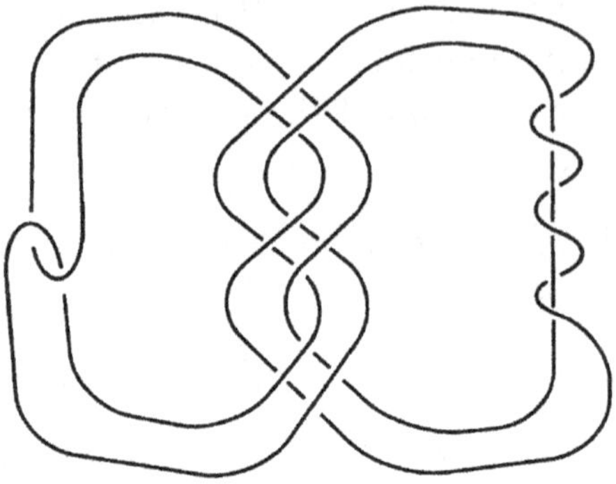

Figure 1: A doubled trefoil.

Monopole classes

The result of the previous subsection is disappointing. The lower bound for the genus of embedded surfaces in terms of basic classes is no better than the lower bound for the genus of a knot which the Alexander polynomial provides: it does not capture the topology. However, the reason for the disappointing result is that too much of the content of the gauge theory has been disposed of in passing from the solution set of the monopole equations (3) to the integer invariant $SW(Y, \mathfrak{c})$ which counts these solutions.

Rather than count the solutions, let us take a step backwards and simply make the following definition:

Definition 3.5. A class $\alpha \in H^2(Y;\mathbb{Z})$ is a *monopole class* if it arises as $c_1(\mathfrak{c})$ for some Spin^c structure $\mathfrak{c}$ on Y for which the equations (3) admit a solution *for every choice of Riemannian metric h on Y.*

This definition ensures that the basic classes are monopole classes. Also, Proposition 3.1 applies to monopole classes, just as it applies to basic classes; so a monopole class α has dual Thurston norm at most 1. (This is simply another step backwards, to Corollary 2.3, from which Proposition 3.1 was deduced.)

However, the monopole classes are in general a *larger* set than the basic classes. The following theorem is our central result. We shall give an outline of the proof in section 5.

Theorem 3.6. *If Y is a closed, irreducible, oriented 3-manifold with a smooth, taut foliation $\mathcal{F}$ by oriented leaves, then $e(\mathcal{F})$ is a monopole class.*

Combining this statement with Corollary 1.6, one obtains

Corollary 3.7. *If Y is a closed, irreducible oriented 3-manifold, then the unit ball $B(Y) \subset H^2(Y;\mathbb{R})$ for the dual Thurston norm is the convex hull of the monopole classes (reduced to real coefficients). In other words, the Thurston norm on H_2 is given by*

$$|\sigma| = \max_{\alpha}\langle \alpha, \sigma\rangle,$$

where the maximum is taken over all monopole classes.

(The extra hypothesis in Corollary 1.6, that there is not a basis for H_2 consisting of tori, was there to ensure that Y had at least one smooth, taut foliation. The hypothesis is unnecessary in Corollary 3.7, for the accidental reason that 0 is always a monopole class, on account of the reducible solution with $\Phi = 0$.)

Thus the monopole classes give us sharp information about the genus of embedded surfaces, while the basic classes (in general) do not. For example, in the 3-manifold $Y(K)$ obtained by zero-surgery on a knot K of genus g, the classes $\pm(2g-2)$ in $H^2 \cong \mathbb{Z}$ are monopole classes. This means that, for the Spin^c structure which we called $\mathfrak{c}_{g-1}$, solutions of the monopole equations always exist, even though the algebraic count of the solutions will be zero if the Alexander polynomial has small degree.

We can also return to our discussion of the scalar curvature, and see that our inequalities there were sharp also. Our previous results stated that the unit ball for the norm

$$4\pi \sup_h \frac{\|\alpha\|_h}{\|s_h\|_h}$$

was sandwiched between the convex hull of the monopole classes and the unit ball for the dual Thurston norm (Lemmas 2.1 and 2.2). Knowing that these last two coincide, we can replace the inequality of Lemma 2.2 with an equality:

Proposition 3.8. *If Y is a closed, irreducible oriented 3-manifold, then the dual Thurston norm on $H^2(Y;\mathbb{R})$ is given by*

$$|\alpha|_* = 4\pi \sup_h \frac{\|\alpha\|_h}{\|s_h\|_h},$$

where the supremum is taken over all Riemannian metrics on Y. □

Remark. Up until this point, it was by no means clear that the supremum on the right hand side was ever finite. Our results now say that the supremum is reached, in the limit, by stretching the metric along a cylinder $[-R, R]\times\Sigma$, where Σ is a minimum-genus representative for a class σ with $\langle\alpha, \sigma\rangle = |\alpha|_*|\sigma|$.

4 Detecting monopole classes

How can one detect that a given class is a monopole class, without it being a basic class? An existence theorem is needed for solutions to the equations. A simple scenario in which one can see that solutions must exist arises when the 3-manifold in question is embedded in a suitable 4-manifold. We therefore turn to dimension four.

The 4-dimensional equations

The equations we have been discussing are a 3-dimensional version of the monopole equations which were first introduced, by Witten [42], in dimension 4. On an oriented Riemannian 4-manifold X, a Spin^c structure $\mathfrak{c}$ consists of a hermitian vector bundle W of rank 4, together with a Clifford multiplication

$$\rho : T^*X \to \mathrm{End}(W)$$

with the property that, if $e^1, \dots, e^4$ are an orthonormal coframe at a point in X, then the endomorphisms $\rho(e^i)$ are skew-adjoint and satisfy the Clifford relations

$$\rho(e^i)\rho(e^j) + \rho(e^j)\rho(e^i) = -2\delta_{ij}.$$

Clifford multiplication is extended to forms of higher degree as before. It is a consequence of this definition that the spin bundle W decomposes into two bundles of rank 2, $W^+ \oplus W^-$, whose determinants are equal. The action of 1-forms maps $W^+ \to W^-$, the action of 2-forms preserves the decomposition, and one can characterize W^- as the subspace annihilated by $\rho(\omega)$ for all *self-dual* 2-forms ω (forms satisfying $*\omega = \omega$). A spin connection is defined as before, and given a spin connection A, one has a Dirac operator D_A acting on sections of W. We write D_A^+ for the restriction of D_A to W^+, which is an operator

$$D_A^+ : \Gamma(W^+) \to \Gamma(W^-).$$

The connections on $\Lambda^2 W^+$ and $\Lambda^2 W^-$ induced by a spin connection A are equal, and we write $\hat{A}$ for either. We write $c_1(\mathfrak{c})$ for the first Chern class of W^+, which is the class represented in de Rham cohomology by the form $(i/2\pi)F_{\hat{A}}$.

Once again, the set of Spin^c structures is acted on transitively by $H^2(X;\mathbb{Z})$, and we have the same rule,

$$c_1(\mathfrak{c}+e) = c_1(\mathfrak{c}) + 2e,$$

which shows that $c_1(\mathfrak{c})$ determines $\mathfrak{c}$ to within a finite ambiguity measured by the 2-torsion subgroup of $H^2(X;\mathbb{Z})$.

The 4-dimensional monopole equations are the following pair of equations for a section Φ of W^+ and a spin connection A:

$$\begin{aligned} \rho(F^+_{\hat{A}}) - \{\Phi \otimes \Phi^*\} &= 0 \\ D^+_A \Phi &= 0. \end{aligned} \tag{13}$$

The first equation is to be interpreted as an equality between endomorphisms of W^+. The curly brackets denote the trace-free part on W^+, not on W, and F^+ denotes the projection of the curvature onto the self-dual forms, as usual.

The *moduli space* $M_{\mathfrak{c}}$ is the space of solutions (A, Φ) modulo the action of the gauge group, $G = \mathrm{Map}(X, S^1)$. We can also perturb the equations, rather as in the 3-dimensional case, by an arbitrary self-dual 2-form η:

$$\begin{aligned} \rho(F^+_{\hat{A}} + i\eta) - \{\Phi \otimes \Phi^*\} &= 0 \\ D^+_A \Phi &= 0. \end{aligned} \tag{14}$$

We write $M_{\mathfrak{c},\eta}$ for the solution space.

In dimension 3, for a generic perturbation, the irreducible solutions are isolated. In dimension 4, the equations have an index. We suppose X is compact and write

$$d(\mathfrak{c}) = \frac{1}{4}\big(c_1(\mathfrak{c})^2[X] - 2\chi(X) - 3\sigma(X)\big),$$

which one can also recognize as the second Chern number, $c_2(W^+)[X]$. The basic facts about the moduli space are these:

Proposition 4.1. *The moduli space $M_{\mathfrak{c},\eta}$ is compact. For an open, dense set of perturbations η, the irreducible part of the moduli space (the locus of solutions with $\Phi \neq 0$) is a smooth manifold of dimension $d(\mathfrak{c})$, cut out transversely by the equations.*

Remark. The proof of compactness runs much as in the 3-dimensional case, beginning with an essentially identical calculation leading to (4).

For the unperturbed equations, a solution with $\Phi = 0$ means a connection $\hat{A}$ in $\Lambda^2 W^+$ with anti-self-dual curvature, and hence an anti-self-dual representative for $c_1(\mathfrak{c})$. An anti-self-dual form has negative square, so there can be no solutions if $c_1(\mathfrak{c})^2[X]$ is positive. If $c_1(\mathfrak{c})^2[X]$ is zero, there can be reducible solutions only if $c_1(\mathfrak{c})$ is a torsion class, so that there is a flat connection. The same is true of the perturbed equations if η is small. Even if $c_1(\mathfrak{c})^2[X]$ is negative, however, there can only be a reducible solution if $2\pi c_1(\mathfrak{c}) - \eta$ is represented by an anti-self-dual form. The real cohomology $H^2(X;\mathbb{R})$ is the direct sum of the self-dual and anti-self-dual harmonic spaces, so the space of η for which such a representative exists is an affine subspace of codimension $b^+(X)$, the dimension of the space of self-dual forms. If $b^+(X)$ is non-zero, there is no reducible solution for generic η, and if $b^+(X)$ is at least 2, there is no solution for all η in a generic path.

Now let $\mathfrak{c}$ be a Spin^c structure with $d(\mathfrak{c}) = 0$, so that $c_1(\mathfrak{c})^2[X] = 2\chi + 3\sigma$. We shall suppose that X either has $b^+ \geq 2$ or has $2\chi + 3\sigma$ non-negative, and if $2\chi + 3\sigma$ is zero we shall also ask that let $c_1(\mathfrak{c})$ is not a torsion class. For a generic η, and also for a generic path of η, the moduli space $M_{\mathfrak{c},\eta}$ then consists of finitely many points which are transverse, irreducible solutions of the equations. The number of solutions, counted with suitable signs, is independent of the choice of perturbation and the choice of metric. We write

$$SW(X, \mathfrak{c}) \in \mathbb{Z}$$

for this number. This is the Seiberg-Witten monopole invariant for X with the Spin^c structure $\mathfrak{c}$ [42]. As before, if $SW(X,\mathfrak{c})$ is well-defined and *non-zero*, we call $c_1(\mathfrak{c})$ a *basic class* of X. Note that if $SW(X,\mathfrak{c})$ is non-zero, then the moduli space $M_{\mathfrak{c}}$ of solutions to (13) is non-empty for *every* choice of Riemannian metric.

The first significant result about basic classes was proved by Witten in [42]. This was the statement that, for a smooth algebraic surface with $b^+ > 1$ (e.g. a hypersurface in $\mathbb{CP}^3$ of degree 4 of more), the first Chern class and its negative (the canonical class) are basic classes. This was soon generalized by Taubes in [36, 37] to symplectic manifolds, in the following form. A symplectic structure ω on a manifold determines an almost-complex structure uniquely up to deformation, and hence has Chern classes $c_i(\omega)$. The *canonical class* K_ω is $-c_1(\omega)$.

Theorem 4.2 (Taubes, [36, 37]). *Let (X, ω) be a compact symplectic 4-manifold. Suppose either that $b^+ > 1$, or that $b^+ = 1$ and $K_\omega \smile [\omega]$ and K_ω^2 are both positive. Then the canonical class is a basic class.*

Remarks. Note that the hypotheses rule out $\mathbb{CP}^2$. (On a Kähler manifold, the sign of $K \smile [\omega]$ is opposite to that of the mean scalar curvature.) The statement for the case $b^+ = 1$ can be sharpened, but not without refining our treatment of the monopole invariants. A full treatment of the monopole invariants in the case $b^+ = 1$ is given in [26].

This version of the theorem is a little careless. As well as determining a canonical class K_ω, a symplectic structure gives rise to a canonical Spin^c structure $\mathfrak{c}_\omega$, with $c_1(\mathfrak{c}_\omega) = K_\omega$. Taubes result asserts that this Spin^c structure is basic, and in fact

$$SW(X, \mathfrak{c}_\omega) = \pm 1 \tag{15}$$

under the hypotheses of the Theorem.

Stretching 4-manifolds

A simple relationship between the 3- and 4-dimensional equations gives the following criterion for a class α on a 3-manifold to be a monopole class.

Proposition 4.3. *Let Y be a closed, oriented 3-manifold embedded in a compact, oriented 4-manifold X. Let α be a basic class on X. Then the restriction $\alpha|_Y$ is a monopole class on Y.*

Proof. Let $\mathfrak{c}$ be the Spin^c structure with $c_1(\mathfrak{c}) = \alpha$ and $SW(X, \mathfrak{c})$ non-zero, so that solutions to (13) exist for all metrics h on X. Let h_Y be a Riemannian on Y, and let h_1 be any metric on X such that a collar neighborhood $[-1,1] \times Y$ carries a product metric $dt^2 + h_Y$. Let h_R be obtained from h_1 by replacing this short cylinder by a longer cylinder, $[-R, R] \times Y$, for $R > 1$. For each h_R, there exists a solution (A_R, Φ_R) on X.

The idea of the proof is to show that, as R approaches ∞, we can find a subsequence such that the corresponding solutions converge to a translation-invariant solution on some portion of the cylindrical piece. A translation-invariant solution can be interpreted as a solution of the 3-dimensional equations on Y, for the metric h_Y, so showing that a solution exists. Since h_Y is arbitrary, the class $c_1(\mathfrak{c})|_Y$ is a monopole class.

To make this work, we need first to understand the relationship between the equations in dimensions three and four. On a cylinder $[-R, R] \times Y$ with a product metric, the action of $\rho(dt)$ gives an isomorphism between W^+ and W^-. Using this isomorphism, Clifford multiplication by 1-forms orthogonal to dt become endomorphisms of W^+. In this way, Y acquires a Spin^c structure (with spin bundle $W_3 \cong W^+|_Y$) which one can call the restriction of $\mathfrak{c}$. Given a solution of the unperturbed 4-dimensional equations on the cylinder, one can apply a gauge transformation to make the dt component of $\hat{A}$ zero. If A is in such a *temporal gauge*, it can be recovered from the path $A(t)$ in space of spin connections on Y, obtained by restricting A to the slices $\{t\} \times Y$. The spinor Φ on the 4-manifold gives a path $\Phi(t)$ in the space of spinors on the 3-manifold.

In a temporal gauge, the 4-dimensional equations (13), become the following equations for the paths $A(t)$ and $\Phi(t)$,

$$\begin{aligned} \rho(\dot{A}) &= -\rho(F_A) + \{\Phi \otimes \Phi^*\} \\ \dot{\Phi} &= -D_A\Phi, \end{aligned} \tag{16}$$

in which D_A now stands for the 3-dimensional Dirac operator, and the dot is differentiation with respect to t. These equations can be recognized as the downward gradient-flow equations for the functional $\mathrm{CSD}(A, \Phi)$.

Having understood this relationship, we can complete the proof. The solution (A_R, Φ_R) on the cylinder $[-R, R] \times Y$ can now be interpreted as a gradient trajectory for the Chern-Simons-Dirac functional. The compactness properties of the equations can be used to show that the change of CSD along these trajectories is bounded, by a constant independent of R. It follows that, when R is large, there is at least some portion of the cylinder in which the change in the functional is small. Passing to a subsequence, one obtains in the limit a translation-invariant solution to the equations on the cylinder, in a temporal gauge. This is a critical point of CSD, or in other words a solution of the 3-dimensional equations. This outline is filled out in [23]. □

Remark. It is worth commenting that, if $\mathfrak{c}$ restricted to Y is trivial, the solution whose existence is established by this argument may only be the trivial solution. With some additional hypotheses however, one can establish a stronger conclusion.

Using Theorem 4.2, we can draw the following simple corollary.

Corollary 4.4. *Let Y be a closed oriented 3-manifold embedded in a closed, symplectic 4-manifold (X, ω). If $b^+(X) = 1$, suppose that the hypotheses of Theorem 4.2 hold. Then $K_\omega|_Y \in H^2(Y)$ is a monopole class on Y.* □

(In fact, solutions on Y exist for the Spin^c structure $\mathfrak{c}_\omega|_Y$.) This observation, with a little ingenuity, is already enough to show that the set of monopole classes can have larger convex hull than the set of basic classes on a 3-manifold. We will need to adapt the corollary, however, before it becomes very useful. It is a puzzling question to characterize the classes $\alpha \in H^2(Y)$ which arise in this way, for a general Y.

Floer homology

There is a well-understood framework in which to place the ideas just discussed, namely the framework of 'Floer homology'. The model in the literature that is closest to what we need is Floer's construction in [11] of an invariant of 3-manifolds, using the gradient-flow of the Chern-Simons functional (of an $SU(2)$ connection).

There appears to be no serious obstacle to adapting [11] to the Chern-Simons-Dirac functional, particularly in the case that $c_1(\mathfrak{c})$ is not torsion, but there is not yet a complete account of such a construction in the literature. Nevertheless, it is clear how to proceed, and we shall content ourselves with some remarks. The starting point of [11] is the basic observation that one can calculate the homology of a compact manifold M by the following recipe. Choose a Morse function f on M whose gradient flow satisfies the additional 'Morse-Smale' condition, that the stable and unstable manifolds of all the critical points meet transversely. This means in particular that the trajectories which run from a critical point a at $t = -\infty$ to a critical point b at $t = +\infty$ form a family of dimension equal to the difference of the indices of a and b. Of these degrees of freedom, one is the freedom to reparametrize the trajectory $\gamma(t)$ as $\gamma(t + c)$. If a and b have index differing by 1, the trajectories are isolated once one forgets the parameterization. Now form the vector space C with a basis e_a indexed by the critical points a, and define a linear transformation ∂ by

$$\partial(e_a) = \sum n_{ab} e_b,$$

where n_{ab} counts the number of trajectories from a to b in the case that $i(b) = i(a) - 1$ and is zero otherwise. To avoid questions of orientation, one can take $\mathbb{Z}/2$ as the field of coefficients for C. Then one shows that $\partial^2 = 0$ and the $\ker \partial / \mathrm{im}\partial$ is the homology of M. In particular it is independent of the choice of f and the choice of Riemannian metric used to define the gradient. Without having an alternative definition of the homology however, one can verify this independence directly.

Floer applied this construction in an infinite-dimensional setting, taking the Chern-Simons functional as f. In the monopole setting, one should use the functional CSD on the space $\mathcal{C}$. The situation is simplest in the case that $c_1(\mathfrak{c})$ is not torsion, so that the monopole equations on Y have no solution with $\Phi = 0$. As mentioned above, for a suitable exact 2-form μ, the perturbed functional CSD_μ has non-degenerate critical points, and these are a finite set. Although there is no well-defined difference of indices between a pair of critical points, we can measure the index difference between a and b along a given trajectory γ, as the spectral flow of the Hessian, as before. If a Morse-Smale condition is satisfied, we can then construct C and ∂ as before, defining n_{ab} as the number of trajectories whose spectral flow is 1.

Unfortunately, we cannot expect to achieve the stronger Morse-Smale condition by such a restricted class of perturbations as the addition of an exact 2-form μ. One must seek a larger class of perturbations. At the same time, it is a particular property of the equations involved that the spaces of trajectories have any reasonable compactness properties (as noted in the previous subsection, the gradient trajectories can be interpreted as solutions of the 4-dimensional monopole equations), and one must choose the perturbations of CSD_μ so as not to upset this feature. We make some remarks in the following subsection about how one might define a suitable larger class of perturbations, and for the moment we shall pass over this point. It is precisely here that more work needs to be done to carry through the Floer program for the monopole equations.

After taking care of perturbations and compactness, one should arrive by this construction at a vector space $HF(Y, \mathfrak{c})$ with $\mathbb{Z}/2$ coefficients and an even-odd grading. It should depend only on Y and $\mathfrak{c}$, not on the choice of Riemannian metric or the perturbation chosen for the equations. The construction makes clear that $HF(Y, \mathfrak{c})$ is zero if the original monopole equations, or their perturbation, have no solution. Also, the euler characteristic of $HF(Y, \mathfrak{c})$ (the difference of the odd and even betti numbers) is equal to the integer invariant $SW(Y, \mathfrak{c})$.

The usefulness of Floer homology in the present context is that Proposition 4.3 and Corollary 4.4 can be strengthened, so as to conclude that the

Floer homology is non-zero. (For the 4-dimensional Yang-Mills invariants, this role for the instanton Floer homology of [11] was first noted by Donaldson.) For example, if Y is embedded in a closed symplectic 4-manifold (X, ω) (satisfying the hypotheses of Corollary 4.4 in the case $b^+(X) = 1$), and if $K_\omega|_Y$ is not torsion, then one would conclude that $HF(Y, \mathfrak{c}_\omega|_Y)$ is non-zero, once the definition of this Floer homology was in place.

Perturbing the gradient flow

It may be worth noting that perturbing the Chern-Simons-Dirac functional so as to achieve a Morse-Smale condition for the trajectories of the gradient flow may not be particularly difficult. Let Y be a closed Riemannian 3-manifold with Spin^c structure $\mathfrak{c}$, and let $\mathcal{S}$ be the space of all pairs (A, Φ), where A is a spin connection and Φ is a section of W. The space $\mathcal{C}$ above is the quotient $\mathcal{S}/\mathcal{G}$.

Let A_0 be a fixed spin connection, so that we can write the general spin connection as

$$A = A_0 + a\,\mathbf{1},$$

so identifying the space of spin connections with the space of imaginary-valued 1-forms a, as before. We have already considered adding to CSD a function of the form

$$\tau_\mu(a) = i \int a \wedge \mu$$

for an exact 2-form μ. Let us now relax the requirement slightly, and suppose only that μ is *closed.* Take a collection $\mu_1, \ldots, \mu_N$ of closed 2-forms, which included a basis for $H^2(Y)$, and let $\tau_1, \ldots, \tau_N$ be the corresponding functions of a. As noted earlier, the functions τ_i on $\mathcal{S}$ are invariant only under the identity component $\mathcal{G}_e \subset \mathcal{G}$. The map

$$(\tau_1, \ldots, \tau_N) : \mathcal{S} \to \mathbb{R}^N$$

commutes with the $\mathcal{G}$ action, however, when $\mathcal{G}$ acts on $\mathbb{R}^N$ through a discrete action of the quotient, $\mathcal{G}/\mathcal{G}_e \cong H^1(Y; \mathbb{Z})$, by translations. So to obtain a $\mathcal{G}$-invariant function on $\mathcal{S}$, we should take a function

$$F : \mathbb{R}^N \to \mathbb{R}$$

which is invariant under these translations, and define

$$f(a,\Phi) = F\big(\tau_1(a),\dots,\tau_N(a)\big).$$

(Functions of this shape include the 'smoothed-out' holonomy maps used in [11] for non-abelian connections.)

Our function f is not yet sufficiently general, for it does not depend on Φ. To incorporate Φ, we can proceed as follows. Let L be the Greens operator for the ordinary Laplacian on $C^\infty(Y)$. Thus L inverts the Laplacian when restricted to functions with zero mean, and the kernel and cokernel of L are the constant functions. Let $H \subset G_e$ be the subgroup

$$H = \{\, e^{i\xi} \mid \xi : Y \to \mathbb{R}, \text{ with } \textstyle\int \xi = 0 \,\}.$$

The quotient G_e/H is the circle, represented by the constant maps, and G/H is represented by the harmonic maps $Y \to S^1$, a group isomorphic to $S^1 \times H^1(Y;\mathbb{Z})$.

For any fixed spinor $\psi \in \Gamma(W)$, let σ_ψ be the complex function on $\mathcal{S}$ defined by the hermitian inner product

$$\sigma_\psi(a,\Phi) = \int \langle \psi^a, \Phi \rangle,$$

where ψ^a denotes the expression

$$\psi^a = e^{-Ld^*a}\psi.$$

The definition of ψ^a is such that it transforms as Φ does under the action of H: if $u = e^{i\xi}$, where ξ has mean zero, then

$$\begin{aligned} \psi^{a-u^{-1}du} &= \psi^{a-id\xi} \\ &= e^{-Ld^*a}e^{iLd^*d\xi}\psi \\ &= u\psi^a. \end{aligned}$$

Thus $\sigma_\psi : \mathcal{S} \to \mathbb{C}$ is invariant under H. Under the circle G_e/H, however, it transforms with weight 1.

Now choose a collection of spinors ψ_i $(i = 1,\dots K)$, and let σ_i be the corresponding complex functions on $\mathcal{S}$. We now have a collection of functions

$$(\tau_1,\dots,\tau_N,\sigma_1,\dots,\sigma_K) : \mathcal{S} \to \mathbb{R}^N \times \mathbb{C}^K.$$

These functions are equivariant for G when G/H is made to act suitably on $\mathbb{R}^N \times \mathbb{C}^K$. The quotient of $\mathbb{R}^N \times \mathbb{C}^K$ by this action is a bundle with fiber $\mathbb{C}^K/S^1$ over the base $T^b \times \mathbb{R}^{N-b}$, where b is the Betti number of Y.

Now choose a smooth function

$$F : \mathbb{R}^N \times \mathbb{C}^K \to \mathbb{R}$$

which is invariant under the action of G/H, and define

$$f(a, \phi) = F(\tau_1, \ldots \tau_N, \sigma_1, \ldots, \sigma_K).$$

Now consider perturbing CSD by the addition of such a function f. With terms of this sort we can C^1-approximate, for example, any smooth function on a compact submanifold of $\mathcal{C}$ (lying in the locus where Φ is non-zero), and quite formally the class of functions is large enough to give the necessary transversality. If the partial derivatives of F are bounded, it seems that the compactness theorems for spaces of trajectories hold up too. Thus, in the crucial calculation (4), one finds amongst other things an additional cubic term in Φ, which is local in the t coordinate, but non-local on Y, involving an expression of the shape

$$\Phi^* \rho\big(d \circ L\langle \psi^a, \Phi \rangle\big)\Phi.$$

But such a term does not break the argument.

5 Monopoles and contact structures

We now turn to the proof of Theorem 3.6. In view of Corollary 4.4, one might hope to prove this by showing that if Y had a taut foliation $\mathcal{F}$, then one could always embed Y in a closed symplectic manifold (X, ω) in such a way that $c_1(\omega)$ restricted to Y was $\pm e(\mathcal{F})$. Perhaps this can be done; such a result would be very interesting, and presumably very hard. A slight modification of this tactic leads to a proof, however. Applying a theorem of Eliashberg and Thurston [8], we shall embed Y in an *open* symplectic manifold whose ends have a cone-like geometry. We shall then extend the 4-dimensional gauge-theory techniques to this setting.

Using the theorem of Eliashberg and Thurston

The following material can be found in [8]. To begin, the following proposition sheds a geometric light on the meaning of taut. (The converse to the proposition is true also, but is rather deeper.)

Proposition 5.1. *If $\mathcal{F}$ is a taut foliation of Y by oriented leaves, then there is a closed 2-form Ω on Y whose restriction to the leaves is positive.*

Proof. Let γ be a closed curve, transverse to the leaves and meeting every leaf. (The existence of γ was our chosen definition of taut.) A small tubular neighborhood $N(\gamma)$ meets the leaves of $\mathcal{F}$ in a foliation by disks, to give a product structure. Using the product structure, pull back to $N(\gamma)$ a 2-form ψ supported in the interior of the disk and non-negative there. The result is a closed form $\Omega(\gamma)$ which is non-negative on the leaves of $\mathcal{F}$. By pushing γ along the leaves, one obtains transverse curves running through any point of Y. By taking a suitable finite collection of such curves γ_i and adding up the corresponding forms $\Omega(\gamma_i)$, one obtains a suitable Ω. □

Now let α be a non-vanishing 1-form on Y whose kernel at each point is the tangent plane to $\mathcal{F}$ and whose orientation is such that $\alpha \wedge \Omega$ is positive. The integrability of the tangents to the foliation means that $\alpha \wedge d\alpha$ is zero. On the cylinder $[-1, 1] \times Y$, consider the closed 2-form

$$\omega = d(t \wedge \alpha) + \Omega.$$

In ω^2, the only term to survive is the terms $dt \wedge \alpha \wedge \Omega$, which is positive. So ω is a symplectic form.

We have succeeded in embedding Y in a symplectic 4-manifold with boundary, $[-1, 1] \times Y$, and it is not hard to see that the first Chern class $c_1(\omega)$ restricts to $e(\mathcal{F})$ on the 3-manifold. But this elementary step is insufficient for our needs.

A *contact structure* on a 3-manifold is a field of 2-planes ξ which strictly fails to be integrable at every point of Y. This means that, if ξ is defined locally as the kernel of a 1-form β, then $\beta \wedge d\beta$ is nowhere zero. If the 3-manifold and the 2-plane field are oriented (as will always be the case for us), then a suitable form β exists globally. Note however that, as a non-vanishing 3-form, the product $\beta \wedge d\beta$ itself determines an orientation of Y. We shall say that the contact structure ξ is compatible with the orientation of Y if the form $\beta \wedge d\beta$ is positive, for some and hence for all choices of β with kernel

ξ. The theorem of Eliashberg and Thurston which we need states that the tangent planes to a foliation can be deformed to give a contact structure, compatible with either orientations:

Theorem 5.2 (Eliashberg-Thurston [8]). *Let $\mathcal{F}$ be a smooth, oriented foliation of an oriented 3-manifold, other than the foliation of $S^1 \times S^2$ by spheres. Then the 2-plane field $T\mathcal{F}$ can be C^0 approximated by contact structures ξ compatible with either the given orientation of Y or its opposite.* □

(An example of this phenomenon arises when Y is a circle bundle over a surface arising as a compact left quotient of $SL(2,\mathbb{R})$. A left-invariant 2-plane field is determined by a 2-plane π in the Lie algebra $\mathfrak{sl}(2,\mathbb{R})$. If π is tangent to the null cone of the Killing form, then the corresponding 2-plane field is a foliation. If the Killing form is either definite or hyperbolic on π, then the 2-plane field is a foliation, compatible with one or other orientation of the 3-manifold.)

Now let us return to the manifold $X = [-1,1] \times Y$ with the symplectic form ω constructed above. The oriented boundary of X is $\bar{Y} \cup Y$ (the bar denotes the opposite orientation), and $\mathcal{F}$ foliates both components. Using the Theorem, one obtains foliations ξ_- and ξ_+ on $\bar{Y}$ and Y, compatible with their respective orientations, and at a small angle from the tangents to $\mathcal{F}$. If the angle is made small enough, we can arrange that these contact structures are *compatible* with ω, in the weak sense that ω is positive on the 2-planes at the boundary:

$$\omega|_\xi > 0 \quad \text{at } \partial X. \tag{17}$$

Indeed, ω is positive on the tangent planes to $\mathcal{F}$, which $\xi_\pm$ approximate. (To clarify the signs involved, this sort of compatibility holds for the standard, Kähler, symplectic form on a pseudo-convex domain in $\mathbb{C}^2$, such as a ball, when the boundary is given the contact structure defined by the complex tangent directions.)

To summarize, starting from a 3-manifold Y with a foliation $\mathcal{F}$, we have constructed a symplectic 4-manifold (X,ω) in which Y is embedded, with $K_\omega|_Y$ equal to the euler class $e(\mathcal{F})$. The 4-manifold has a contact structure $\xi = \xi_- \cup \xi_+$ on its boundary (compatible with the boundary orientation, in a 'convex' direction), and the symplectic form is compatible with ξ, in the sense described by (17).

Four-manifolds with contact boundary

Although we have not embedded Y in a closed manifold, the convex contact structure on the boundary of X is all we shall need, because we can extend the monopole invariants $SW(X, \mathfrak{c})$ for closed manifolds so as to define similar invariants for 4-manifolds with contact boundary.

We give an account of the construction from [21]. Let X be a compact, connected, oriented 4-manifold with non-empty boundary ∂X, and let ξ be an oriented contact structure on ∂X, compatible with the boundary orientation. In the presence of a metric, any oriented 2-plane field such as ξ determines a Spin^c structure on ∂X, and hence a 4-dimensional Spin^c structure on a collar of the boundary. One can think of this in various ways. For example, define the spin bundle W on ∂X can be defined as the sum $\mathbb{C}\oplus\xi^c$, where the second summand means that the oriented 2-planes of ξ are being regarded as complex lines; then define Clifford multiplication at a point y by picking a basis e_1, e_2, e_3 of tangent vectors at y, with e_1 the positive normal to ξ and declaring that these act on $(\mathbb{C}\oplus\xi^c)_y$ by the Pauli matrices

$$\begin{pmatrix} i & 0 \\ 0 & -i \end{pmatrix}, \quad \begin{pmatrix} 0 & -1 \\ 1 & 0 \end{pmatrix}, \quad \begin{pmatrix} 0 & i \\ i & 0 \end{pmatrix},$$

using the basis vector e_2 to trivialize ξ. Alternatively, one can think of this as a special case of the way in which an almost-complex structure determines a Spin^c structure in even dimensions. Note that the spin bundle comes with a canonical section $\Phi_0 = (1, 0)$.

Now let $\mathfrak{c}$ be any extension of $\mathfrak{c}_\xi$ to the interior of X. Given such an extension, we shall define a monopole invariant

$$SW(X, \xi, \mathfrak{c}) \in \mathbb{Z},$$

which is a diffeomorphism invariant of the triple (no condition on $b^+(X)$ is needed).

The invariant $SW(X, \xi, \mathfrak{c})$ is defined as follows. First we enlarge the manifold X by adding expanding cones to the boundary components. In more detail, if Y is a 3-manifold with a positive contact structure defined as the kernel of a 1-form β, then there is a symplectic form on the cone $(0, \infty) \times Y$, given as

$$\omega = d(f(t)\beta)$$

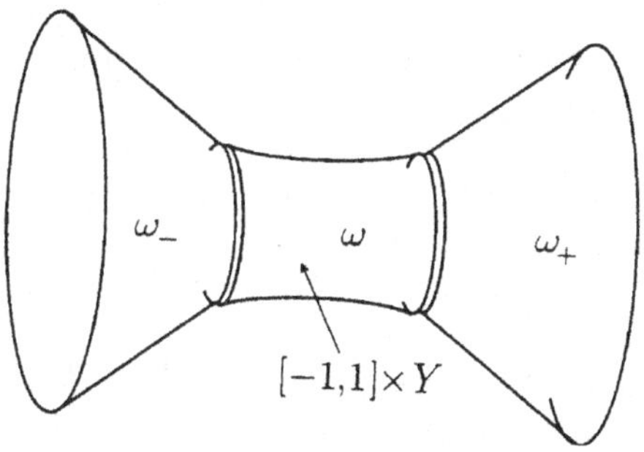

Figure 2: The geometry of Z.

for any monotone increasing function f of $t \in (0, \infty)$. To reproduce the way in which $\mathbb{R}^4$ with its standard symplectic structure arises from S^3, we prefer to set

$$\omega = (1/2)d(t^2\beta).$$

We apply this standard construction to the components of ∂X, and attach conical pieces $[1, \infty) \times \partial X$ to the boundary, to obtain an open 4-manifold Z (diffeomorphic to the interior of X):

$$Z \;=\; X \,\cup\, [1, \infty) \times \partial X.$$

On Z we choose a Riemannian metric h compatible with ω on the conical pieces. This means that there are local orthonormal coframes in which ω can be expressed as $e^1 \wedge e^2 + e^3 \wedge e^4$. Figure 2 shows an illustration of Z in the case that X is the manifold $[-1, 1] \times Y$ from the previous subsection. In this case, the contact structures ξ_- and ξ_+ give symplectic forms $\omega_\pm$ on the two conical ends.

The symplectic structure ω on the conical pieces determines a canonical Spin^c structure $\mathfrak{c}_\omega$ there, essentially the same as the Spin^c structure determined by ξ on the boundary of X. The choice of $\mathfrak{c}$ on X gives an extension of $\mathfrak{c}_{omega}$ to all of Z. The spin bundle $W^+ = W^+_\mathfrak{c}$ has a canonical section Φ_0 of unit length on the conical pieces of Z, and there is a unique spin connection there, with the property that $D^+_{A_0}\Phi_0 = 0$. (Such a connection is determined by any non-vanishing spinor on a 4-manifold.) We extend Φ_0 and A_0 arbitrarily over the remained of Z.

Motivated by the constructions of [36, 37], we now consider a modified version of the Seiberg-Witten monopole equations on the Riemannian manifold Z. The equations are

$$\begin{aligned} \rho(F^+_{\hat{A}}) - \{\Phi \otimes \Phi^*\} &= \rho(F^+_{\hat{A}_0}) - \{\Phi_0 \otimes \Phi_0^*\} \\ D^+_A \Phi &= 0. \end{aligned} \tag{18}$$

We can also consider, as before, perturbing these equations by the addition of a self-dual 2-form η, which should decay on the ends of Z:

$$\begin{aligned} \rho(F^+_{\hat{A}} + i\eta) - \{\Phi \otimes \Phi^*\} &= \rho(F^+_{\hat{A}_0}) - \{\Phi_0 \otimes \Phi_0^*\} \\ D^+_A \Phi &= 0. \end{aligned}$$

The unperturbed equations are set up so that (A_0, Φ_0) satisfies the equations on the conical ends, though not necessarily in the interior, since Φ_0 may not satisfy the Dirac equation there. We seek solutions (A, Φ) in general which are asymptotic to (A_0, Φ_0) at infinity. It turns out that, even if only mild decay is required of $A - A_0$ and $\Phi - \Phi_0$, any solution of the unperturbed equations will approach the canonical pair exponentially fast after adjustment by a suitable gauge transformation.

We again write $M_\mathfrak{c}$ (or $M_{\mathfrak{c},\eta}$ in the perturbed case) for the set of such solutions, considered up to the equivalence relation defined by the gauge transformations. Note that there can be no 'reducible' solutions with $\Phi = 0$, because Φ is required to approach the unit-length spinor Φ_0 at infinity. The main facts about the moduli space are these:

Proposition 5.3 ([21]). *The space $M_{\mathfrak{c},\eta}$ is compact, and for generic η it is a smooth manifold cut out transversely by the equations. In this case, the dimension of the moduli space is given by*

$$d(\mathfrak{c}) = c_2(W^+_\mathfrak{c}, \Phi_0)[Z, Z \setminus X],$$

(which is the relative euler class of the bundle $W^+_\mathfrak{c}$ on Z, relative to the non-vanishing section Φ_0 on the ends).

(The formula for the dimension coincides with the alternative formula $c_2(W^+[X]$ which we gave in the closed case.)

In the case $d(\mathfrak{c}) = 0$, we can again count the number of solutions, either with suitable signs, or just modulo 2, to obtain a definition of the invariant $SW(X, \xi, \mathfrak{c})$. It is again independent of the choices made, such as the 1-form β, the metric h, and the perturbation η.

Symplectic filling

In the definition of $SW(X, \xi, \mathfrak{c})$ as just described, no symplectic form on X is involved and none is needed in the construction. When one has a symplectic structure on X compatible with ξ at the boundary, then there is a non-vanishing theorem for the invariant. The symplectic form ω determines a canonical Spin^c structure $\mathfrak{c}_\omega$ on X, and the compatibility condition with ξ means in particular that $\mathfrak{c}_\omega$ and $\mathfrak{c}_\xi$ are the same at the boundary, so and invariant $SW(X, \xi, \mathfrak{c}_\omega)$ is defined. The theorem is then:

Theorem 5.4 ([21]). *If X is a compact 4-manifold with contact structure ξ on the boundary, and ω is a symplectic form on X compatible with ξ, then*

$$SW(X, \xi, \mathfrak{c}_\omega = \pm 1.$$

□

Once the analytic framework of the previous subsection is in place, the proof of this result is very much as the same as the proof of (15) from [37]. There is one point to note, however. The proof begins by constructing a symplectic form ω_Z on Z, the manifold with conical ends. Although Z is a union of pieces each of which carries a symplectic form (as illustrated in Figure 2 for the case of $[-1, 1] \times Y$), these forms do not necessarily agree at the joins, and even their cohomology classes may be different. Nevertheless, one can patch the forms together, in that one can find a symplectic form ω_Z which is *asymptotic* to the conical form on the ends, and agrees with the given form ω on X except in a small neighborhood of ∂X. See [21], for example.

The stretching argument used in Theorem 4.3 and Corollary 4.4 works just as well in the present setting of 4-manifolds with contact boundary. From the above theorem, we can therefore deduce:

Corollary 5.5. *Let (X, ω) be a symplectic 4-manifold with a compatible contact structure ξ on the boundary. If Y is an oriented 3-manifold embedded in X, then $K_\omega|_Y$ is a monopole class.* □

Using the construction of Eliashberg and Thurston, with $X = [-1,1] \times Y$, we deduce Theorem 3.6:

Corollary 5.6. *If Y is a closed, irreducible, oriented 3-manifold with a smooth, taut foliation $\mathcal{F}$ by oriented leaves, then $e(\mathcal{F})$ is a monopole class.* □

(Again, the hypothesis of irreducibility excludes $S^1 \times S^2$.) Slightly more generally, one can apply Corollary 5.5 to the case that Y is parallel to ∂X or a component of ∂X. In Eliashberg's terminology, [8], a contact 3-manifold Y is *symplectically fillable* if Y arises as the correctly oriented boundary of a 4-manifold X carrying a compatible symplectic form ω. If Y arises as a union of components of such a boundary, then it is symplectically *semi*-fillable.

Corollary 5.7. *If a contact 3-manifold (Y, ξ) is symplectically semifillable, then $e(\xi)$ is a monopole class.* □

One useful feature of this corollary is one can form a connected sum of semifillable contact structures, and so obtain results about monopole classes on reducible 3-manifolds also.

All these statements can be sharpened a little, because one knows which Spin^c structure is involved, not just the Chern class. More significantly perhaps, one should draw the stronger conclusion that the Floer homology is non-zero in these cases. For example, if (Y, ξ) is symplectically semifillable and e_ξ is not torsion, one would conclude that $HF(Y, \mathfrak{c}_\xi)$ is non-zero. (A more refined statement could be made in the case of a torsion class.) In particular then, one should say:

5.8. *If Y carries a taut foliation, then the Seiberg-Witten Floer homology of Y is non-zero for the corresponding* Spin^c *structure.*

The obstruction to proving such statements is no larger than the problem of verifying a suitable construction of Floer homology.

Invariants of contact structures

There is a way to rephrase part of the construction just described. Given an oriented 3-manifold Y and a contact structure ξ compatible with the orientation, one can form a symplectic cone $[1, \infty) \times Y$ with symplectic form ω_+ as before, and attach to it a cylinder $(-\infty, 1] \times Y$, as shown in Figure 3.

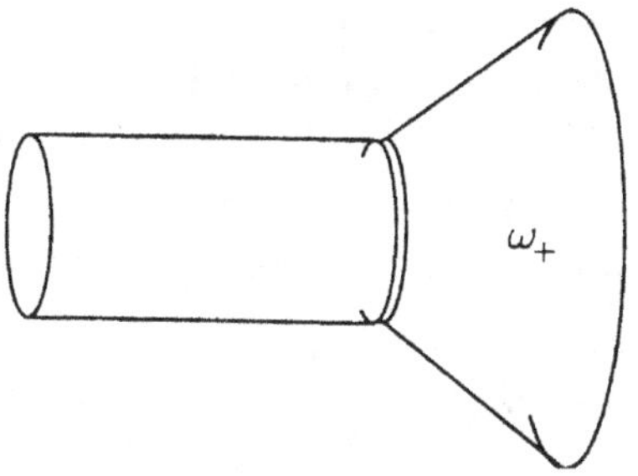

Figure 3: Defining an invariant of (Y, ξ).

Using the Spin^c structure given by ξ, one can then write down a version of the monopole equations on this 4-manifold which resemble the deformed equations (18) on the conical piece and resemble the usual equations (13) on cylindrical end, where they can be interpreted as the gradient flow equations for a trajectory of CSD.

For each critical point a of the functional CSD on Y (or of the perturbed functional CSD_μ considered before), one can look at the moduli space of solutions (A, Φ) which are asymptotic to (A_0, Φ_0) on the cone, and which descend from the critical point a on the cylinder. After perturbation, these moduli spaces are smooth manifolds. For each α, let $n(a, \xi)$ be the number of solutions belonging to zero-dimensional moduli spaces, counted with signs as usual, and consider the expression

$$\sum_a n_a e_a$$

as an element of the chain group C which defines Floer homology. If we suppose that $e(\xi)$ is not torsion, to eliminate the reducible solutions, then the above sum is closed under ∂ (by arguments that are familiar in other applications of Floer homology), and the resulting homology class

$$[\sum_a n_a e_a] \in HF(Y, \mathfrak{c}_\xi) \tag{19}$$

is an invariant of the contact 3-manifold.

In this way, one can define an invariant of contact structures, once Floer homology is in hand. These seems the most natural setting in which to place the constructions of [28], where collections of contact structures are exhibited which are homotopic as 2-plane fields but not isotopic as contact structures. (The manifolds Y in [28] are homology spheres, so one needs to tackle Floer homology for the case that $c_1(\mathfrak{c})$ is zero.)

The non-vanishing result, Corollary 5.7, should be rephrased so as to say that the invariant (19) of (Y, ξ) is a non-zero element of HF if the contact structure is semi-fillable.

6 Potential applications

When the Seiberg-Witten equations were introduced as an alternative gauge theory tool to replace the self-dual Yang-Mills equations exploited by Donaldson, an important link with topology was temporarily lost. The 3-dimensional companions of Donaldson's Yang-Mills invariants are, as we have mentioned, the instanton Floer homology $I(Y)$ introduced in [11], and the Casson invariant $\lambda(Y)$ [1]. These play the roles of $HF(Y, \mathfrak{c})$ and $SW(Y, \mathfrak{c})$ from the monopole theory. Floer defined $I(Y)$ for homology 3-spheres Y by studying the gradient-flow of the $SU(2)$ Chern-Simons functional, whose critical points correspond to flat $SU(2)$ connections, or representations of $\pi_1(Y)$ in $SU(2)$. There is also a version $I_w(Y)$ for the case that Y is a homology $S^1 \times S^2$, where the chain group is built from flat $SO(3)$ connections with non-zero Stiefel-Whitney class [3]. Defining $I(Y)$ in other situations presents technical difficulties related to reducible connections.

In the definition of $I(Y)$, there is an immediate connection with the fundamental group. Thus, for example, Floer's instanton homology vanishes if Y is a homotopy sphere (Floer's definition did not use the trivial representation of π_1). No such statement can be made very easily for the monopole Floer homology (our tentatively defined $HF(Y, \mathfrak{c})$).

On the other hand, in the Seiberg-Witten version, we have a handle on anon-vanishing result: irreducible 3-manifolds with $b_1 \neq 0$ admit taut foliations by Gabai's results, and we have argued that a foliation forces $HF(Y, \xi)$ to be non-zero. If one could establish even a weak relationship between the monopole $HF(Y, \mathfrak{c})$ and instanton Floer homology, then there would be a useful payoff.

Surgery and property 'P'

To elaborate on the last remark, the application we have in mind is in line with Casson's application of the invariant $\lambda(Y)$ to the question of 'property P'. This is the question of whether one can make a simply connected 3-manifold by non-trivial surgery on a non-trivial knot K. (If one cannot, then K is said to have property P.) Work of Gordon an Luecke [18] showed that one cannot make S^3 this way. So if one did manufacture a homotopy-sphere by surgery on a knot, it would be a fake 3-sphere (a counterexample to the Poincaré conjecture).

The basic example to consider is $+1$ surgery, in which a neighborhood $N(K)$ is removed and sewn back in so that the meridian on $N(K)$ is attached to a curve in the class of the meridian plus longitude. Let us call this manifold Y_1, or $Y_1(K)$. We also have the manifold Y_0 obtained by zero-surgery, which is a homology $S^1 \times S^2$. To prove property P, we would like to know that Y_1 has non-trivial fundamental group, and to this end we could seek to show that $\pi_1(Y_1)$ had non-trivial representations in $SU(2)$, or that $I(Y_1)$ is non-zero.

There is a powerful tool at hand, in the following theorem of Floer. (This is a special case of his 'exact triangle' in instanton homology [3]).

Theorem 6.1 (Floer, [3]). *In the above situation the instanton Floer homology $I(Y_1)$ of the homology sphere Y_1 is isomorphic to the instanton Floer homology $I_w(Y_0)$ of the homology $S^1 \times S^2$ obtained by zero-surgery.* □

In this form, the theorem is already hard to prove. But the special case we are interested is actually very *easy* to prove:

Corollary 6.2. *If the instanton Floer homology $I_w(Y_0)$ is non-trivial, then Y_1 is not a homotopy 3-sphere.* □

(In fact, if Y_1 is a homotopy 3-sphere, then one can easily use the holonomy perturbations introduced by Floer to deform the equations $F_A = 0$ on Y_0 so that they have no solutions for $SO(3)$ connections A on Y with non-zero w_2.)

Suppose now that we could prove:

Conjecture 6.3. *If the monopole Floer homology $HF(Y, \mathfrak{c})$ of a homology $S^1 \times S^2$ is non-trivial for some Spin^c structure with $c_1(\mathfrak{c})$ not torsion, then the instanton Floer homology $I_w(Y)$ is non-trivial also.*

Then we would be home, at least in the case that the genus of K is 2 or more. Indeed, we have learned from the foliation theory that in this case, Y_0

admits a smooth, taut foliation having a genus 2 surface has a compact leaf. The euler class of this foliation is non-trivial, and modulo the verification of the definitions, the monopole Floer homology of Y_0 for the corresponding Spin^c structure has been shown to be non-zero. The conjecture would imply that instanton Floer homology is non-trivial also, and it would follow that Y_1 was not a homotopy sphere, by Floer's result. (In the case of genus 1, there are two additional difficulties. The first is that Gabai's foliation may not be smooth. The second is that the relevant Spin^c structure on Y_0 has $c_1 = 0$, so one must treat reducibles with respect. The second point is moot, perhaps, because one would need to consider all Spin^c structures together, most likely, to prove the conjecture above.)

Note that, without using any connection between the instanton and monopole Floer homologies, one could try and reprove the theorem of Gordon and Luecke by establishing an exact triangle for the monopole Floer homology.

The Pidstrigatch-Tyurin program

Conjecture 6.3 does not stand unsupported. Indeed, there is evidence for a closer relationship. In introducing the monopole invariants of 4-manifolds, Witten [42] conjectured a very specific relationship between the monopole invariants and the older instanton invariants, for a large class of 4-manifolds. (This conjecture is extended in [32].) A mathematical approach to proving Witten's conjecture was proposed by Pidstrigatch and Tyurin [33], and although their program does raise some technical challenges at the time of writing, it seems most likely that a proof will be along the lines suggested. There is work in this direction in [9]. The proposed method is to study a larger moduli space of solutions to the $PU(2)$ monopole equations. These equations have the same shape as the equations (13) but involve a non-abelian connection A. The equations contain both the instanton equations and the usual monopole equations.

If one pursues this line in three dimensions instead of four, one can arrive at an elegant proof that the 3-manifold invariants we have called $SW(Y, \mathfrak{c})$ are related to the 'odd' Casson invariant. For example, if Y is obtained by zero surgery on a knot K, there is a Casson-type invariant $\lambda(Y)$ which is the euler characteristic of the instanton Floer homology $I_w(Y)$, and without encountering any of the technical difficulties of the 4-dimensional case, one can prove that

$$\lambda(Y) = \sum_{s} SW(Y, \mathfrak{c}).$$

Of course, one can verify this relationship externally, because both sides can be reduced to the Alexander invariant. (In the case of the left-hand side, this is due to Casson.) But the 'internal' proof using the $PU(2)$ equations seems to show that the relationship extends beyond the euler characteristics. It may well be that one can relate the Floer homologies using this approach.

7 Surfaces in 4-manifolds

Where the theory succeeds

We turn now to the question of representing a 2-dimensional homology class σ in a smooth, oriented 4-manifold X by a smoothly embedded, oriented surface Σ of minimal complexity. There are many statements here that can be made to look very much like their 3-manifold counterparts, particularly when Σ has trivial normal bundle. (This condition implies that the self-intersection number $\sigma \cdot \sigma$ is zero, but the converse is false, because we still allow that Σ may be disconnected.) For example, if $\sigma \cdot \sigma$ is zero, then $\mathrm{PD}[\sigma] \in H^2(X)$ is the pull-back of the generator of $H^2(S^2)$ by some map $f : X \to S^2$, and one can find a representative surface Σ as the inverse image of a regular value of f, so establishing that representative surface exists. Every embedded surface with trivial normal bundle arises this way, just as a surface in a 3-manifold Y arises from a map to S^1.

We have already introduced the 4-dimensional monopole equations and the basic classes, at least in the case that $b^+(X) > 1$. For surfaces with trivial normal bundle, the basic classes provide a lower bound for the genus, just as in dimension three (compare Proposition 3.1):

Proposition 7.1. *Let X be a smooth, oriented, closed 4-manifold with $b^+(X) > 1$, let α be a basic class and Σ an embedded surface representing a class σ. Suppose that the normal bundle of Σ is trivial. Then*

$$\chi_-(\Sigma) \geq \langle \alpha, \sigma \rangle.$$

Proof. Being a basic class means for α that there is a Spin^c structure $\mathfrak{c}$ with $c_1(\mathfrak{c}) = \alpha$ for which the invariant $SW(X, \mathfrak{c})$ is non-zero. All we need, however, is that the moduli space $M_{\mathfrak{c}}$ of solutions to the equations (13) is non-empty, for every choice of Riemannian metric on X. Just as in the 3-dimensional case (see (5)), if (A, Φ) is a solution for a given metric, then the Weitzenbock formula leads to

$$\int_X |\Phi|^4 d\mathrm{vol} \leq \int_X s^2 d\mathrm{vol}.$$

In dimension 4, the first of the two Seiberg-Witten equations leads to the relationship $|F_{\hat{A}}^+|^2 = (1/8)|\Phi|^4$, so there is a relationship between the L^2 norms,

$$||F_{\hat{A}}^+||^2 \leq ||s||^2/8. \tag{20}$$

Now let $||\alpha||$ stand for the L^2 norm of the harmonic representative, as before. We have

$$||\alpha|| \leq (1/2\pi)||F_{\hat{A}}||$$

On the other hand, alpha is the orthogonal sum of its self-dual and anti-self dual parts, and

$$||\alpha^+||^2 - ||\alpha^-||^2 = (\alpha \smile \alpha)[X],$$

so from the inequality (20) we obtain the bound

$$||\alpha||_h^2 \leq ||s_h||_h^2/(4\pi)^2 + (1/2)\alpha^2[X],$$

in which we have once again adjusted our notation to indicate the dependence on a Riemannian metric h. We can write

$$||\alpha||_h \leq ||s_h||_h/(4\pi) + C, \tag{21}$$

for a constant C which is independent of h, to reach an inequality that we can compare with (7).

Now let Σ be an embedded surface representing σ, as in the statement. We may assume that Σ is connected, because the general case follows by linearity. Since σ^2 is zero, the normal bundle of Σ is trivial, and we can find embedded in X a product region $[0,1] \times S^1 \times \Sigma$, as a collar on the boundary of a tubular neighborhood of Σ. Let h_1 be a metric on X whose restriction to this region is a product metric, with the metric on the Σ factor being of unit area and constant non-negative scalar curvature. (We can suppose Σ is not a sphere.) Then we can do just what we did in dimension three (see Lemma 9). Let h_r be a metric on X which contains a product region $[0, r] \times S^1 \times \Sigma$ and coincides with h_1 outside this region. Exactly as before, we can calculate for $h = h_r$,

$$\|s_h\|_h = 4\pi r^{1/2}(2g-2) + O(1)$$

as $r \to \infty$, while

$$\|\alpha\|_h \geq r^{1/2}\langle \alpha, [\Sigma]\rangle.$$

These two estimates are inconsistent with the inequality (21), unless $2g-2 \geq \langle \alpha, \sigma \rangle$. □

One can pass from the proposition above to a statement about surfaces with non-negative normal bundle (that is, surfaces Σ such that each component has non-negative self-intersection number):

Proposition 7.2. *Let X be a smooth, oriented, closed 4-manifold with $b^+(X) > 1$, let α be a basic class and Σ an embedded surface representing a class σ. Suppose that the normal bundle of Σ is non-negative. Then*

$$\chi_-(\Sigma) \geq \sigma \cdot \sigma + \langle \alpha, \sigma \rangle.$$

Proof. The statement can be deduced from the case of trivial normal bundle by using the 'blow-up formula' for the 4-dimensional monopole invariants. It is enough once more to consider the case that Σ is connected. Let $k = \sigma \cdot \sigma$ and et $X_k = X \# k\bar{\mathbb{CP}}^2$ be the connected sum of X and k copies of $\mathbb{CP}^2$ with reversed orientation. Put

$$\alpha_k = \alpha + e_1 + \cdots + e_k, \tag{22}$$

where e_i is the generator of H^2 in the ith copy of $\bar{\mathbb{CP}}^2$. The blow-up formula ([19], Proposition 2) says that α_k is a basic class on X_k if α is a basic class on X, because the monopole invariants of corresponding Spin^c structures are equal. Let $\tilde{\Sigma}$ be the embedded surface in X_k formed by an internal connected sum of Σ with the spheres representing the generators of homology in the $\bar{\mathbb{CP}}^2$ summands. (We orient these spheres so that the class e_i evaluates as $+1$ on the ith sphere, and we form the sum respecting these orientations.)

The surface $\tilde{\Sigma}$ has trivial normal bundle and the same genus as Σ. So the previous proposition gives

$$\begin{aligned}\chi_-(\Sigma) &= \chi_-(\tilde{\Sigma}) \\ &\geq \langle \alpha_k, [\tilde{\Sigma}] \rangle \\ &= k + \langle \alpha, \sigma \rangle \\ &= \sigma \cdot \sigma + \langle \alpha, \sigma \rangle,\end{aligned}$$

which is the desired result. □

Remarks. In some applications of this proposition, a proof of the blow-up formula is not needed. One may know that α is a basic class by an application of Taubes' result, Theorem 15, when α arises as K_ω for some symplectic form, in which case one knows that α_k is a basic class also, because there is a symplectic structure on X_k with this class as its canonical class. Without using the blow-up formula, one can therefore deduce

$$\chi_-(\Sigma) \geq \sigma \cdot \sigma + \langle K_\omega, \sigma \rangle. \tag{23}$$

This inequality is an equality for symplectic submanifolds, or smooth algebraic curves in a complex surface, where it is usually referred to as the adjunction formula. The inequality in the proposition is often called the 'adjunction inequality'.

Unlike the 3-dimensional version of this statement, which led only to such basic facts as the lower bound for the genus of a knot in terms of the degree of its Alexander polynomial, the 4-dimensional version provides information which we can reach in no other way at present. Even with the 4-torus, where the Spin^c structure $\mathfrak{c}$ with $c_1 = 0$ has a monopole invariant $SW(T^4, \mathfrak{c}) = 1$, we learn that embedded surfaces in T^4 satisfy

$$\chi_-(\Sigma) \geq |\Sigma \cdot \Sigma|,$$

which is a significant result. For example, it leads to a proof of Milnor's conjecture [31] on the unknotting number of torus knots [22]. (The absolute value appears on the right-hand side because T^4 looks the same with either orientation.) By contrast, the 3-dimensional result, applied to the 3-torus, is entirely contentless.

The result for T^4 above is sharp: every class σ can be represented by a surface of complexity $|\sigma \cdot \sigma|$. More generally, if X carries a symplectic form ω, then Proposition 7.2 is sharp for at least a significant range of classes σ, by Donaldson's theorem [7] on the existence of symplectic submanifolds. That is, since non-degeneracy of a closed 2-form is an open condition, any rational cohomology class $\Omega' \in H^2(X;\mathbb{Q})$ sufficiently close to the class $\Omega = [\omega]$ in $H^2(X;\mathbb{R})$ is represented by a symplectic form ω', and the theorem of [7] says that for some large k (depending on Ω') an integer class $\mathrm{PD}(k\Omega')$ is represented by a symplectic submanifold Σ, for which the inequality (23) is inevitably an equality.

Where the theory hesitates

One should not, however, expect too much of Proposition 7.2. Symplectic 4-manifolds are, perhaps, close cousins of the 3-manifolds Y which fiber over the circle, such as manifolds obtained by zero-surgery on a fibered knot in S^3. For these knots, the degree of the Alexander polynomial and the genus of the knot are equal, and the lower bounds coming from the basic classes are sharp. For general 3-manifolds, the basic classes (as opposed to the monopole classes) give us information which is rather less than sharp, as explained in section 3, and the 4-dimensional situation is probably no better. In dimension three, we recovered much better information by looking at monopole classes detected using the stretching arguments from one dimension higher, or what is essentially the same, the non-vanishing of Floer homology. In dimension four, we have no analogous tool: if we defined a 'monopole class' to be a class $c_1(\mathfrak{c})$ for a Spin^c structure $\mathfrak{c}$ with the property that the equations had solutions for all metrics, as we did in dimension three, then we would have (at present) no tools for detecting monopole classes, other than the observation that basic classes are monopole classes. We have no Floer homology in dimension four.

The relationship between dimensions three and four is clarified somewhat by focusing on 4-manifolds of the form $X = S^1 \times Y$, where Y is a closed oriented 3-manifold. If $\mathfrak{c}$ is a Spin^c structure on X which is pulled back from Y, then it is quite easy to see that the monopole invariants are related:

$$SW(X,\mathfrak{c}) = SW(Y,\mathfrak{c}).$$

Note that, as long as $c_1(\mathfrak{c})$ is non-zero, the invariant $SW(X,\mathfrak{c})$ is well-defined according to our exposition above, because even if $b^+(X) = 1$ (which occurs when $b_1(Y) = 1$), the class $c_1(\mathfrak{c})$ has square zero.

Thus if Y is the manifold $Y(K)$ obtained by zero-surgery on a knot K, then the invariants $SW(X,\mathfrak{c})$ are determined by the Alexander polynomial of K, through a formula like (12). If K has Alexander polynomial 1, like the unknot, then the monopole invariants of X are all zero. A similar example with $b^+ > 1$ is the 4-manifold $S^1 \times Y_2$, where Y_2 is the connected sum of two copies of the same $Y(K)$. For such 4-manifolds, the inequality of Proposition 7.2 tells us nothing.

For the special manifolds $S^1 \times Y$, this set-back is temporary. There is actually a simple device which gives excellent information (sharp in many cases) concerning the minimum genus problem, despite the failure of the basic classes. It is a variant of the stretching argument.

Proposition 7.3. *Let Y be an oriented 3-manifold embedded in a closed, oriented 4-manifold Z with $b^+(Z) > 1$, and suppose that the image of $H_1(Y)$ in every component of $H_1(Z \setminus Y)$ is zero. Let α be a basic class on Z, and let α' be the class on $S^1 \times Y$ obtained by restricting α to Y and pulling back to the product. Then if Σ is an oriented embedded surface in the 4-manifold $S^1 \times Y$, we have*

$$\chi_-(\Sigma) \geq \sigma \cdot \sigma + \langle \alpha', \sigma \rangle,$$

where σ is the class represented.

Remark. Thus α' behaves like a basic class on $X = S^1 \times Y$, even though the monopole invariants of this 4-manifold may all be zero.

Proof. Let Σ be a surface in $S^1 \times Y$. We may take it that Σ is connected. Cut open $S^1 \times Y$ to form the manifold with boundary $[0,1] \times Y$, and make the cut transverse to Σ, so that we obtain a surface with boundary Σ_1. Let $\Sigma_n \subset [0,n] \times Y$ be the surface in $[0,n] \times Y$ obtained by concatenating n copies of Σ_1. Regard Σ_n as a surface with boundary in Z, by embedding $[0,n] \times Y$ as a collar neighborhood N of Y. The boundary of Σ_n, as a subset of ∂N in Z, is independent of n, and by the hypothesis on H_1 it is the boundary of a surface S in $Z \setminus N$.

Consider now the closed surface $S_n \subset Z$ formed as the union of $\Sigma_n \subset N$ and $S \subset Z \setminus N$. We apply the basic class inequality, Proposition 7.2, to the surface S_n, and we find that all three terms are linear in n:

$$\begin{aligned} \chi_-(S_n) &= n\chi_-(\Sigma) + C_1 \\ S_n \cdot S_n &= n(\sigma \cdot \sigma) + C_2 \\ \langle \alpha, S_n \rangle &= n\langle \alpha', \sigma \rangle + C_3, \end{aligned}$$

for constants C_i independent of n. Taking n sufficiently large, we deduce the required inequality. □

As in section 5, one can apply the same device to the case that Z is not closed but has a boundary carrying a contact structure ξ compatible with the orientation. If Z carries a symplectic structure compatible with ξ, then K_ω is a basic class, and we have a revision of the above proposition:

Proposition 7.4. *Let Z be a compact 4-manifold with boundary carrying a contact structure ξ on the boundary, compatible with the boundary orientation, and having a compatible symplectic form ω. Let Y be a 3-manifold embedded in Z and suppose that the image of $H_1(Y)$ in each component of $Z \setminus Y$ is zeero. Let $K' \in H^2(S^1 \times Y)$ be obtained from the canonical class K_ω as before. Then for any embedded surface Σ in $S^1 \times Y$, one has*

$$\chi_-(\Sigma) \geq \sigma \cdot \sigma + \langle K', \sigma \rangle,$$

where σ is the class represented. □

Using the theorem of Eliashberg and Thurston, we deduce:

Corollary 7.5. *Let Y be a closed, oriented 3-manifold carrying a taut foliation $\mathcal{F}$ by oriented leaves. Let e be the euler class of $T\mathcal{F}$ pulled back to $S^1 \times Y$. Then for any embedded surface Σ in $S^1 \times Y$ representing a class σ, one has*

$$\chi_-(\Sigma) \geq \sigma \cdot \sigma + \langle e, \sigma \rangle,$$

where σ is the class represented.

Proof. Using the result of Eliashberg and Thurston, Theorem 5.2, we construct a 4-manifold $Z = S^1 \times Y$ with contact structure ξ on the boundary and a symplectic structure ω, for which K_ω is $e(\mathcal{F})$. The only thing missing is the condition on $H_1(Y)$. This can be put right by the Legendrian surgery of [41]. That is, we choose Legendrian curves γ_i in the contact 3-manifold ∂Z so as to represent a basis for the homology of each component. We can then form a 4-manifold Z^* by adding 2-handles to these curves. According to [41], if the framing of the 2-handles is correctly chosen, the manifold Z^* has contact boundary and carries a compatible form ω^* extending ω. In the larger manifold Z^*, the condition on H_1 is satisfied. □

Corollary 7.6. *Let Σ be an embedded surface in $S^1 \times Y$ representing a class σ. Suppose that Y does not have a basis for H_2 represented by tori. Then the complexity of Σ satisfies the lower bound*

$$\chi_-(\Sigma) \geq |\sigma \cdot \sigma| + |\pi_*(\sigma)|,$$

in which the last term denotes the Thurston norm of the image of σ under the map $\pi : S^1 \times Y \to Y$.

Proof. Under the given hypotheses, the classes $e(\mathcal{F})$ as $\mathcal{F}$ runs through (smooth) taut foliations have the Thurston polytope $B(Y)$ as their convex hull, by Gabai's result, Theorem 1.5. (Recall also that, even without the hypothesis on tori, enough smooth foliations may well exist.) The absolute value sign is appropriate because the 4-manifold has orientation-reversing diffeomorphisms. □

The last corollary usually gives sharp information. To represent a class σ by a surface Σ whose complexity is as given on the right-hand side, one can proceed as follows. Write

$$\sigma = i_* \pi_*(\sigma) + \tau,$$

where i_* is the inclusion of $\{1\} \times Y$ and τ has the form $[S^1] \times [\gamma]$ for a closed curve γ in Y. Let S be a surface representing $\pi_*(\sigma)$ in Y, whose complexity is as small as possible (so given by the Thurston norm). Try to arrange that S meets γ transversely and without excess intersection in Y, so that the geometric and algebraic intersection numbers coincide. If this can be done, then we form a singular surface

$$\Sigma' = i(S) \cup (S^1 \times \gamma)$$

representing σ, and by smoothing the double points which occur at $i(S \cap \gamma)$ we arrive at a smooth surface Σ in the 4-manifold having the right complexity. The condition on excess intersection easily fulfilled if S is connected or if S is a union of parallel surfaces. Thus we can always apply this construction in the case that $b_1(Y)$ is 1.

This corollary seems to draw more from the 3-dimensional world than it does from gauge theory. Our proof has used the existence theorem for taut foliations, as well as the contact perturbations of foliations obtained from [8]. And yet the gauge theory remains an essential part of the story.

An interesting corollary of the lower bound above is to the question of whether the 4-manifold $S^1 \times Y$ can be symplectic. As we said in the previous subsection, if $S^1 \times Y$ is symplectic, then the lower bounds coming from basic classes of $S^1 \times Y$ are sharp, at least for some homology classes σ of positive square. If Y is obtained by zero-surgery on a knot K, then this lower bound is

$$\chi_-(\Sigma) \geq \sigma \cdot \sigma + (2r - 2)\beta \cdot \pi_*(\sigma),$$

where r is the degree of the symmetrized Alexander polynomial and β is the generator of $H_1(Y)$. On the other hand, the corollary above gives the potentially stronger lower bound

$$\chi_-(\Sigma) \geq \sigma \cdot \sigma + (2g - 2)\beta \cdot \pi_*(\sigma),$$

where g is the genus of the knot, at least if g is at least 2. (If g is 1, we need to know that the taut foliation can be made smooth.)

Corollary 7.7. *Let K be a knot of genus g and let r be the degree of its symmetrized Alexander polynomial. Let Y be obtained by zero-surgery on K. If g is 1, suppose in addition that Y carries a smooth taut foliation. Then a necessary condition for the existence of a symplectic structure on $S^1 \times Y$ is that g and r are equal.* □

Where the theory fails

An interesting construction of 4-manifolds was described recently by Fintushel and Stern in [10]. The building brick of this construction is a 4-manifold with boundary, of the form $S^1 \times Y$, where Y is a knot complement, $S^3 \setminus N(K)$. The boundary of $S^1 \times Y$ is a 3-torus. Let X_0 be a $K3$ surface, obtained for example by the Kummer construction which resolves the sixteen double points in $T^4/\pm 1$. Let $T \subset X_0$ be a standard 2-torus. In the Kummer model, T might be the image of a standard 2-torus in T^4. Let X_K be a closed 4-manifold obtained by removing a neighborhood $N(T)$ from X_0 and replacing it with $S^1 \times Y$, attaching the 3-torus boundaries. This should be done in such a way as to leave the homology of X_K the same as that of X_0. (Note that $S^1 \times Y$ and $N(T)$ look the same at the level of homology.)

In [10] it is shown that one can read off the Alexander polynomial of K from the monopole invariants of X_K. In general, X_K has the same homotopy type as X_0, and if K has trivial Alexander polynomial we have no invariants to distinguish the two manifolds. The question raised in [10] is whether X_K can be diffeomorphic to $X_{K'}$ if K and K' are different knots. In particular, if K is a knot with trivial Alexander polynomial, is X_K diffeomorphic to the $K3$ surface X_0?

It is tempting to guess that the *genus* of K is visible in X_K. For example, let $S \subset X_0$ be a sphere meeting T orthogonally in one point and having normal bundle of degree 2. (One can see such a sphere in the Kummer construction.) The cohomology class σ carried by S in X_0 corresponds to a class σ' in X_K. What is the smallest possible genus for a representative of σ'? The simplest surface that is easily visible is the surface Σ obtained from S by removing the disk in which S meets $N(T)$ and replacing the disk with a spanning surface of the knot in $S^1 \times Y$. This representative has

$$\chi_-(\Sigma) = 2g - 2,$$

where g is the genus of the knot. The basic classes give us the lower bound

$$\chi_-(\Sigma') \geq 2r - 2,$$

for any other surface Σ' representing this class, but it seems most likely that Σ is already best possible.

Note that if one carries out this construction using simply the 4-torus for X_0 rather than $K3$, then one has 4-manifolds X_K with an S^1 factor, and the results of the previous subsection can be applied. In particular, the genus of the knot *is* reflected in the minimum genus of embedded surfaces in the 4-manifold, even if it is not reflected in the monopole invariants. This lends some support to the conjecture that $2g - 2$ cannot be bettered in the $K3$ examples, but we seem to have no tools.

Unanswered questions

There are many more questions than answers concerning the minimum genus problem in dimension 4. We know very little about classes of negative square in algebraic surfaces, for example. There are manifolds such as $\mathbb{CP}^2 \# \mathbb{CP}^2$, where the gauge theory invariants have told us very little.

To illustrate the lack of present understanding in another direction, consider the following question.

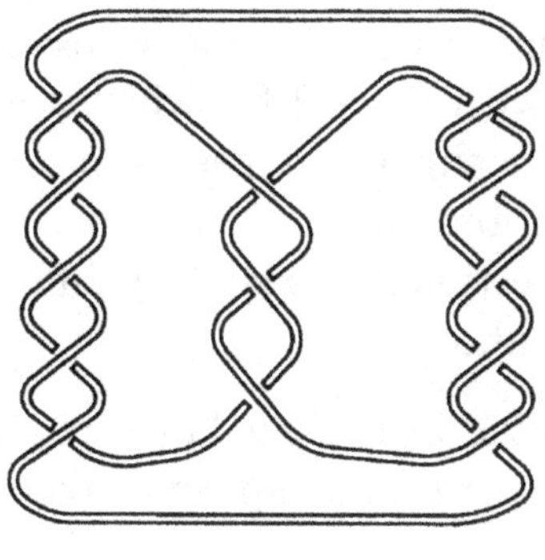

Figure 4: The pretzel knot $(5, -3, 5)$.

Question 7.8. Let $\tau : \tilde{X} \to X$ be a finite covering of a closed 4-manifold X. If Σ is a surface representing the smallest possible complexity for a class $\sigma \in H_2(X; \mathbb{Z})$, does its inverse image $\tilde{\Sigma} = \tau^{-1}(\Sigma)$ have least complexity in its homology class? Is there any large class of 4-manifolds (not simply connected) for which the answer is yes?

This question was posed by Thurston in dimension 3, and answered by Gabai in [13]: since one can lift a taut foliation to a cover, Theorem 1.5 and Corollary 1.3 give an affirmative answer in this case, at least if the manifold is irreducible. There is no comparable tool in dimension 4, which is why the question arises. In a similar spirit, one can ask about branched covers. For example:

Question 7.9. Let $\tau : \tilde{X} \to X$ be a branched double covering of a closed 4-manifold X, branched along a surface $B \subset X$. Let $\tilde{B}$ be its inverse image. If B has least complexity in its homology class, is the same true of $\tilde{B}$? Is there a class of 4-manifolds X for which the answer is yes?

One can also ask whether $\chi_{\min}$ is linear on the ray generated by a class σ in the case that the normal bundle of a representative is trivial:

Question 7.10. Let Σ be a representative of least complexity for the class $\sigma \in H_2(X; \mathbb{Z})$. Suppose that the normal bundle of Σ is trivial, and let Σ_n be the surface consisting of n parallel copies of Σ in the tubular neighborhood. Is Σ_n a representative of least complexity for the class $n\sigma$? Is there a class of 4-manifolds X for which the answer is yes?

It is easy to invent other questions whose answers seem out of reach. The list of 4-manifolds with $b_2 > 0$ for which the function $\chi_{\min}(\sigma)$ is known is very short. There is a slightly longer list if one only asks about surfaces with non-negative normal bundle.

The following question is of a slightly different nature, but arises naturally from the discussion above.

Question 7.11. Let Y be obtained by zero-surgery on a knot K in S^3. If the 4-manifold $S^1 \times Y$ carries a symplectic structure, is the knot K necessarily fibered?

That K be a fibered knot is a sufficient condition for $S^1 \times Y$ to be symplectic. This is an observation of Thurston's. On the other hand, we have seen that a *necessary* condition is that the genus of K coincide with the degree of its symmetrized Alexander polynomial, at least if the genus is 2 or more. Another necessary condition is that the Alexander polynomial have leading coefficient ± 1 (see [10]). The pretzel knot shown in Figure 4 satisfies both of these necessary conditions (though with genus 1), but is not a fibered knot [5].

One might speculate that there is a role for codimension-two foliations of 4-manifolds, somewhat akin to the role of codimension-one foliations of 3-manifolds, though perhaps without the powerful existence theorem. Let $\mathcal{F}$ be an oriented foliation of a closed, oriented 4-manifold X by 2-dimensional leaves. Let us say that $\mathcal{F}$ is *taut* if there is closed 2-form ω which is positive on the leaves. Let $\kappa(\mathcal{F})$ be the 'canonical class' of the almost complex structure on X which $\mathcal{F}$ determines. That is, κ is minus the sum of the euler classes of the tangential and normal 2-plane fields. If Σ is a compact leaf of $\mathcal{F}$, its genus is determined by an adjunction formula,

$$2g - 2 = \sigma \cdot \sigma + \langle \kappa(\mathcal{F}), \sigma \rangle,$$

where σ is the class carried in homology.

Question 7.12. Is it true that a compact leaf of a taut foliation $\mathcal{F}$ of this sort is always genus-mimizing? More generally, is it possible that embedded surfaces in such a foliated 4-manifold satisfy an adjunction inequality, which bounds there genus from below by a formula such as

$$\chi_{-}(\Sigma) \geq \sigma \cdot \sigma + \langle \kappa(\mathcal{F}), \sigma \rangle,$$

at least in the case that $\langle \omega, \sigma \rangle$ is positive?

(Some extra hypothesis of the sort indicated is needed to deal with the case that $\kappa(\mathcal{F})$ is negative, as the example of $S^2 \times S^2$ shows.) Note that $S^1 \times Y$ has codimension-two foliations arising from the foliations of Y, so the above question includes Corollary 7.5 as a special case. Another special case arises when the foliation is a fibration. The quesion then has an affirmative answer, because the fibration carries a symplectic form.

In the version given above, the question may not be very useful, because of a lack of examples of foliations. The situation looks more interesting if one allows the foliation to have singularities of the sort that arise in holomorphic foliations (taking due account of orientations). What we have in mind is the sort of foliation $\mathcal{F}$ that one obtains in a complex surface when $T\mathcal{F}$ is defined by the vanishing of a holomorphic 1-form, or more generally a 1-form with values in some holomorphic line bundle. One could extend the previous question to singular foliations based on this model.

In this form, the question encompasses the question of mimimum genus in the Fintushel-Stern fake $K3$ surfaces (the spaces we called X_K earlier). The 4-manifolds X_K carry singular codimension-two foliations $\mathcal{F}$ formed by combining a holomorphic foliation of $K3$ with Gabai's foliation of the knot comlement $S^3 \setminus K$ (extended trivially to the product with the circle). The canonical class of this foliation $\mathcal{F}$ is given by $2g\,\text{P.D.}[T]$, where $[T]$ is the class of the torus on which the modification was performed and g is the genus of the knot. Thus an affirmative answer to the last question would imply that the genus of K is visible in X_K, as we speculated earlier. Note that the manifolds X_K provide interesting examples of taut, singular foliations on a manifold admitting no symplectic structure. (It is pointed out in [10] that X_K is not symplectic if the Alexander polynomial of K is not monic.)

References

[1] S. Akbulut and J. D. McCarthy, *Casson's invariant for oriented homology 3-spheres*, Mathematical Notes, vol. 36, Princeton University Press, Princeton, NJ, 1990.

[2] D. Auckly, *The Thurston norm and three-dimensional Seiberg-Witten theory*, Osaka J. Math. **33** (1996), no. 3, 737–750.

[3] P. J. Braam and S. K. Donaldson, *Floer's work on instanton homology, knots and surgery*, The Floer memorial volume, Progr. Math., vol. 133, Birkhäuser, Basel, 1995, pp. 195–256.

[4] W. Chen, *Dehn surgery formula for Seiberg-Witten invariants of homology 3-spheres*, preprint.

[5] R. H. Crowell and H. F. Trotter, *A class of pretzel knots*, Duke Math. J. **30** (1963), 373–377.

[6] S. K. Donaldson, *Polynomial invariants for smooth four-manifolds*, Topology **29** (1990), no. 3, 257–315.

[7] ———, *Symplectic submanifolds and almost-complex geometry*, J. Differential Geom. **44** (1996), no. 4, 666–705.

[8] Y. M. Eliashberg and W. P. Thurston, *Confoliations*, to appear in Mem. Amer. Math. Soc.

[9] P. M. N. Feehan and T. G. Leness, *Non-abelian monopoles, I: regularity, transversality and compactness*, preprint.

[10] R. Fintushel and R. Stern, *Knots, links and four-manifolds*, preprint.

[11] A. Floer, *An instanton-invariant for 3-manifolds*, Comm. Math. Phys. **118** (1988), no. 2, 215–240.

[12] K. A. Frøyshov, *The Seiberg-Witten equations and four-manifolds with boundary*, Math. Res. Lett. **3** (1996), no. 3, 373–390.

[13] D. Gabai, *Foliations and the topology of 3-manifolds*, J. Differential Geom **18** (1983), no. 3, 445–503.

[14] ———, *Foliations and genera of links*, Topology **23** (1984), no. 4, 381–394.

[15] ———, *Foliations and the topology of 3-manifolds. II*, J. Differential Geom. **26** (1987), no. 3, 461–478.

[16] ———, *Foliations and the topology of 3-manifolds. III*, J. Differential Geom. **26** (1987), no. 3, 479–536.

[17] ———, *Foliations and 3-manifolds*, Proceedings of the International Congress of Mathematicians (Kyoto 1990), vol. I, Math. Soc. Japan, 1991, pp. 601–609.

[18] C. McA. Gordon and J. Luecke, *Knots are determined by their complements*, J. Amer. Math. Soc. **2** (1989), no. 2, 371–415.

[19] D. Kotschick, J. W. Morgan, and C. H. Taubes, *Four-manifolds without symplectic structures but with nontrivial Seiberg-Witten invariants*, Math. Res. Lett. **2** (1995), no. 2, 119–124.

[20] P. B. Kronheimer, *An obstruction to removing intersection points in immersed surfaces*, Topology **36** (1997), no. 4, 931–962.

[21] P. B. Kronheimer and T. S. Mrowka, *Monopoles and contact structures*, Inventiones Mathematicae, to appear.

[22] ———, *Gauge theory for embedded surfaces, I*, Topology **32** (1993), no. 4, 773–826.

[23] ———, *The genus of embedded surfaces in the projective plane*, Math. Res. Lett. **1** (1994), no. 6, 797–808.

[24] ———, *Embedded surfaces and the structure of Donaldson's polynomial invariants*, J. Differential Geom. **41** (1995), no. 3, 573–734.

[25] ———, *Gauge theory for embedded surfaces, II*, Topology **34** (1995), no. 1, 37–97.

[26] T. J. Li and A. Liu, *General wall crossing formula*, Math. Res. Lett. **2** (1995), no. 6, 797–810.

[27] Y. Lim, *Seiberg-Witten invariants for 3-manifolds and product formulae*, preprint.

[28] P. Lisca and G. Matić, *Tight contact structures and the Seiberg-Witten invariants*, Inventiones Mathematicae, to appear.

[29] M. Marcoli and B.-L. Wang, *Equivariant Seiberg-Witten Floer homology*, preprint.

[30] G. Meng and C. H. Taubes, SW = *Milnor torsion*, Math. Res. Lett. **3** (1996), no. 5, 661–674.

[31] J. Milnor, *Singular points of complex hypersurfaces*, Princeton University Press, Princeton, N.J., 1968, Annals of Mathematics Studies, No. 61.

[32] G. Moore and E. Witten, *Integration over the u-plane in Donaldson theory*, preprint.

[33] V. Y. Pidstrigach and A. N. Tyurin, *Localisation of Donaldson invariants along Seiberg-Witten basic classes*, preprint.

[34] R. Roussarie, *Plongements dans les variétés feuilletées et classification de feuilletages sans holonomie*, Inst. Hautes Études Sci. Publ. Math. (1974), no. 43, 101–141.

[35] C. H. Taubes, *Casson's invariant and gauge theory*, J. Differential Geom. **31** (1990), no. 2, 547–599.

[36] ______, *The Seiberg-Witten invariants and symplectic forms*, Math. Res. Lett. **1** (1994), no. 6, 809–822.

[37] ______, *More constraints on symplectic forms from Seiberg-Witten invariants*, Math. Res. Lett. **2** (1995), no. 1, 9–13.

[38] W. P. Thurston, *Foliations of three-manifolds which are circle bundles*, Ph.D. thesis, Berkeley, 1972.

[39] ______, *A norm for the homology of 3-manifolds*, Mem. Amer. Math. Soc. **59** (1986), no. 339, i–vi and 99–130.

[40] R. G. Wang, *On Seiberg-Witten Floer invariants and the generalized Thom problem*, preprint.

[41] A. Weinstein, *Contact surgery and symplectic handlebodies*, Hokkaido Math. J. **20** (1991), no. 2, 241–251.

[42] E. Witten, *Monopoles and four-manifolds*, Math. Res. Lett. **1** (1994), no. 6, 769–796.

The Geometry of the Seiberg-Witten Invariants

Clifford H. Taubes*

Department of Mathematics
Harvard University
Cambridge MA 02138

The Seiberg-Witten invariants were defined by Witten [W1] for any compact, oriented 4-manifold; and they are determined by the underlying differentiable structure when the Betti number b^{2+} is larger than 1. After the choice of orientation for the real line $\det^+ = H^0 \otimes \det(H^1) \otimes \det(H^{2+})$, the Seiberg-Witten invariants constitute a map from the set, $\mathcal{S}$, of $\mathrm{Spin}^{\mathbb{C}}$ structures on the 4-manifold to the integers. There is also an extension of SW in the case where the Betti number b^1 is positive to a map SW: $\mathcal{S} \to \Lambda^* H^1(X;\mathbb{Z})$. (Here, $\Lambda^* H^1(X;\mathbb{Z}) = \mathbb{Z} \oplus H^1 \oplus \Lambda^2 H^1 \oplus \cdots \oplus \Lambda^{b_1} H^1$. Note that the projection of the image of SW on the summand $\mathbb{Z}$ reproduces the original map as defined from $\mathcal{S}$ to $\mathbb{Z}$ in [W1].) In either guise, this map, SW, is computed by an algebraic count of solutions to a certain non-linear system of differential equations on the manifold. The invariant SW and the Seiberg-Witten equations were introduced to the mathematical community by Witten [W1] after his ground breaking work with Seiberg in [SW1], [SW2]. See also [KM], [Mor], [KKM] and [T1].

*Supported in part by the National Science Foundation.

It is hard to deny that Seiberg and Witten's equation has revolutionized the study of smooth, 4-dimensional manifolds. This advance is due primarily to the fact that the Seiberg-Witten invariants are eminently computable, much more so than the Donaldson's invariants. (Witten has conjectured that the information coded in the two invariants is equivalent.) This is to say that the basic strategies for computing Donaldson invariants can be applied with comparitively little effort (not no effort, mind you) to computing the Seiberg-Witten invariants. Thus, our knowledge has expanded greatly due to the relative ease of the calculations involved. (See, e.g. [KM], [MST], [FS1].)

However, a second impetus for the Seiberg-Witten revolution comes directly from the observation of an intimate relationship between the Seiberg-Witten invariants and symplectic geometry. This last observation has brought symplectic geometry into the 4-manifold picture, with surprising gains both for 4-manifold topology and for symplectic geometry.

As remarked in [T1], a symplectic manifold has a natural orientation as does the line $\det^+$. Furthermore, there is a canonical identification of the set $\mathcal{S}$ with $H^2(X;\mathbb{Z})$. Thus, on a symplectic 4-manifold, SW can be viewed as a map from $H^2(X;\mathbb{Z})$ to $\mathbb{Z}$, or, more generally, from $H^2(X;\mathbb{Z})$ to $\Lambda^* H^1(X;\mathbb{Z})$. Meanwhile, a compact symplectic 4-manifold has a second natural map from $H^2(X;\mathbb{Z})$ to $\mathbb{Z}$, its Gromov invariant, Gr. The map Gr also extends on a $b^1 > 0$ symplectic 4-manifold to a map from $H^2(X;\mathbb{Z})$ into $\Lambda^* H^1(X;\mathbb{Z})$; the extension is sometimes called the Gromov-Witten invariant, but it will be denoted here by Gr as well. In either guise, Gr, assigns to a class e a certain weighted count of compact, symplectic submanifolds whose fundamental class is Poincare dual to e.

The Gromov invariant was introduced initially by Gromov in [Gr] and then generalized by Witten [W2] and Ruan [Ru]. See also [T2] and [T5]. (Note that Gr here does not count maps from a fixed complex curve. It differs in this fundamental sense from the Gromov-Witten invariant introduced in [W2].) The precise definitions of SW and Gr are provided in the first section of this paper.

Here is the main theorem which relates SW to Gr:

Theorem 1. *Let X be a compact, symplectic manifold with $b^{2+} > 1$. Use the symplectic structure to orient X and the line* $\det^+$; *and use the symplectic structure to define* SW *as a map from $H^2(X;\mathbb{Z})$ to $\Lambda^* H^1(X;\mathbb{Z})$. Also, use the symplectic structure to define* Gr: $H^2(X;\mathbb{Z}) \to \Lambda^* H^1(X;\mathbb{Z})$. *Then* SW=Gr.

Theorem 1 is proved in [T5].

The equivalence between the Gromov invariant and the original SW map into Z was announced by the author in [T1]. The proof of Theorem 1 can be divided into three main parts. The first part explains how a non-zero Seiberg-Witten invariant implies the existence of symplectic submanifolds. The second part explains how a symplectic submanifold can be used to construct a solution to a version of the Seiberg-Witten equations. The final part compares the counting procedures for the two invariants. The first and second parts of the proof can be found in [T3] and [T4], respectively and the final part (together with an overview of the whole strategy) is in [T5]. Some of the early applications of Theorem 1 are also described in [Ko1].

A restricted version of Theorem 1 holds in the case when $b^{2+} = 1$. Here, a fundamental complication is that the Seiberg-Witten invariant depends on more than the differentiable structure. This is to say that there is a dependence on a "so called" choice of chamber. However, the symplectic form selects out a unique chamber, and with this understood, one has:

Theorem 2. *Let X be a compact, oriented 4-manifold with $b^{2+} = 1$ and with a symplectic form. Then the symplectic form canonically defines a chamber in which the equivalence* SW = Gr *holds for classes $e \in H^2(X;\mathbb{Z})$ which obey $\langle e, s\rangle \geq -1$ when ever $s \in H^2(X;\mathbb{Z})$ is represented by an embedded, symplectic sphere with self-intersection number -1.*

(Here, $\langle , \rangle$ denotes the pairing between cohomology and homology.) Theorem 2 is also proved in [T5].

By the way, when X is a $b^{2+} = 1$ symplectic manifold and $e \in H^2(X;\mathbb{Z})$ is a class for which the conditions of Theorem 2 do not hold, it is still the case that the Seiberg-Witten invariant $\mathrm{SW}(e)$ still counts pseudo-holomorphic subvarieties of some sort; Li and Liu [LL] have very recently sorted out the details. See also Theorem 1.6 below. (The Gromov invariant as defined below is not the correct symplectic invariant for such e since the subspaces involved can have singularities.)

The remainder of this essentially expository article is devoted to three related topics, one for each numbered section. The first topic is Theorems 1 and 2 and related results about the Seiberg-Witten invariants on a symplectic manifold. To be more precise, the first section below starts with a summary of the definitions of both the Seiberg-Witten invariant and the Gromov-Witten invariant. This first section also lists other special properties of the Seiberg-Witten invariants on symplectic manifolds. (See Theorems 1.6-1.8, below.) And, the section ends with an instructive example.

The second topic concerns the role of symplectic manifolds in the broader arena of 4-dimensional differential topology. And, the final topic concerns a possible geometric interpretation of the Seiberg-Witten invariants in the non-symplectic case. In particular, the final section contains, for the most part, speculations about future avenues of research.

1 The Seiberg-Witten and the Gromov-Witten invariants.

The purpose of this first section is to give a precise definition of the Seiberg-Witten invariants and also the Gromov invariants for a symplectic 4-manifold. The former is considered in Subsections 1a-c, and the latter in Subsections 1d,e. Then, Subsection 1f describes the basic geometric underpinning of the SW= Gr theorems and Subsection 1g summarizes some of other properties of SW on a symplectic manifold. The final subsections here discuss interpretations, applications and examples of the SW=Gr theorems.

a) The Seiberg-Witten equations.

The Seiberg-Witten equations were first introduce by Seiberg and Witten in [SW1], and [SW2] and see also [W1]. A purely mathematical approach to these equations was first taken in [KM]. The book by Morgan [Mor] is a more complete reference, as is the exposition by Kotschick, Kronheimer and Mrowka [KKM].

In this subsection, X is a compact, connected, oriented, 4-dimensional manifold. Let $b^1 = \dim(H^1(X))$ denote the first Betti number of X and let b^{2+} denote the dimension of a maximal subspace, $H^{2+}(X;\mathbb{R}) \subset H^2(X)$ where the cup product form is positive.

Fix a smooth Riemannian metric on X. The metric defines the principle SO(4) bundle of orthornormal frames, Fr$\to X$. Of the various associated bundles to this frame bundle, two in particular play central roles. These are the bundles Λ_+ of self-dual 2-forms and Λ_- of anti-self dual 2-forms. Note that $\Lambda^2 TX \approx \Lambda_+ \oplus \Lambda_-$.

By definition, a Spin$^{\mathbb{C}}$ structure on X is an equivalence class of lifts of Fr to a principal Spin$^{\mathbb{C}}(4)$ bundle $F \to X$. In this regard, recall that the group Spin$^{\mathbb{C}}(4)$ is the group $(SU(2) \times SU(2) \times U(1))/\{\pm 1\}$, this being a central extension of $SO(4) = (SU(2) \times SU(2))/\{\pm 1\}$ by the circle $U(1)$. The homomorphism Spin$^{\mathbb{C}} \to (SU(2) \times SU(2))/\{\pm 1\}$ simply forgets the factor of $U(1)$.

A Spin$^{\mathbb{C}}$ lift F of Fr has two canoncial associated $\mathbb{C}^2$ bundles, $S_\pm \to$

X which are defined using the two evident homomorphisms of $\mathrm{Spin}^{\mathbb{C}}$ to $U(2) = (SU(2) \times U(1))/\{\pm 1\}$. Note that S_+ is distinguished by the fact that the projective bundle is the unit 2-sphere bundle in Λ_+. There is, of course, an analagous relationship between S_- and Λ_-.

With the preceding understood, the original version of Seiberg and Witten's equations can now be defined. These are equations for a pair (A, ψ), where A is a connection on $\det(S_+)$ and where ψ is a section of S_+. The equations read:

$$\begin{aligned} D_A\psi &= 0, \\ P_+F_A &= \frac{1}{4}\tau(\psi \oplus \psi^*) + \mu. \end{aligned} \tag{1.1}$$

In the first line above, D_A is the Dirac operator, a first order differential operator which maps sections of S_+ to sections of S_-. This D_A is defined as the composition of Clifford multiplication (a homomorphism from $S_+ \otimes T^*X$ to S_-) with covariant differentiation using the connection on S_+ which comes from the Levi-Civita connection on Fr and the connection A on $\det(S_+)$. In the second line of (1.1), P_+ denotes the orthogonal projection from Λ^2T^*X to Λ_+ and F_A denotes the curvature 2-form of A. Meanwhile, τ is the adjoint of the Clifford multiplication endomorphism from $\Lambda_+ \otimes \mathbb{C}$ into $\mathrm{End}(S_+)$ and μ is a fixed, imaginary valued, anti-self dual 2-form on X. Any choice for μ will do.

There is a natural action of the group of smooth maps from X to $U(1)$ on the set of solutions to (1.1). The action sends a map g and a pair (A, ψ) to $(A + 2gdg^{-1}, g\psi)$. Use $\mathcal{M}$ to denote the set of orbits under this group action. Typically, notational distinctions will not be made between a pair (A, ψ) and its orbit in $\mathcal{M}$.

Topologize $\mathcal{M}$ as follows: First, introduce the manifold $\mathrm{Conn}(\det(S_+))$ of Hermitian connections on $\det(S_+)$. This is an affine Frechet manifold modelled on $i\cdot\Omega^1$. Here, Ω^1 denotes the vector space of smooth 1-forms on X. With $\mathrm{Conn}(\det(S_+))$ understood, introduce the space $\mathrm{Conn}(\det(S_+)) \times C^\infty(S_+)$. The group $C^\infty(X; S^1)$ acts smoothly on the latter (as indicated above), and the space of orbits of this group action, $(\mathrm{Conn}(\det(S_+)\times C^\infty(S_+))/C^\infty(X; S^1)$, is given the quotient topology. The space $\mathcal{M}$ sits in this quotient, and the implicit topology on $\mathcal{M}$ is the subspace topology inherited from the orbit space $(\mathrm{Conn}(\det(S_+) \times C^\infty(S_+))/C^\infty(X; S^1)$.

Here are some basic properties of $\mathcal{M}$ (see, [W1] or [KM], [Mor], [KKM]):

- $\mathcal{M}$ is always compact.

- If $b^{2+} > 0$, then there is a Baire set of $\mathcal{U} \subset C^\infty(X; i\Lambda_+))$ of choices for μ in (1.1) whose corresponding $\mathcal{M}$ has the structure of a smooth manifold of dimension

$$2d = -\frac{1}{4}(2\chi + 3\tau) + \frac{1}{4}c_1 \bullet c_1.$$

Here, χ is the Euler characteristic of X and τ is the signature of X. Also, "$\bullet$" signifies the pairing on $H^2(X;\mathbb{Z})$ which is cup product composed with evaluation in the fundamental class. Furthermore, when $\mu \in \mathcal{U}$, then

a) There are no points in $\mathcal{M}$ where the corresponding ψ is zero.

b) $\mathcal{M}$ is orientable, and an orientation of $\det^+$ canonically orients $\mathcal{M}$.

c) The subspace of orbits $(p, (A, \psi)) \in (\det(S_+) \times \mathrm{Conn}(\det(S_+)) \times C^\infty(S_+))/C^\infty(X; S^1)$ where $(A, \psi) \in \mathcal{M}$ naturally defines a smooth, principal S^1 bundle $\mathcal{E} \to X \times \mathcal{M}$.

(1.2)

A Baire set is a countable intersection of open and dense sets and so is dense. The Baire set in question is characterized by the condition that a certain family of first order, elliptic differential operators that is parameterized by the points in $\mathcal{M}$ has, at each point in $\mathcal{M}$, trivial cokernel.

Here are some additional comments about (1.2):

- The number $2d$ in (1.2) can be even or odd. Its parity is the same as that of $\frac{1}{2}(\chi + \tau) = 1 - b^1 + b^{2+}$.

- Equation (1.2) implies the following: When $d < 0$ and $\mu \in \mathcal{U}$, then $\mathcal{M} = \varnothing$. This is because there are no negative dimensional manifolds.

- In the case $d = 0$ and $\mu \in \mathcal{U}$, then $\mathcal{M}$ consists of a finite set of points. In this case, an orientation on $\mathcal{M}$ simply assigns either $+1$ or -1 to each point. (This is because $H_0(\text{point}; \mathbb{Z})$ already has a canonical generator, which is the point itself. The orientation assigns a fundamental class which is either the point, or - the point.)

- Let $c_1(\mathcal{E})$ denote the first Chern class of the principal S^1 bundle $\mathcal{E} \to X \times \mathcal{M}$. Then slant product with $c_1(\mathcal{E})$ defines a map, $\phi : H_*(X;\mathbb{Z}) \to H^{2-*}(\mathcal{M};\mathbb{Z})$.

(1.3)

b) The Seiberg-Witten invariant.

Let $\mathcal{S}$ denote the set of SpinC structures on X. Although $\mathcal{S}$ requires a choice of Riemannian metric for its definition, there is a natural identification between such sets defined by any two metrics. (Remember that the space of metrics on X is convex.) Thus, one can speak unambiguously about $\mathcal{S}$ with out reference to a particular metric. Note that $\mathcal{S}$ is an affine space modelled on $H^2(X;\mathbb{Z})$.

Likewise, the definition of SW requires a choice of Riemannian metric; and it also requires a choice of perturbing form μ in the set $\mathcal{U}$ of (1.2). Here is the definition of SW:

Definition 1.1. Fix an orientation for the line $\det^+$. Fix a Riemannian metric on X and the fix a $\mathrm{Spin}^{\mathbb{C}}$ structure in $\mathcal{S}$. Also, fix $\mu \in \mathcal{U}$ so that the conclusions of (1.2) and (1.3) are valid. Let d be as defined in (1.2). Then, the value of $\mathrm{SW} \in \Lambda^* H^1(X;\mathbb{Z})$ on the given $\mathrm{Spin}^{\mathbb{C}}$ structure is defined as follows:

- If $d < 0$, then SW$= 0$.

- If $d = 0$, then $\mathcal{M}$ is a finite set of points and the chosen orientation for $\det^+$ defines a map $\varepsilon : \mathcal{M} \to \{\pm 1\}$. With this understood, then
$$\mathrm{SW} \equiv \Sigma_{\Xi \in \mathcal{M}}\, \varepsilon(\Xi) \tag{1.4}$$
which is an element in the $\mathbb{Z}$ summand of $\Lambda^* H^1$.

- In general, $\mathrm{SW} \in \mathbb{Z} \oplus H^1 \oplus \cdots \oplus \Lambda^{2d} H^1$ with non-zero projection in $\Lambda^p H^1$ only if p has the same parity $1 - b^1 + b^{2+}$. In this case, SW is defined by its values on the set of decomposable elements in $\Lambda^p(H_1(X;\mathbb{Z})/\,\mathrm{Torsion})$; and
$$\mathrm{SW}(\gamma_1 \wedge \cdots \wedge \gamma_p) = \int_{\mathcal{M}} \phi(\gamma_1) \wedge \cdots \wedge \phi(\gamma_p) \wedge \phi(*)^{d-p/2}, \tag{1.5}$$
where $*$ is the class of a point generating $H_0(X)$.

The next proposition asserts that the apparent dependence of SW on the choice of metric and μ is spurious when $b^{2+} > 1$.

Proposition 1.2. *Let X be a compact, connected, oriented 4-manifold with $b^{2+} > 1$. Then the value of* SW *is independent of the choice* *of Riemannian metric and form* μ. *In fact,* SW *depends only on the diffeomorphism type of* X. *Furthermore,* SW *pulls back naturally under orientation preserving diffeomorphisms. This is to say that if* $\varphi : X \to X'$ *is a smooth, orientation preserving diffeomorphism, then* $SW_X(\varphi^*\eta) = \varphi^* SW_{X'}(\eta)$. *Finally,* SW *changes sign when the orientation of the line* $\det^+$ *is switched.*

(Note that $\mathrm{Spin}^{\mathbb{C}}$ structures pull-back under diffeomorphisms because metrics do.)

See, e.g. [Mor] or [KKM] for a proof of this proposition.

The preceding proposition does not hold in general in the case where the 4-manifold X has $b^{2+} = 1$. However, the failure of this proposition can be readily analyzed, and the results are summarized in Proposition 1.3, below. To state the proposition precisely, it is convenient to make a short digression to consider some special features of $b^{2+} = 1$ manifolds.

To begin the digression, introduce $\mathrm{Met}(X)$ to denote the Frechet space of smooth, Riemannian metrics on X. Given a metric g on X, let ω_g denote the unique (up to multiplication by $\mathbb{R}^*$), non-trivial, self-dual, harmonic 2-form on X. With ω_g understood, then each $c \in H^2(X;\mathbb{Z})$ defines a "wall" in $\mathrm{Met}(X) \times i \cdot \Omega^{2+}$ whose elements consist of pairs (g, μ) where $2\pi \cdot [\omega_g] \bullet c = i \cdot \int_X \omega_g \wedge \mu$. The wall divides $\mathrm{Met}(X) \times i \cdot \Omega^{2+}$ into two open sets, each of which is called a "c-chamber".

With the preceding, then Proposition 1.2 has the following $b^{2+} = 1$ version:

Proposition 1.3. *Let* X *be a compact, connected, oriented 4-manifold with* $b^{2+} = 1$. *Let* s *be a* $\mathrm{Spin}^{\mathbb{C}}$ *structure on* X. *Then the value of* $\mathrm{SW}(s) \in \Lambda^* H^1(X;\mathbb{Z})$ *is constant on any* $c = c_1(\det(S_+))$ *chamber in* $\mathrm{Met}(X) \times i \cdot \Omega^{2+}$.

See, e.g. [Mor] or [KKM] for a proof. However, the point is that the arguments for Proposition 1.2 work on an open set in $\mathrm{Met}(X) \times i \cdot \Omega^{2+}$ where the corresponding $\mathcal{M}$ contains no elements where the corresponding ψ vanishes identically. Indeed, the count for SW can change along a path in $\mathrm{Met}(X) \in i \cdot \Omega^{2+}$ only when the path intersects elements in $\mathcal{M}$ where the corresponding ψ vanishes identically. And, such elements occur if and only if (g, μ) lies in the wall. The change in SW as the wall is crossed can be computed. See [KM], [LL2], [OO].

c) The Seiberg-Witten invariants on symplectic manifolds.

As remarked in the introduction, a symplectic 4-manifold has a natural orientation, a natural orientation for the line $\det^+$ and a natural identification between $\mathcal{S}$ and $H^2(X;\mathbb{Z})$. The introduction also asserted that a symplectic manifold with $b^{2+} = 1$ also has a natural chamber. The purpose of this subsection is to explain these assertions.

The orientation of X. A symplectic 4-manifold is, by definition, a pair (X, ω), where X is a smooth 4-manifold and where ω is a closed 2-form on X with $\omega \wedge \omega$ nowhere zero. (The characteristic number $\frac{1}{2}(\chi + \tau) = 1 - b^1 + b^{2+}$ must be even for X to admit a symplectic form.) Because $\omega \wedge \omega$ is nowhere zero, this form orients X. This is the orientation that the introduction referred to. It will be assumed throughout.

The orientation of $\det^+$. The description of the orientation for the line $\det^+$ is conveniently divided into five steps.

Step 1. A choice of orientation for $\det_+$ is equivalent to a choice of orientation for the virtual vector space $H^1(X;\mathbb{R}) - H^0(X;\mathbb{R}) - H^{2+}(X;\mathbb{R})$. After a metric on X is chosen, the latter can be viewed using Hodge theory as the formal difference between the kernel and the cokernel of the operator $\delta_0 = (P_+d, d^*) : \Omega^1 \to \Omega^0 \oplus \Omega^{2+}$. Here, Ω^1 is the space of smooth 1-forms, Ω^0 is the space of smooth functions and Ω^{2+} is the space of smooth, self dual 2-forms.

Step 2. Every symplectic manifold admits almost complex structures, endomorphisms J of TX with square -1. As noted by Gromov [Gr], one can find almost complex structures with the property that the bilinear form

$$g = \omega(\cdot, J(\cdot)) \tag{1.6}$$

defines a Riemannian metric on TX. Such a J will be called *ω-compatible.*

The almost complex structure J decomposes $TX \otimes \mathbb{C} \approx T_{1,0} \oplus T_{0,1}$ into a sum of complex 2-plane bundles such that J has eigenvalue i on the former and $-i$ on the latter. The complexified cotangent bundle decomposes analogously as $T^{1,0} \oplus T^{0,1}$.

Thus, the endomorphism J acts, by definition on the domain of the operator δ_0.

Step 3. If the metric g is chosen as in (1.6), then there is also a natural almost complex structure (call it J_R) which acts on the range of δ_0. The latter is induced from a square -1 endomorphism (also called J_R) on the vector bundle $\varepsilon_{\mathbb{R}} \oplus \Lambda_+$ whose sections define δ_0's range. Here,

$\varepsilon_{\mathbb{R}} \to X$ denotes the product bundle $X \times \mathbb{R}$. Likewise, $\varepsilon_{\mathbb{C}}$, below, will denote the product complex line bundle.

To define J_R, remark first that the metric in (1.6) splits $\Lambda^2 T^* X$ as $\Lambda_+ \oplus \Lambda_-$. The form ω is self dual with respect to this splitting and has norm $\sqrt{2}$ everywhere. Conversely, if g is any metric for which ω is self-dual and has norm $\sqrt{2}$, then $J \equiv g^{-1}\omega$ defines an almost complex structure J on TX such that (1.6) holds.

Note that J induces an endomorphism of $\Lambda^2 T^* X$ with square 1 which preserves Λ_+. The $+1$ eigenspace of this endomorphism on Λ_+ is the span of ω. The orthogonal compliment is the -1 eigenspace. The latter is an oriented, 2-plane bundle over X which is the underlying real bundle of the complex line bundle $K^{-1} = \Lambda^2 T^{0,1}$.

With the preceding understood, view $\varepsilon_{\mathbb{R}} \oplus \Lambda_+$ as a complex 2-plane bundle by writing the latter as $\varepsilon_{\mathbb{C}} \oplus K^{-1}$, where $x + y \cdot \omega \in \varepsilon_{\mathbb{R}} \oplus \Lambda_+$ is identified with $x + \sqrt{-1} \cdot y \in \varepsilon_{\mathbb{C}}$. Multiplication by $\sqrt{-1}$ on $\varepsilon_{\mathbb{C}} \oplus K^{-1}$ defines the endomophism J_R on $\varepsilon_{\mathbb{R}} \oplus \Lambda_+$.

Step 4. In general, $\delta_0 J - J_R \delta_0 \neq 0$. However, this difference is always a zero'th order operator. (The symbol of δ_0 intertwines J with J_R; and δ_0 itself intertwines J with J_R when J is an integrable almost complex structure.)

The fact that $\delta_0 J - J_R \delta_0$ is zero'th order implies that there is a relatively compact perturbation of δ_0 which does intertwine J and J_R. For example, $\delta_1 = 2^{-1} \cdot (\delta_0 - J_R \cdot \delta_0 \cdot J)$ has this property.

Since δ_1 differs from δ_0 by a zero'th order operator, both its kernel and cokernel are finite dimensional. Furthermore, because δ_1 intertwines J with J_R, its kernel and cokernel have natural structures as complex vector spaces. And, since complex vector spaces have canonical orientations, the virtual vector space kernel(δ_1) – cokernel(δ_1) has a canonical orientation.

Step 5. The complex orientation for kernel(δ_1) – cokernel(δ_1) canonically orients the line $\det^+ =$ kernel(δ_0) – cokernel(δ_0). The argument here is standard K-theory since the family of operator $\{\delta_t = t \cdot \delta_0 + (1-t) \cdot \delta_1\}_{t \in [0,1]}$ defines a continuous map of Fredholm operators with respect to appropriate Hilbert space completions of Ω^1 and $\Omega^0 \oplus \Omega^{2+}$. (The Sobolev spaces L^2_1 for the range and L^2 for the domain will suffice.) The point is that the association of the virtual vector space kernel(δ_t)– cokernel(δ_t) to $t \in [0,1]$ defines an element in the real K-theory of the interval (see the Appendix in [At].) Since the interval is contractible, this element is trivial. In particular, it has vanishing first Stieffel-Whitney class, so it is orientable and an orientation at $t = 1$ induces one at $t = 0$.

Note that the purpose of orienting the line det+ is to obtain a reasonably canonical orientation for the moduli space $\mathcal{M}$. In this regard, the symplectic orientation of $\det^+$ induces an orientation on $\mathcal{M}$ which is described directly in Section 4.

The identification of S with $H^2(X;\mathbb{Z})$. As remarked, the set S has the natural structure of an affine space modelled on $H^2(X;\mathbb{Z})$. This implies that the identification in question arises immediately with the specification of a "canonical" $\text{Spin}^{\mathbb{C}}$ structure. And, as observed in [T1], there is a canonical $\text{Spin}^{\mathbb{C}}$ structure on a symplectic manifold.

With the metric chosen from an ω-compatible J, the canonical $\text{Spin}^{\mathbb{C}}$ structure is characterized by the identifications

$$S_+ = \mathbb{I} \oplus K^{-1} \quad \text{and} \quad S_- = T^{0,1}, \tag{1.7}$$

where $K^{-1} = \Lambda^2 T^{0,1}$. This splitting of S_+ is defined as follows: Clifford multiplication defines an endomorphism from Λ_+ into the bundle of skew hermitian endomorphisms of S_+. With the preceding understood, the splitting of S_+ in (1.7) is the decomposition of S_+ into eigenbundles for the action of ω; here ω acts with eigenvalue $-2i$ on the trivial summand $\mathbb{I}$, and it acts with eigenvalue $+2i$ on the K^{-1} summand.

As just remarked, the identification in (1.5) of a canonical element in s identifies

$$\mathcal{S} \approx H^2(X;\mathbb{Z}). \tag{1.8}$$

Under this identification, a class $e \in H^2(X;\mathbb{Z})$ is sent to the $\text{Spin}^{\mathbb{C}}$ structure whose $S_\pm$ bundles are given by

$$S_+ = E \oplus (K^{-1} \otimes E) \quad \text{and} \quad S_- = T^{0,1} \otimes E, \tag{1.9}$$

where E is a complex line bundle whose first Chern class is isomorphic to e. Once again, this splitting of S_+ is into eigenbundles for the action of ω on S_+; and the convention is that the bundle where ω acts as $-2i$ is written first.

By the way, after the identification in (1.8), the dimension $2d$ of the Seiberg-Witten moduli space (as given in (1.2)) can be rewritten as follows: If $e \in H^2(X;\mathbb{Z})$ and if e is used to determine the $\text{Spin}^{\mathbb{C}}$ structure as in (1.9), then the formal dimension of the moduli space $\mathcal{M}$ is

$$2 \cdot d = e \bullet e - c \bullet e \tag{1.10}$$

where $c = c_1(K)$ with $K = \Lambda^2 T^{1,0}$. (The number $e \bullet e - c \bullet e$ is even because the class c is characteristic: Its mod 2 reduction is the second Stieffel-Whitney class of X.)

The symplectic chamber when $b^{2+} = 1$. Suppose now that X is a compact, oriented 4-manifold with $b^{2+} = 1$ and with a symplectic form ω. The latter defines a canonical c-chamber for each $c \in H^2(X;\mathbb{Z})$ by requiring μ in (1.1) to obey $i \cdot \int_X \omega \wedge \mu > 2\pi \cdot [\omega] \bullet c$. This last chamber will be called the "symplectic chamber".

Note, by the way, that two symplectic forms ω and ω' on X define the same chamber when $[\omega] \bullet [\omega'] > 0$. Thus, the symplectic chamber depends only on the form ω up to continuous deformations through closed forms ν with $[\nu] \bullet [\nu] > 0$.

In the subsequent discussions, the Seiberg-Witten invariant for such a pair (X, ω) will always denote the map SW from Proposition 1.3 as defined in the symplectic chamber. This $b^{2+} = 1$ definition of SW will be implicit in the subsequent discussions.

d) Pseudo-holomorphic submanifolds.

As noted in the introduction, the Gromov-Witten invariant is defined by counting (in a suitable sense) pseudo-holomorphic submanifolds on the symplectic manifold X. Thus, a more complete description of this invariant must start with a digression to discuss pseudo-holomorphic submanifolds. There are four parts to this discussion.

Part 1. The complex line bundle $K = \Lambda^2 T^{1,0}$ is called the canonical bundle. Note that the isomorphism class of K, and thus its first Chern class $c \in H^2(X;\mathbb{Z})$, are independent of the choice of ω-compatible almost complex structure J. Furthermore, this isomorphism class and also c are both unchanged if ω is changed through a continuous family of symplectic forms. (Note the sign convention here: $c \bullet [\omega] < 0$ when $X = \mathbb{CP}^2$.)

Part 2. A submanifold Σ in X is called pseudo-holomorphic when J preserves $T\Sigma$. It follows from the non-degeneracy of (1.6) that ω is non-degenerate on $T\Sigma$ and so orients Σ. Infact, J induces the structure of a complex curve on Σ. Then, the inclusion map of Σ into X is pseudo-holomorphic in the sense of Gromov [Gr].

If Σ is a connected and compact pseudo-holomorphic submanifold, then the genus of Σ is constrained by the adjunction formula to equal

$$\text{genus} = 1 + \frac{1}{2}(e \bullet e + c \bullet e) \tag{1.11}$$

where e is the Poincare' dual to the fundamental class $[\Sigma]$ of Σ.

Henceforth, all pseudo-holomorphic submanifolds in this Section 1 should be assumed to be compact unless stated to the contrary.

Part 3. Fix a pseudo-holomorphic submanifold Σ. Since J preserves $T\Sigma$, it must also preserve the orthogonal compliment in TX of $T\Sigma$.

The latter is the normal bundle, N, of Σ. Thus, N has a natural structure as a complex line bundle over Σ. The metric from TX defines a connection on $N \to \Sigma$, and thus endows N with a holomorphic structure as a bundle over the complex curve Σ. With this understood, one can introduce the associated d-bar operator, $\bar{\partial}$, to map sections of N to sections of $N \otimes T^{0,1}C$. (Here, $T^{0,1}C$ is the usual anti-holomorphic summand of $T^*C \otimes_{\mathbb{R}} \mathbb{C}$.)

One's first guess is that the kernel of $\bar{\partial}$ corresponds to the vector space of deformations of Σ in X which are pseudo-holomorphic to first order. However, this guess is wrong, in general. Rather, this vector space corresponds to the kernel of certain canonical, zero'th order deformation of $\bar{\partial}$. This deformation is an $\mathbb{R}$ linear operator, D, which also maps sections of N to sections of $N \otimes T^{0,1}C$ and which is defined as follows: The 1-jet off of Σ of the almost complex structure defines a pair (ν, μ) of section of $T^{0,1}C$ and $N^{\otimes 2} \otimes T^{0,1}C$. (See (2.3) in [T4].) Then

$$Dh = \bar{\partial}h + \nu \cdot h + \mu \cdot \bar{h}. \tag{1.12}$$

Part 4. Note that the index of D is given by the Riemann-Roch formula, which is to say that it equals $2d$ in (1.10) in the case where $e \in H^2(X;\mathbb{Z})$ is Poincare' dual to $[\Sigma]$. As the index is, by definition, the difference between the dimensions (over $\mathbb{R}$) of the kernel and the cokernel of D, a necessary condition for the triviality of cokernel(D) is that $2 \cdot d$ be non-negative. In general, this condition is not sufficient. However, if $2d \geq 0$, all pseudo-holomorphic submanifolds have trivial cokernel if the almost complex structure is chosen from a certain Baire subset of ω compatible almost complex structures. (This fact is proved in, e.g. [MS].)

e) The Gromov-Witten type invariants.

Fix $e \in H^2(X;\mathbb{Z})$. This subsection defines $\mathrm{Gr}(e) \in \Lambda^* H^1(X;\mathbb{Z})$. (The reader is referred to [T2] for the proofs of the assertions below.) The discussion here is broken into seven parts.

Part 1. Introduce the integer $d = d(e)$ as defined by (1.10). Then $\mathrm{Gr}(e)$ lies in the summand $\mathbb{Z} \oplus \Lambda^2 H^1 \oplus \cdots \oplus \Lambda^{2d} H^1$. Its projection into $\Lambda^{2p} H^1$ (for $0 \leq p \leq d$) can be determined by evaluating $\mathrm{Gr}(e)$ on a decomposable element in $\Lambda^{2p}(H^1(X;\mathbb{Z})/\,\mathrm{Torsion})$. Of course, when $p = 0$, the corresponding component of $\mathrm{Gr}(e)$ is simply an integer.

With the preceding understood, make the following choices when $d > 0$: First, choose $p \in \{0, \ldots, 2d\}$ and if $p > 0$, choose an element $\gamma_1 \wedge \cdots \wedge \gamma_{2p} \in \Lambda^{2p} H^1$. Then, for each $j \in \{1, \ldots, 2p\}$, choose an oriented, embedded circle in X to represent the class γ_j. To simplify notation, the chosen circle will be denoted by γ_j also. Make these choices of $\Gamma = \{\gamma_j\}_{1 \le j \le 2p}$ so that the distinct circles are disjoint. With Γ chosen, choose a set $\Omega \subset X$ of $d - p$ distinct points which miss each circle in Γ.

Part 2. Let $\mathcal{H} \equiv \mathcal{H}(e, J, \Gamma, \Omega)$ denote the set whose typical element is an unordered set, h, of pairs $\{(C_k, m_k)\}$ where each C_k is a compact, oriented, pseudo-holomorphic submanifold in X, and where the corresponding m_k is a positive integer. The elements in h should be constrained as follows:

1. For each k, introduce e_k to denote the Poincare' dual to $[C_k]$, and $d_k = e_k \bullet e_k - c \bullet e_k$. Require $d_k \ge 0$.
2. Require that $m_k = 1$ unless $d_k = 0$ and the genus of C_k is also 0. (Thus, C_k is a torus with trivial normal bundle.)
3. Require that $\Sigma_k m_k e_k = e$.
4. There is a partition $\Gamma = \cup_k \Gamma_k$, where each Γ_k contains some even number $2 \cdot p_k$ elements with $0 \le p_k \le d_k$. Furthermore, C_k intersects precisely once each $\gamma \in \Gamma_k$; and no $\gamma \in \Gamma_k$ is tangent to C_k at their intersection point. Moreover, C_k has empty intersection with the elements of $\Gamma - \Gamma_k$.
5. Each C_k contains precisely $d_k - p_k$ points of Ω.
6. Require that the $C_k \cap C_{k'} = \varnothing$ when $k \ne k'$.

(1.13)

(The final condition implies that $e_k \bullet e_{k'} = 0$ when $k \ne k'$. And this implies that $\mathcal{H}$ is empty whenever d (from (1.10)) is negative.)

Part 3. Suppose that $h \in \mathcal{H}$, and that $(C_k, m_k) \in h$ is such that $d_k > 0$. This data can be used to define a real vector space V_k of dimension $2 \cdot d_k$ as follows: First of all, each $z \in C_k \cap \Omega$ contributes a summand $N|_z$ to V_k, where $N \to C_k$ is the normal bundle to C_k in X. Meanwhile, each $\gamma \in \Gamma_k$ contributes a real line summand to V_k; the latter being the line $N|_z / p(T\gamma|_z)$, where $z = \gamma \cap C_k$ and where $p : TX|_z \to N|_z$ is the tautological projection.

Note that N is naturally oriented, as is each $\gamma \in \Gamma_k$. This means that each of the summands of V_k has a natural orientation. Thus, V_k inherits an orientation with the choice of an ordering for the set Γ_k. For, this ordering gives the order of the oriented real line summands in

V_k. (The summands which are indexed by the points in $C_k \cap \Omega$ are each naturally complex, and so their order in V_k is immaterial.)

With V_k understood, note that any section α of the normal bundle N defines a tautological element in V_k by restricting α to the points in $C_k \cap \Omega$ and to the points where the elements of Γ_k intersect Ω. The preceding defines a tautological map from $C^\infty(N)$ to V_k whose restriction to the kernel of the operator D in (1.12) will be denoted by G_k.

Part 4. Here are some salient properties of the set $\mathcal{H}$:

- Each $h \in \mathcal{H}$ is a finite set.

- There is a Baire set $\mathcal{W}$ of triples (J, Γ, Ω) for which J is ω compatible and for which the corresponding set $\mathcal{H}$ is finite. Furthermore when $h \in \mathcal{H}$ and when $(C_k, m_k) \in h$, then

  a) The operator D in (1.12) has trivial cokernel.

  b) If $d_k > 0$, the homomorphism G_k : kernel$(D) \to V_k$ is an isomorphism.

  c) If $d_k = 0$ and $m_k > 1$, then the pull-back of D to any finite cover of the torus C_k also has trivial cokernel (and kernel).

(1.14)

These facts are proved in [T2], see also [Mc1].

Part 5. Assume now that the data (J, Γ, Ω) is chosen from the set $\mathcal{W}$ in (1.14). Let $h \in \mathcal{H}$ and let $(C, m) = (C_k, m_k) \in h$. The purpose of this part of the discussion is to associate to such a pair an integer, $r(C, m)$. There are three cases to consider.

If $m = 1$ and $d_k = 0$. Here, $r(C, 1) \in \{\pm 1\}$ and it counts (mod 2) the spectral flow for a path of zero'th order deformations of D which starts with D and ends with an operator $D_1 = \bar{\partial} + \nu' : C^\infty(C; N) \to C^\infty(C; N \otimes T^{0,1}C)$ whose kernel and cokernel are also trivial. The path $t \to D_t$ can be chosen so that

- The set of t where cokernel$(D_t) \neq \{0\}$ is a finite number, N.

- At such t where cokernel$(D_t) \neq \{0\}$, the dimension of this cokernel is 1.

- At such t where cokernel$(D_t) \neq \{0\}$, the restriction of the t-derivative of D_t to kernel(D_t) composes with projection onto cokernel(D_t) as an isomorphism.

(1.15)

(Because D' is $\mathbb{C}$-linear, the set of ν' where kernel$(D') \neq \{0\}$ is a codimension 2 variety in $C^\infty(C; T^{0,1}C)$. This insures that $r(C,1)$ depends only C and, in particular, not on the details of D's deformation.)

If $m = 1$ and $d_k > 0$. The integer $r(C,1) \in \{\pm 1\}$ again. However, the definition in this case requires the choice of an ordering of the elements of Γ_k. As remarked above, the latter serves to orient the vector space V_k. Next, choose a continuous path $t \to D_t$ so that:

- For each $t \in [0,1]$, D_t is a zero order deformation of D.
- $D_0 = D$.
- D_t has trivial cokernel for all t.
- $D_1 = \bar{\partial} + \nu'$ is $\mathbb{C}$-linear.

(1.16)

With the preceding understood, the association of the kernel(D_t) to $t \in [0,1]$ defines a $2 \cdot d_k$ dimensional, real vector bundle over $[0,1]$. The fiber of this vector bundle over $t = 1$ is complex, so naturally oriented; and the latter induces an orientation of the kernel(D), the fiber over $t = 0$. With this orientation for kernel(D), the linear map G_k is then an isomorphism between two oriented vector spaces. Now define $r(C,1) = +1$ if G_k preserves orientation, and otherwise define $r(C,1) = -1$.

Note that $r(C,1)$ is independent of the choice of the path $\{D_t\}$, but it will change sign if the ordering of Γ_k is changed by a permutation with odd parity.

If $m > 1$. As noted above, this requires C to be a torus (and N to be topologically trivial). There are three distinct isomorphism classes of non-trivial real line bundles over C, and by tensoring (over $\mathbb{R}$) the range and domain of D with any one of these, one obtains a suite of 3 new operators. Agree to call any one of these a "twisted version" of D. Note that the index of D and any of its twisted versions is zero. This is because d_k is zero when C is a torus with trivial normal bundle.

With the preceding understood, the value of $r(C,m)$ depends only on the various possibilities for the mod(2) spectral flow for the operator

D and for its twisted versions. (Once again, the genericity assumptions on J are such as to insure that these spectral flows are well defined.) This is to say, that $r(C,m)$ depends only on the mod 2 spectral flow for D and on the number of D's twisted version which have non-trivial spectral flow. (And, of course, it depends on m.) In this regard, once m is fixed, there are eight possibilities for $r(C,m)$; and it is convenient to label the possibilities with a tag, $\pm k$, where the $\pm$ indicates whether the spectral flow for D is $+1$ or -1, and where $k \in \{0,1,2,3\}$ indicates the number of the twisted versions of D which have non-trivial spectral flow.

For a fixed tag, $\pm k$, it proves convenient to present the data $\{r(C,m)\}_{m=1,2,\dots}$ with the help of a "generating function", $f_{\pm k}(t)$. This is to say that $f_{\pm k}$ is, by definition, that formal power series for which the coefficient of t^m is $r(C,m)$. This sort of presentation is convenient here only because $f_{\pm k}(t)$ is, in all cases, a fairly simple analytic function of t. Here are the eight generating functions:

- $f_{+0}(t) = \frac{1}{1-t}$.
- $f_{+1}(t) = 1+t$.
- $f_{+2}(t) = \frac{1+t}{1+t^2}$.
- $f_{+3}(t) = \frac{(1+t)(1-t^2)}{1+t^2}$.
- $f_{-0}(t) = 1-t$.
- $f_{-1}(t) = \frac{1}{1+t}$.
- $f_{-2}(t) = \frac{1+t^2}{1+t}$.
- $f_{-3}(t) = \frac{1+t^2}{(1+t)(1-t^2)}$.

(1.17)

End the digression.

Part 6. Suppose $e \in H^2(X;\mathbb{Z})$ has been chosen and suppose that $d = e \bullet e - c \bullet e \geq 0$. Let $p \in \{0,\dots,d\}$ and let $\gamma_1 \wedge \cdots \wedge \gamma_{2p} \in \Lambda^{2p}H^1(X;\mathbb{Z})$. Fix $(J,\Gamma,\Omega) \in \mathcal{W}$ so that the conclusions of (1.14) hold. Then, let $h = \{(C_k,m_k)\} \in \mathcal{H}$. The preceding step defined an integer weight $r(C_k,m_k)$ for each $(C_k,m_k) \in h$ from the given data and from the choice of an ordering on the corresponding Γ_k. The purpose of this step is to use $\{r(C_k,m_k)\}$ to define an integer weight, $q(h)$, to the set h. The definition of $w(h)$ is simplest when $p=0$, whence

$$q(h) = \Pi_k r(C_k,m_k). \qquad (1.18)$$

In the case where $p > 0$, each $r(C_k, m_k)$ depends on the choice of an ordering for the corresponding Γ_k. This dependence is compensated for in the definition of $q(h)$ as follows: The chosen orderings of the Γ_k also induce an ordering of Γ which differs from the given labeling by a permutation, σ, of the set $\{1, \dots, 2p\}$. The latter has a parity, which will be denoted by $\varepsilon(\sigma) \in \{\pm 1\}$. Note that $\varepsilon(\sigma)$ is insensitive to the choice of ordering for $\{(C_k, m_k)\}$ as each Γ_k has an even number of elements.

With the preceding understood, associate the weight

$$q(h) = \varepsilon(\sigma) \cdot \Pi_k r(C_k, m_k) \tag{1.19}$$

to each $h \in \mathcal{H}$ in the case when $p > 0$. (Note that $q(h)$ in (1.19) is insensitive to the chosen orderings of each of the Γ_k's.)

Part 7. Here is the definition of Gr:

Definition 1.4. Define Gr: $H^2(X; \mathbb{Z}) \to \Lambda^* H^1(X; \mathbb{Z})$ as follows: Set $\mathrm{Gr}(0) = 1$. For $e \in H^2(X; \mathbb{Z}) - \{0\}$, set $d = e \bullet e - c \bullet e$. Then

- $\mathrm{Gr}(e) = 0$ if $d < 0$.
- If $d \geq 0$,

  a) Fix $J \in \mathcal{W}$, and use J to define $\mathcal{H} = \mathcal{H}(e, J, \varnothing, \varnothing)$. With $\mathcal{H}$ understood, define the projection of $\mathrm{Gr}(e)$ in the $\mathbb{Z}$ summand of $\Lambda^* H^1(X; \mathbb{Z})$ to equal $\Sigma_{h \in \mathcal{H}} q(h)$ where $w(h)$ is given by (1.18).

  b) Fix $p \in \{1, \dots, d\}$ and then fix $\gamma_1 \wedge \cdots \wedge \gamma_{2p} \in \Lambda^{2p}(H^1(X; \mathbb{Z})/$ Torsion). Then, choose $(J, \Gamma, \Omega) \in \mathcal{W}$ and use this data to define $\mathcal{H} = \mathcal{H}(e, J, \Gamma, \Omega)$. With $\mathcal{H}$ understood, define $\mathrm{Gr}(e)(\gamma_1 \wedge \cdots \gamma_{2p}) \equiv \Sigma_{h \in \mathcal{H}} q(h)$ where $q(h)$ is given by (1.19).

The following proposition describes the salient properties of the preceding definition:

Proposition 1.5. *Let (X, ω) be a pair consisting of a smooth, compact, connected 4-manifold X with a symplectic form ω. If $e \in H^2(X; \mathbb{Z})$, then the value of Gr(e) as given in Definition 1.4 is independent of the precise choice for the data (J, Γ, Ω) and thus depends only on the symplectic form ω. Furthermore, Gr$(\cdot)$ is constant if ω is changed through a continuous path of symplectic forms. Finally, Gr behaves naturally with respect to diffeomorphisms of X in the following sense: Let $\varphi : X \to X$ be a diffeomorphism and let Gr_ω and $Gr_{\varphi^*\omega}$ denote the respective Gromov invariants as defined by ω and $\varphi^*\omega$. Then $Gr_{\varphi^*\omega}(\varphi^* e) = \varphi^*(Gr_\omega(e))$.*

A proof of this proposition can be found in [T2].

f) A geometric relationship between SW and Gr.

The relationship between SW and Gr in Theorems 1 and 2 is ultimately a consequence of a certain geometric property of the zero set of a component of the spinor ψ when a pair (A, ψ) solves a certain version of (1.1). To make this more precise, it is necessary to first rewrite (1.1) on a symplectic manifold X which exploits the decomposition in (1.9). This rewriting of (1.1) requires a preliminary, two part digression. Part 1 of the digression observes that the bundle K^{-1} comes equipped with a canonical connection (up to the action of $C^\infty(X; S^1)$ (see, e.g. [T6]). To define this canonical connection, remember first that for any fixed $\mathrm{Spin}^{\mathbb{C}}$ structure, the choice of a connection on $\det(S_+)$ and the Levi-Civita connection on the bundle Fr defines a connection on the $\mathrm{Spin}^{\mathbb{C}}$ lift F. Thus, the choice of a connection (say A) on $\det(S_+)$ gives a covariant derivative, ∇_A, on sections of S_+. Now consider the canonical $\mathrm{Spin}^{\mathbb{C}}$ structure in (1.7). Restriction of ∇_A to a section of the trivial summand $\mathbb{I}$ and projection of the resulting covariant derivative onto $\mathbb{I} \otimes T^*X$ defines a covariant derivative ∇_A on the trivial complex line bundle. With the preceding understood, remark that there is a unique choice of connection A_0 (up to the afore-mentioned gauge equivalence) on $\det(S_+) = K^{-1}$ for which the corresponding covariant derivative on the trivial line bundle admits a non-trivial, covariantly constant section.

For Part 2 of the digression, consider the general $\mathrm{Spin}^{\mathbb{C}}$ structure in (1.7). Since $\det(S_+) = E^2 \otimes K^{-1}$, the choice of the connection A_0 on K^{-1} allows any connection A on $\det(S_+)$ to be written uniquely as

$$A = A_0 + 2a, \tag{1.20}$$

where a is a connection on the complex line bundle E. Thus, with A_0 chosen, the Seiberg-Witten equations in (1.1) can be thought of as equations for a pair (a, ψ) where a is a connection on E and where ψ is a section of S_+ in (1.9).

End the digression. With this reinterpretation of (1.1) understood, remark now that it proves useful to "renormalize" the form μ in (1.1) by writing

$$\mu = \frac{ir}{4}\omega + P_+F_{A_0} + i\mu_0. \tag{1.21}$$

Here, r can be any non-negative number and μ_0 can be any section of Λ_+. (In practice, think of μ_0 as being close to 0.) Furthermore, in the case where $r > 0$, it also proves useful to write

$$\psi = r^{1/2}(\alpha, \beta) \tag{1.22}$$

to correspond with the splitting in (1.9). Then, with the preceding understood, the Seiberg-Witten equations in (1.1) read

$$D_A(\alpha,\beta) = 0.$$
$$P_+F_a + \frac{ir}{8}(1 - |\alpha|^2 + |\beta|^2)\omega - \frac{r}{4}(\alpha\beta^* - \alpha^*\beta) - i\mu_0 = 0. \qquad (1.23)$$

Here, $\alpha\beta^*$ and $\alpha^*\beta$ are sections of K and K^{-1} where the latter are naturally identified as the orthogonal compliment of the span of ω in $\Lambda_+ \otimes \mathbb{C}$. (Note that this last equation differs from the analogous equations in [T1], [T3] and [T6] in that the β used here is $-i$ times that used in the previous papers. The insertion of this factor of $-i$ here avoids numerous factors of i later on.)

Rewriting (1.1) as in (1.23) realizes the Seiberg-Witten equations as equations for data $(a, (\alpha, \beta))$ where a is a connection on the line bundle E, α is a section of E, and β is a section of $K^{-1} \otimes E$.

End the digression. The following theorem provides the geometric underpinnings of the theorems in the introduction:

Theorem 1.6. *Let X be a compact, oriented, symplectic manifold. Fix an ω-compatible almost complex structure on X, and use the resulting metric to define the Seiberg-Witten equations. Fix $e \in H^2(X;\mathbb{Z})$ and use e to define the* Spin$^{\mathbb{C}}$ *structure in (1.9). Also, fix a finite (maybe empty) collection $\{\varsigma_k\}$ of closed subsets of X. Given $\varepsilon > 0$, and then given r sufficiently large, the following is true: If $(a, (\alpha, \beta))$ solves the r and $\mu_0 = 0$ version of (1.23) and has $\alpha^{-1}(0)$ intersect each ς_k, then there is a compact (not necessarily connected), complex curve C with a pseudo-holomorphic map $\varphi : C \to X$ with*

- $\varphi_*[C]$ *equal to the Poincare dual of* e;
- $\varphi(C) \cap \varsigma_k \neq 0$ *for each* k.
- $\sup_{x:\alpha(x)=0} \mathrm{dist}(x, \varphi(C)) + \sup_{x\in\varphi(C)} \mathrm{dist}(x, \alpha^{-1}(0)) < \varepsilon.$

(This theorem follows from Theorem 1.3 in [T3].)

g) Symplectic manifold constraints on SW.

Apart from Theorems 1 and 2 of the introduction and Theorem 1.6, above, the following summarizes the main results about the Seiberg-Witten invariants for a symplectic manifold X with $b^{2+} > 1$.

Theorem 1.7. *Let X be a compact 4-manifold with $b^{2+} > 1$ and with symplectic form ω. Then SW, as a map from $H^2(X;\mathbb{Z})$ to $\Lambda^* H^1(X;\mathbb{Z})$, has the following properties:*

- *If* $SW(e) \neq 0$, *then* $e \bullet e = c \bullet e$ *so* d *in (1.5) is zero.*
- SW *maps into the* $\Lambda^0 = \mathbb{Z}$ *summand of* $\Lambda^* H^1(X; \mathbb{Z})$.
- $SW(0) = 1$ *and* $SW(c) = \pm 1$, *where* $c = c_1(K)$.
- $SW(e) = \pm\,\mathrm{SW}(c - e)$.
- *If* $SW(e) \neq 0$, *then* $0 \leq e \bullet [\omega] \leq c \bullet [\omega]$. *Furthermore, the lower bound is strict except for* $e = 0$, *and the upper bound is strict except for* $e = c$.

(1.24)

Remark that third assertion is the main theorem in [T6]. (Note that an easier proof is had by using the analysis in [T3].) The fourth assertion follows from Witten's observation that the set $\mathcal{S}$ admits an involution which changes the Seiberg-Witten invariant by a sign, at most. The fifth assertion is proved in [T7], and the first assertion in [T3]. (The second assertion says that X has "simple type".) The second assertion follows from the first one.

In the case when $b^{2+} = 1$, similar arguments yield:

Theorem 1.8. *Let* X *be a compact, symplectic 4-manifold with* $b^{2+} = 1$. *Then the invariant* SW *obeys:*

- $SW(0) = 1$.
- *If* $SW(e) \neq 0$, *then* $0 \leq e \bullet \omega$ *with equality if and only if* $e = 0$.

h) An interpretation of the classes e with $\mathrm{SW}(e) \neq 0$.

There is a sense in which SW for symplectic X measures solely properties of the class c. Indeed, if e is any class with $\mathrm{Gr}(e) \neq 0$, then any connected, pseudo-holomorphic submanifold which contributes to the formula for $\mathrm{Gr}(e)$ also contributes to that for $\mathrm{Gr}(c)$. This comment is justified by the following observation: If e has $\mathrm{Gr}(e) \neq 0$, then the same is true for $c - e$ using the fourth point in (1.24). Thus, both $\mathcal{H}(e)$ and $\mathcal{H}(c - e)$, as defined in (1.14), are non-empty. Pick arbitrary elements $h \in \mathcal{H}(e)$, and let $h' \in \mathcal{H}(c - e)$. If there are no tori of square zero which appear both in h and h', then the union of these two sets is in $\mathcal{H}(c)$. More generally, let $\{C_a\}$ denote the list of tori which are shared between the sets h and h'. Let $h^0 \subset h$ denote the subset which is obtained by deleting any (C, m) where $C \in \{C_a\}$ is a shared torus. Define h'^0 analagously. Then $h^0 \cup h'^0 \cup \{(C_a, m_a + m'_a)\}$ is in $\mathcal{H}(c)$. This last point follows from the condition $e \bullet (c - e) = 0$ (due to the first point in (1.24)), since the latter implies that except for the sharing of tori, any representation of e by a union of disjoint, pseudo-holomorphic submanifolds is disjoint from any representation of $c - e$ by such a set. (Here, remember that distinct pseudo-holomorphic submanifolds intersect with locally positive intersection number. Thus, the only way e and $c - e$ can be have intersecting representatives is if e and $c - e$ have pseudo-holomorphic representatives which share some number of tori with self intersection zero and/or spheres of self-intersection -1. The case where e and $c - e$ share a sphere of self intersection -1 can be ruled out using the adjunction formula and the $d = 0$ condition.)

Note that the preceding picture was more or less conjectured by Witten in the case where X is Kahler.

i) Applications.

The main application to date of Theorems 1 and 2 and 1.7-1.8 has been to find certain kinds of pseudo-holomorphic or symplectic submanifolds in X. In particular, in the $b^{2+} > 1$ case, Theorems 1 and 1.7 imply that every class e with SW$\neq$ 0 is Poincare' dual to the fundamental class of a symplectic submanifold of X. Furthermore, for generic choice of metric (for which ω is still self dual with norm $\surd 2$), every class e with SW$\neq$ 0 will be Poincare' dual to the fundamental class of a pseudo-holomorphic submanifold. For example, one corollary states that if X has $b^{2+} > 1$ and $c \bullet c < 0$, then $X = Y\#(-\mathbb{CP}^2)$, where Y is a symplectic manifold. (Note that in every case, this existence theorem holds for the class c.)

As noted in Theorem 2, when $b^{2+} = 1$, the Seiberg-Witten equations typically yield invariants of X with some additional structure chosen. In any event, Theorem 2 again produces (under the correct hypothesis) pseudo-holomorphic submanifolds in $b^{2+} = 1$ manifolds. Of particular interest are pseudo-holomorphic spheres with non-negative square, because the existence of such spheres can be used to determine X. The point is that spheres with non-negative square come in positive dimensional families, and thus sweep out "coordinates" on X. This surprising observation was first made in the symplectic context by Gromov [Gr]; and then McDuff [Mc2] considerably extended Gromov's original application. In any event, the game to date in the $b^2_+ = 1$ case has been to find pseudo-holomorphic spheres. This game has been played with much cleverness by Li and Liu [LL3], [Liu], and also by [OO], [Bi], [Mc3] to the extent that a great deal is now known about the classification of symplectic manifolds with $b^2_+ = 1$. McDuff and Salamon [MS2] have a nice review of the $b^2_+ = 1$ story.

By the way, Donaldson [D1] has a remarkable existence theorem for symplectic submanifolds of very high degree ($[\omega] \bullet e \gg 0$) on any symplectic manifold. Donaldson's existence theorem is proved by a very different sort of argument.

j) Examples from 3-manifolds.

Let M be a compact, oriented 3-manifold which admits a fibering $\phi : M \to S^1$. Then, one can choose a metric on M so that ϕ is a harmonic map. This implies that the ϕ-pull back, ν, of the standard volume form on the circle is a harmonic 1-form. That is, $d\nu = 0$ and also $d * \nu = 0$. Here, $*$ is the Hodge star operator on M. With the preceding understood, let $X = S^1 \times M$. Then X has a symplectic form, namely $\theta \wedge \nu + *\nu$, where θ is the pull-back to X of the standard volume form on S^1 via the projection on the first factor of S^1. Note that $b^2_+(X)$ is equal to $b^1(M)$.

Let $F \subset M$ denote a typical fiber of ϕ. The manifold M can be realized as the quotient of $[0,1] \times F$ by the equivalence relation which identifies $(0, x)$ with $(1, \varphi(x))$, where $\varphi : F \to F$ is a diffeomorphism. In this guise, the map φ is induced by the tautological projection from $[0,1] \times F$ to $[0,1]$. Let $A : H^1(F) \to H^1(F)$ denote the action of φ^*. The characteristic polynomial of A is defined to be $\det(1 - t \cdot A)$.

It turns out that the characteristic polynomial of A can be computed from the Seiberg-Witten invariants of X by as described by:

Proposition 1.9. *Let X be as described above. For each integer $n \geq 1$, let $sw_n \in \mathbb{Z}$ denote the sum of the Seiberg-Witten invariants from the set of* Spin$^{\mathbb{C}}$ *structures on X which have the property that $\langle c_1(\det(S_+)), F\rangle = n$. Let $sw_0 = 1$. Then*

$$\Sigma_n sw_n t^n = \frac{\det(t^{-1} - tA)}{(t^{-1} - t)^2}. \tag{1.25}$$

Remark that the right side of (1.25) is equal to $t^{-2g+2} \cdot \det(1 - t^2 \cdot A)/(1 - t^2)^2$ where $g = \text{genus}(F)$, and so defines a formal power series in t and t^{-1}.

(Meng-Taubes [MT] and also Salamon [Sa] have arrived at (1.25) independently, and by different techniques.)

Here are some explanatory remarks: In the case where $b^2_+(X) > 1$, there are only finitely many Spin$^{\mathbb{C}}$ structures with non-zero Seiberg-Witten invariants. This implies that the left hand sum is finite, and so $(t^{-1} - t)^2$ should divide $\det(t^{-1} - tA)$. That such is the case can be argued as follows: As remarked, $b^2_+(X) = b^1(M)$, and the latter is 1 more than the number of zero eigenvalues of the matrix $A - 1$. Furthermore, the number of such zero eigenvalues is always even because A preserves the symplectic cup product pairing on $H^1(F)$. Thus, if $b^2_+(X) > 1$, then $A - 1$ has at least two zero eigenvalues. And, if $A - 1$ has n zero eigenvalues, then $\det(t^{-1} - tA)$ is divisible by $(t^{-1} - t)^n$. (Note that the determinant in question is equal to t^{-g} times the characteristic polynomial of A.)

In the case when $b^2_+(X) = b^1(M) = 1$, the formula in (1.25) should be interpreted in the symplectic chamber.

By the way, examples of such fibering 3-manifolds with $b^1 = 1$ can be obtained from a so-called fibered knot $K \subset S^3$. Indeed, if K is any knot in S^3, take out a solid torus neighborhood of the knot, and then glue the latter back in so that the meridian and the zero framed longitude are switched. The result will be a compact, oriented 3-manifold $M = M_K$ with $b^1 = 1$. For fibered knots, $S^3 - K$ fibers over S^1, and for these knots, the manifold M_K will also fiber over S^1. In this case, the polynomial $\det(t^{-1} - tA)$ can be identified (up to sign) with a certain normalization of the Alexander polynomial of the knot, $\Delta_K(t)$. With this understood, then (1.25) becomes a special case of a general formula (valid for any knot in S^3) of the theorem of Meng and the author [MT] which is discussed in Section 3b, below. (The Alexander polynomial Δ_K which appears here is normalized so that $\Delta_K(t) = \Delta_K(t^{-1})$. This version of Δ_K is determined [Co] by the skein relations

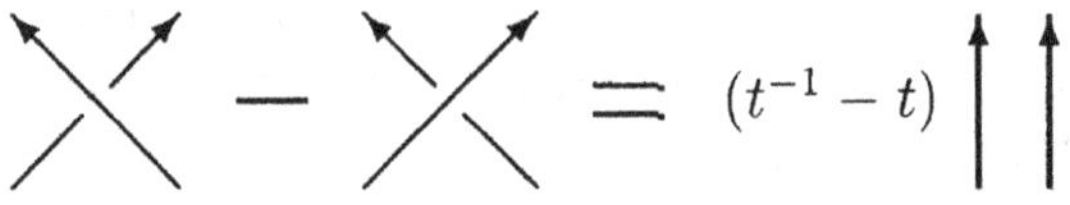

(1.26)

with the normalization that its value on the unknot is 1.)

Here is an outline of the proof of Proposition 1.9: The equivalence between SW and Gr turns the problem into one of counting pseudo-holomorphic submanifolds in the appropriate homology class. This process is simplified when the metric for X is taken to be the obvious product metric. In this case, the connected, pseudo-holomorphic submanifolds that are relevant to the computation are tori of square zero which have the form $S^1 \times \gamma$, where $\gamma \subset M$ is a closed, integral curve of the vector field on M which is metrically dual to the 1-form ν. Furthermore, for the calculation of sw_n, the curve γ should intersect the fiber F exactly n times. If the metric on M is chosen appropriately, then such a curve γ is determined by a fixed point of the diffeomorphism $\varphi^n : F \to F$; thus the problem comes down to counting fixed points of φ^n with appropriate weights. To see the results of this count, it proves convenient to introduce the set $S(n) \subset F$ of points which are fixed by φ^n, but not by φ^m for any positive $m < n$. From the SW=Gr equivalence, the left hand side of (1.14) is equal to

$$Q(t) = t^{-2g+2}\Pi_n \Pi_{q\in S(n)}(1 - \varepsilon'(q) \cdot t^{2n})^{-\varepsilon(q)/n} \tag{1.27}$$

Here, $\varepsilon(q)$ is the sign of the determinant of the identity minus the differential of φ^n at q and $\varepsilon'(q) = 1$ if $\varepsilon = -1$, but ε' can be either 1 or -1 when $\varepsilon = 1$. (This ε' is 1 if ε also gives the sign of the determinant of the identity minus the differential of φ^{2n}, otherwise ε' is -1.) Note that the factor of $1/n$ in the exponent arises for the following reason: A point $x \in S(n)$ determines the same γ as does $\varphi(x)$, and so counting fixed points in $S(n)$ will over count by a factor of n the number of embedded, pseudo-holomorphic tori in X which have intersection number n with the fiber F. The factor of t^{-2g+2} arises because the canonical class for the symplectic structure evaluates as $2g - 2$ on the fiber F. Here, one must remember that when the S_+ bundle of the $\mathrm{Spin}^{\mathbb{C}}$ structure is given by (1.9), then $c_1(\det(S_+)) = 2e - c$, where $e = c_1(E)$ and where c is the first Chern class of the canonical bundle.

With (1.27) understood, the proof of Proposition 1.9 is completed by identifying (1.27) with $\det(t^{-1} - tA)/(t^{-1} - t)^2$. The argument here (as shown to the author by R. Bott) uses the Lefschetz fixed point formula on F. Indeed, consider that the natural log of Q has the expansion:

$$\begin{aligned} \ln(Q) + (2g-2)\cdot\ln(t) &= \Sigma_{n\geq 0}\Sigma_{q\in S(n)}(-\varepsilon n^{-1}\ln(1-\varepsilon' t^{2n})) \\ &= -\Sigma_{n\geq 0}\Sigma_{q\in S(n)}\Sigma_{k\geq 0}\ \varepsilon\cdot(\varepsilon')^k\frac{t^{2nk}}{nk}. \end{aligned} \tag{1.28}$$

Now, rewrite this last expression by reordering the summation and introducing $m = nk$ to obtain

$$\ln(Q) + (2g-2)\cdot\ln(t) = -\Sigma_{m\geq 0}\ a_m \cdot \frac{t^{2m}}{m}, \tag{1.29}$$

where $a_m = \Sigma_{k:k \text{ is a root of } m}(\Sigma_{q\in S(k)}\ \varepsilon(q)\cdot\varepsilon'(q)^k)$. This sum for a_m is simply the sum over all fixed points of φ^m weighted by the sign of the determinant of the differential of φ^m. And, according the the Lefschetz fixed point theorem, the latter is nothing but $2 - \mathrm{trace}(A^m)$, the Lefschetz number of φ^m. Thus,

$$\begin{aligned} \ln(Q) + (2g-2)\cdot\ln(t) &= -2\cdot\Sigma_{m\geq 0}\frac{t^{2m}}{m} + \Sigma_{m\geq 0}\mathrm{trace}(A^m)\frac{t^{2m}}{m} \\ &= -\ln((1-t^2)^2) + \ln(\det(1-t^2A)). \end{aligned} \tag{1.30}$$

(The author is indebted to E. Ionel and T. Parker for pointing out an omission from (1.27) and (1.28) that was present in an earlier version of this article. Note as well that Ionel and Parker can compute Gromov invariant of $S^1 \times M$ directly, see [IP2].)

2 The smooth classification of 4-manifolds.

Up until recently, one could conjecture that a compact, orientable, simply connected 4-manifold was a connect sum of parts which were symplectic with one or the other orientation. This conjecture was proved false in the summer of 1996 by Zoltan Szabo [Sz] who exhibited indecomposable manifolds whose Seiberg-Witten invariants (for either orientation) violate the constraints in Theorem 1.7 or 1.8. (They are proved indecomposable by virtue of the fact that their Seiberg-Witten invariants for one orientation are non-zero.) Subsequently, Fintushel and Stern [FS2] have found a vast array of such examples. For example, Fintushel and Stern find infinitely many manifolds with the homotopy type of the K3 surface which are not symplectic and not mutually diffeomorphic.

Prior to Szabo's constructions, Shugang Wang [Sw] (see also [Ko2]) constructed (with the help of the Sieberg-Witten invariants) non-symplectic but irreducible 4-manifolds with $\pi_1 = \mathbb{Z}/2$. Wang's manifolds are quotients of algebraic surfaces by anti-holomorphic involutions. (On the other hand, Gompf [Go1] has shown that every finitely presentable group is the fundamental group of a symplectic manifold.)

Note that the issue in all of the above is the C^∞ classification of X. Mike Freedman [Fr], [FQ], has completed the classification up to homeomorphism for simply connected, topological 4-manifolds. The topological classification data consists simply of a triple $(V, Q, \pm 1)$ up to equivalence, where V is a finite $\mathbb{Z}$-module (identified with $H_2(X;\mathbb{Z})$) and $Q : V \otimes V \to \mathbb{Z}$ is a unimodular, symmetric pairing (identified with the intersection pairing on $H_2(X;\mathbb{Z})$). The final $\mathbb{Z}/2$ data is the Kirby-Seibenman invariant (which is automatically 1 when Q is even). (The equivalence here is that of the pair (V, Q) under the action of $Gl(V;\mathbb{Z})$.)

Donaldson's work shows that the smooth and topological classification problems have radically different solutions. In particular, in [D2], topological manifolds with no stable smoothing obstruction are shown not to be smoothable, and in [D3], a simply connected, topological manifold is exhibited with at least two inequivalent smoothings. (With the help of the Seiberg-Witten invariants, a vast number of examples of non-smoothable manifold have now been found; see e.g. [Fu].)

By the way, the C^∞ classification of compact symplectic 4-manifolds is completely up in the air. For example, as of January 1997, the following questions have no answer:

- Fix a finitely presentable group π and list all compact, oriented, smooth 4-manifolds (up to diffeomorphism) which are symplectic and have $\pi_1 = \pi$.
- Let X be a compact, oriented 4-manifold with symplectic forms ω and ω'. Does there exist a diffeomorphism φ of X, and a continuous family $\{\omega_t\}_{t\in[0,1]}$ of symplectic forms such that $\omega_0 = \omega$ and $\omega_1 = \varphi^*\omega'$?

(2.1)

Note that there is a classical obstruction to a manifold having a symplectic form, this being the parity of $b^{2+} - b^1$. Odd parity is required. (The parity here is the obstruction to finding a nowhere vanishing section of the bundle of self-dual 2-forms. This is the same obstruction as that of reducing the structure group from the SO(4) to U(2).)

The Seiberg-Witten invariants and in particular, Theorem 1.6 give additional constraints. For example, the Seiberg-Witten invariants prove that the multiple connect sum $\#_{2n+1}\mathbb{CP}^2$ has no symplectic structure when $n > 0$. (The even connect sums violated the parity condition for b^{2+}.)

This last non-existence result follows from Theorem 1.7 and:

Proposition 2.1. *Let Z be a compact, oriented, $b^{2+} > 1$ four-manifold which smoothly decomposes as a connect sum $X\#Y$ with both X and Y having $b^{2+} \geq 1$. Then all Seiberg-Witten invariants of Z are zero.*

The proof of this last statement is very much like the proof of a similar statement about the Donaldson invariants (see, e.g [DK] for the Donaldson invariant assertion, and [KKM] for the Seiberg-Witten statement).

Note that Proposition 2.1 suggests the following very small subcase of (2.1):

Let Z be a $b^{2+} > 1$, symplectic 4-manifold with a smooth decomposition $Z = X\#Y$. As one summand, say Y, must have $b^{2+} = 0$, must X be symplectic and $Y = \#_n(-\mathbb{CP}^2)$?

(2.2)

(It follows from the main theorem in [D4] that the intersection form of Y is diagonalizable over $\mathbb{Z}$. It is also known [KMT] that Y can have no non-trivial finite covering spaces. Furthermore, using Theorem 1, Kotschick proved [Ko2] that Z must also decompose smoothly as $X'\#_n(-\mathbb{CP}^2)$ for some n with X' being symplectic.)

Vis a vis (2.1), potentially interesting cases also come from algebraic surfaces of general type. Perhaps the most well known examples are the so called Horikawa surfaces [Ho]. These are minimal algebraic surfaces whose geometric genus p and first Chern class c satisfy the equality $c \bullet c = 2p - 4$. There are two deformation types when $c \bullet c$ is divisable by 16, but it is not known whether the two types are diffeomorphic. Or, if diffeomorphic, whether the two symplectic structures are equivalent in the manner noted in the second point of (2.1).

Also note that Morgan and Szabo [MSz] have a construction which yields infinite families of pairs of manifolds which

1. are homeomorphic,
2. may not be diffeomorphic,
3. if they are diffeomorphic, may have inequivalent symplectic structures

(2.3)

As long as C^∞ classifications are under discussion, remark that as of January 1997, the smooth dimension four Poincare' conjecture has yet to be settled. (The conjecture posits a unique smooth structure on the topological 4-sphere.) Of course, as of January 1997, the three dimensional Poincare' conjecture is also up in the air, but unlike the case of dimension three, there are a slew of potentially fake 4-spheres. A family of such examples was constructed by Cappell and Shaneson [CS] along the following lines: Let A be an integer values, 3×3 matrix with determinent 1 such that $\det(A - \mathbb{I}) = 1$ also. Because A has determinant 1, A defines a self diffeomorphism of the 3-torus, $T^3 = \mathbb{R}^3/\mathbb{Z}^3$. The mapping torus of A is the 4-manifold Y which is obtained as the quotient space $(T^3 \times [0,1])/\sim$, where the equivalence is $(x,0) \sim (A \cdot x, 1)$. Because $\det(A - \mathbb{I}) = 1$, the space Y has the homology of $S^3 \times S^1$. Let $B \subset T^3$ be a small ball about the origin. Surger out $B \times S^1$ from Y and glue in $S^2 \times D^2$ in its stead. (Here, D^2 is the standard 2-disk.) There are two inequivalent ways to make this surgery; they differ by a "Gluck twist", which is to say that the gluing map for the second differs from that for the first by the fact that as one moves around the S^1, the S^2 is rotated once around its axis relative to a fixed identification of $S^2 \times S^1$ with the boundary of $S^2 \times D^2$. In any event, let $X_\pm$ denote the two manifolds so constructed. Both have the homotopy type of S^4. Some of these are known to be standard [AK], [Go2], but the identity of a large class has not been determined. (Note that when A is a square, then these manifolds have involutions whose quotients give fake $\mathbb{RP}^4$'s. See [CS].)

3 Can a symplectic form be made?

As remarked previously, a symplectic form is characterized as being both closed and non-degenerate. Lacking an existence theorem for symplectic forms, one can try to study 4-manifolds with a non-degenerate, but not closed form, or with a closed form which is degenerate in places. This section discusses some preliminary steps along the second path.

The second of the approaches above is based on the observation that a compact, oriented, Riemannian manifold X has, thanks to Hodge theory, a b^2_+ dimensional space of closed, self dual forms. Such a form is symplectic where it is not zero. Indeed, when ω is a self dual form, then $\omega \wedge \omega = |\omega|^2 \cdot d\mathrm{vol}$. Furthermore, if the metric is suitably generic, then there will be closed, self dual forms ω which vanish transversely as sections of the $\mathbb{R}^3$-bundle L_+. In this case, $Z = \omega^{-1}(0)$ is a disjoint union of embedded circles. Thus, any compact, oriented 4-manifold has a symplectic form on the compliment of a union of embedded circles.

Meanwhile, on the compliment of Z,

$$J = \frac{\sqrt{2}}{|\omega|} g^{-1} \cdot \omega \tag{3.1}$$

defines an ω-compatible, almost complex structure. This allows one to talk unambiguously about pseudo-holomorphic submanifolds on $X - Z$.

Needless to say, the circles in Z embody (in some mysterious sense) the obstructions to making a symplectic form on X.

At this juncture, there are two self-evident courses of action:

1. Pseudo-holomorphic curves in $X - Z$ are studied with the object of using them (as done in the symplectic case) to understand the differential topology of X.

2. The form ω is manipulated (keeping $d\omega = 0$) so as to remove (or otherwise simplify) components of Z.

(3.2)

The examples below indicated a possible close relation between these two courses of action.

a) A non-compact example.

Here is a (non-compact) example: The manifold in question is $\mathbb{R}^4$, with its standard flat metric. To write down the form, introduceEuclidean coordinates (x, y, z, t) and then the complex coordinates $\eta = x + i \cdot y$ and $\xi = z + i \cdot t$. Now consider the following 2-form:

$$\omega = \frac{i}{2}\Big\{(1 - |\eta|^2 + |\xi|^2) \cdot (d\eta \wedge d\bar{\eta} + d\xi \wedge d\bar{\xi}) \\ - \eta \cdot \xi \cdot d\bar{\eta} \wedge d\bar{\xi} + \bar{\eta} \cdot \bar{\xi} \cdot d\eta \wedge d\xi\Big\} \quad (3.3)$$

It is left as an exercise to verify that ω is both closed and self-dual.

The set Z for ω is the circle $|\eta| = 1$ in the plane where $\eta = 0$. Notice that the disk where $\eta = 0$ and $|\eta| < 1$ is a pseudo-holomorphic submanifold in $X - Z$.

b) Examples from 3-manifolds.

Let M be a compact, oriented, Riemannian 3-manifold with $b^1 \geq 1$. Via Hodge theory, every class in $H^1(M; \mathbb{Z})$ is represented by a harmonic, $\mathbb{R}/\mathbb{Z}$-valued function on M. This is an $\mathbb{R}/\mathbb{Z}$ valued function f which obeys $d * df = 0$, where $*$ is the Hodge star of the Riemannian metric.

Now, let $X = S^1 \times M$, and use the product metric to give X a Riemannian structure. Then

$$\omega = d\theta \wedge df + *df. \quad (3.4)$$

In particular, note that ω is symplectic when f has no critical points. (In the notation of Section 1i, the fibration ϕ is given by $e^{2\pi i f}$ and the 1-form df is denoted by ν. As noted in Section 1i, if $\phi : M \to S^1$ is a fibration, then M has a metric for which $f = (2\pi i)^{-1} \cdot \ln(\phi)$ is harmonic.) In general,

$$Z = S^1 \times \text{ crit}(f) \quad (3.5)$$

where crit(f) is the set of critical points of the $\mathbb{R}/\mathbb{Z}$ valued function f. By the way, it is not hard to prove that f has only non-degenerate critical points when the metric on M is suitably generic.

Any submanifold of the form $\{\text{Point in } S^1\} \times \{f^{-1}(\text{constant})\}$ is a pseudo-holomorphic submanifold of X. A second family of pseudo-holomorphic submanifolds is given by $S^1 \times \{\text{Gradient flow line of } f\}$.

This last example is instructive for a number of reasons. First, as indicated in the previous section, there is something about the fundamental group of X that may give obstructions to making X symplectic.

When such is the case, a first guess is that the elements in Z define generators of this group. However, (3.5) suggests that the relationship between Z and the fundamental group of X is much more subtle than this.

Furthermore, this example indicates that the problem of manipulating ω to cancel elements in Z is, in the S^1-invariant context, the same problem as that of cancelling critical points of an $\mathbb{R}/\mathbb{Z}$-valued function on a 3-manifold M. (This, of course, "explains" how Z sees the fundamental group of M.) This correspondence between a critical points of f and a component of Z is a manifestation of the afore mentioned fact that when $\gamma : \mathbb{R} \to M$ is a gradient flow line of f, then $S^1 \times \gamma(\mathbb{R})$ is a pseudo-holomorphic submanifold of X (and any S^1-invariant pseudo-holomorphic submanifold of X has this form.)

Thus, in the S^1-invariant context, there is a dictionary which translates Morse theoretic constructions into constructions with self-dual forms and pseudo-holomorphic submanifolds:

1. Critical points of $f \leftrightarrow$ zeros circles of ω.

2. Gradient flow lines of $f \leftrightarrow$ pseudo-holomorphic cylinders.

3. Cancelling critical points of $f \leftrightarrow$ removing components of Z.

(3.6)

Furthermore, recent work of the author and G. Meng [MT] plus work of Hutchings and Lee [HL] have extended the dictionary in (3.6) to include the Seiberg-Witten invariants. This is to say that the Seiberg-Witten invariants of $S^1 \times M$ can be interpreted in Morse theoretic terms on M. The simplest case for this interpretation has M given by zero-framed surgery on a knot, K, in the 3-sphere and so has the homology of $S^1 \times S^2$. Here, $X = S^1 \times M$ has $b^{2+} = 1$, but there are only two chambers, and a choice of generator $o \in H^1(M; \mathbb{Z})$ determines a chamber as follows: Let θ be any closed one form on M whose cohomology class gives o and consider (1.1) in the case where $\mu = -i \cdot r \cdot P_+(dt \wedge \theta)$ and r is positive and very large.

With the preceding understood, introduce the formal Laurent series

$$\underline{SW} = \Sigma_n sw_n \cdot t^n, \tag{3.7}$$

where sw_n is the sum of the Seiberg-Witten invariants for $\text{Spin}^{\mathbb{C}}$ structures for which the cup product of $c_1(\det(S_+))$ and o evaluates as n on the fundamental class. Then, [MT] prove that the formal series $\underline{SW} = \Delta_K(t)/(t - t^{-1})^2$, where Δ_K is the symmetrized, Alexander-Conway polynomial of K (see (1.26)). Meanwhile, Hutchings and Lee

[HL] show how to compute the invariant $\Delta_K(t)/(t-t^{-1})^2$ as a suitable, weighted count of gradient flow lines of a circle valued Morse function $\phi : M \to \mathbb{R}/\mathbb{Z}$ which generates H^1.

In the general case, [MT] relates Seiberg-Witten invariants of $X = S^1 \times M$ in the case where $b^1(M) \geq 1$ to a purely topological 3-manifold invariant known as the Milnor torsion (see, e.g. [Mi1], [Tu]). And, [HL] show how to compute the latter using Morse theory. The reader is referred to [MT] and [HL] for the details. See also the very recent [Tu2].

By the way, Theorems 1.7 and 1.8 and the results in [MT] give new obstructions to the existence of a symplectic form on $X = S^1 \times M$. For example, when M is obtained by zero surgery on a knot K in S^3, then it follows from Theorem 1.8 that X has a symplectic form only if Δ_K is a monic polynomial. (This is to say that the highest order term in t begins with 1.) For a fibered knot, the fact that Δ_k is monic follows from (1.25). On the other hand, there are knots K with monic Δ_K which are not fibered. This leads to the following question:

- Can $S^1 \times M$ have a symplectic structure when K is not fibered?
- If K is fibered, is the S^1 invariant symplectic structure (as described in Section 1j, above) unique up to deformation and diffeomorphism?

(3.8)

Kronheimer [K] has the most recent result on these questions.

c) Does Morse theory generalize?

It is not known at present how many of these Morse theoretic notions survive the transition to the non-S^1-invariant world. For example, there are well known techniques (i.e. Morse theory) for manipulating the critical points of a function (see, e.g [Mi2]), and so a relevant question is: Which Morse theory techniques have analogs in the world of self-dual 2-forms on an arbitrary, compact and oriented 4-manifold?

In particular, here is the famous cancellation lemma from Morse theory: Supppose that f is an $\mathbb{R}/\mathbb{Z}$-valued function with a pair of non-degenerate critical points. Suppose further that this pair of critical points is joined by a single, minimal gradient flow line along which the Morse index drops by one. Then there is a new $\mathbb{R}/\mathbb{Z}$ valued function on M with two less critical points. (A gradient flow line between two critical points is minimal when the drop in the value of f along the flow line is no greater than the drop along any other gradient flow line between the two critical points.)

This Morse theoretic cancellation lemma suggests the following question: Is it possible to use pseudo-holomorphic cylinders (or higher genus surfaces) in X to modify the given closed form so as to "simplify" its zero set? In this regard, it is not clear precisely what one means be "simplify". However, one can imagine the following (perhaps naive) version of the simplification question: Suppose that $C \subset X - Z$ is a pseudo-holomorphic submanifold. Define the energy of C to be the integral over C of the form ω. Say that C is minimal if C has least energy amongst all closed, pseudo-holomorphic submanifolds in $X - Z$ which have these components of Z as boundary. Suppose further that C is stable under perturbations of the given metric. Now, assume that C is a cylinder and is the only minimal, stable pseudo-holomorphic submanifold in $X - Z$ with its boundary. Does it then follow that ω can be changed so as to remove the two components of Z which are boundary circles of C, but without modifying either the remaining components of Z or their bounding pseudo-holomorphic curves?

By the way, an analogy with Morse theory is supported by the following observations: Let X be a compact, oriented 4-manifold with vanishing H^1. Suppose that ω is a closed, self dual form on X with transversal zero set. Let $\mathcal{T}$ denote the Frechet manifold of compact, oriented, dimension 1 submanifolds of X whose fundamental class is zero in $H^1(X;\mathbb{Z})$. Define an $\mathbb{R}/\mathbb{Z}$-valued function f on $\mathcal{T}$ as follows: Given a submanifold γ, choose an oriented surface $\Sigma \subset X$ with boundary γ. Now define

$$f(\gamma) = \int_\Sigma \omega. \tag{3.8}$$

Here are some simple observations about this function f: Let Z' denote the set of pairs consisting of a component of $\omega^{-1}(0)$ with an orientation. Thus, Z' sits naturally in $\mathcal{F}$. With this understood, consider:

- Z' coincides with the critical points of f.
- A gradient flow line of f defines a map of a cylinder into X which intersects $X - Z$ as a pseudo-holomorphic subvariety.

(3.9)

Thus (3.9) provides a second indication that the pseudo-holomorphic cylinders in $X - Z$ carry some of the obstruction to making X symplectic.

d) Pseudo-holomorphic submanifolds in $X - Z$.

The preceding indicates that pseudo-holomorphic submanifolds in $X - Z$ play some sort of role in the obstruction problem to making

X symplectic. However, even without this as motivation, one might study these objects for the interesting analytical issues that arise. In this regard, the author offers two results (with proofs to appear in a forthcoming article) on this subject. One result is an existence assertion, and the other is a regularity assertion. The statement of both results requires the following definition:

Definition 3.1. A subset $C \subset X - Z$ will be called a *pseudo-holomorphic subvariety* when C is the image of a complex curve C' under a proper, pseudo-holomorphic map

$$\varphi : C' \to X$$

with the following two properties:

- φ is 1-1 except for at most a countable set of points.
- $\int_{C'} \varphi^* \omega < \infty$.

The statement of the first result also requires the notion of a homological boundary. Here is the definition: Say that Z is the homological boundary of a pseudo-holomorphic subvariety $C \subset X$ if C has intersection number ± 1 with every linking 2-sphere of Z. (As an embedded circle in X, each component of Z has a tubular neighborhood whose boundary is diffeomorphic to $S^1 \times S^2$. Any 2-sphere of the form (point) $\times S^2$ in this boundary is a linking 2-sphere.)

Proposition 3.2 ([T8]). *Let X be a compact, oriented, Riemannian 4-manifold with $b^2_+ \geq 2$ and with non-trivial Seiberg-Witten invariants. Let ω be a closed, self-dual form on X which vanishes transveraly. Then the set Z is the homological boundary of a pseudo-holomorphic subvariety in $X - Z$.*

(The set of circles Z with appropriate orientations is always null-homologous in $H^1(X; \mathbb{Z})$. This can be seen as follows: Choose a $\mathrm{Spin}^{\mathbb{C}}$ structure for X, and choose a section ψ of the corresponding S_+ bundle which vanishes at isolated points. Then $\tau(\psi)$ is a section of Λ_+ which vanishes only where ψ does. A generic one parameter family of sections which interpolates between ω and $\tau(\psi)$ will define, by 1-parameter family of zero sets, a null-homology for Z.)

The second assertion is a regularity theorem for pseudo-holomorphic subvarieties. In this regard, remark that the "classical" regularity theory (see, for example [MS], [Ye], [PW]) holds away from Z. (Basically, the singularities are no worse than those which appear in the complex holomorphic case.) The issue here is the behavior near to Z. Here is the fundamental conjecture:

Conjecture 3.3. *Let $C \subset X$ be a pseudo-holomorphic subvariety. If some subset $\{Z_a\} \subset Z$ form the homological boundary of C, then there are arbitrarily small perturbations of C which result in a smooth, submanifold with boundary, $C' \subset X$, whose interior is symplectic and whose boundary is $\cup_a Z_a$.*

At the time of this writing, a good part of this last conjecture can be proved. In particular, one has

Proposition 3.4 ([T9]). *Suppose the metric and form ω obey certain special, but generally realizable (unobstructed) constraints near Z. Then, let C be as in Conjecture 3.3. There is at most a finite set of points in $\cup_a Z_a$, and given a neighborhood of this set of points, there are arbitrarily small perturbations of C which obey the conclusions of Conjecture 3.3, except in the given neighborhood.*

Note that the behavior of C near the finite set in Proposition 3.4 can be characterized precisely and it is likely that the singularities at these points can be shown to be removable after a further perturbation. The constraints on the metric and form near Z are described in detail in [T9] to which the reader is referred.

With regard to the behavior of C' where the latter is smooth near a point in Z, one can be quite a bit more precise. For this purpose, let ε be a small positive number and let $B^1 \subset \mathbb{R}$ and $B^3 \subset \mathbb{R}^3$ denote the balls of radius ε. Now, let $p \in Z$ be the point in question. Then, there exists $\varepsilon > 0$ and a neighborhood $U \subset X$ of p in X with coordinates $(t, x) \in B^1 \times B^3$ which has the following properties:

1. $(0, 0)$ corresponds to p.
2. $\{x = 0\}$ corresponds to $Z \cap U$ and the 1-form dt canonically orients $Z \cap U$.
3. $\omega = (dt \wedge x^{\dagger} A dx + x^{\dagger} A * dx) + \mathcal{O}(|x|^2)$ where $A = A(t)$ is a symmetric, trace zero, invertible 3×3 matrix. Here, $*$ is the Hodge star on $\mathbb{R}^3$.
4. $ds^2 = dt^2 + dx^t dx + \mathcal{O}(|x|)$

(3.10)

Some comments are in order: First, as pointed out to me by M. Hutchings, the submanifold Z has a canonical orientation since both TX and Λ_+ are oriented. Second, the matrix A is everywhere invertible because of the assumption that ω vanishes transversely along Z. Third A is symmetric and trace zero to insure that $d\omega = 0$ along Z.

Because A has trace zero everywhere, there is, at each $t \in B^1$, either two negative eigenvalues of A or else two positive eigenvalues. This is to say that $A(t)$ defines an orthogonal decomposition of $\mathbb{R}^3$ as $\mathbb{R}_t \oplus W_t$, where $\mathbb{R}_t$ is an eigenspace of $A(t)$, and where $A(t)$ is definite on the 2-plane W_t. Of course, this decomposition varies smoothly with $t \in B^1$.

With the preceding understood, remark that when U has small radius, then it should be possible to choose C' so that

$$C' \cap U = \{t, s \cdot v(t) : (t, s) \in B^1 \times [0, \varepsilon)\}, \tag{3.11}$$

where $v(t)$ is a unit length vector which depends smoothly on t and which either lies in W_t or in $\mathbb{R}_t$. Moreover, if the metric on X is generic, the latter case will not arise.

References

[AK] S. Akbulut and R. Kirby, *A potential smooth counter example in dimension 4 to the Poincare' conjecture, and the Andrews-Curtis conjecture*, Topology, **24** (1985), 375-390.

[At] M.F. Atiyah, *K-Theory*, Benjamin, New York, 1967.

[Bi] P. Biran, *Symplectic packing in dimension 4*, preprint.

[CS] S. Cappell and J. Shaneson, *Some new 4-manifolds*, Ann. Math., **104** (1976), 61-72.

[Co] J.H. Conway, *An enumeration of knots and links and some of their algebraic properties*, in 'Computational Problems in Abstract Algebra', Pergamon Press, New York, (1970), 329-358.

[D1] S.K. Donaldson, *Symplectic submanifolds and almost complex geometry*, J. Diff. Geom., **44** (1996), 666-705.

[D2] S.K. Donaldson, *An application of gauge theory to four dimensional topology*, J. Diff. Geom., **18** (1983), 279-315.

[D3] S.K. Donaldson, *Irrationality and the h-cobordism conjecture*, J. Diff. Geom., **26** (1987), 141-168.

[D4] S.K. Donaldson, *Connections, cohomology and the intersection form of 4-manifolds*, J. Diff. Geom., **24** (1986), 275-341.

[DK] S.K. Donaldson and P.B. Kronheimer, *The Geometry of Four-Manifolds*, Oxford University Press, 1990.

[FQ] M.H. Freedman and F. Quinn, *Topology of 4-Manifolds*, Princeton University Press, 1990.

[Fr] M.H. Freedman, *The topology of four-dimensional manifolds*, J. Diff. Geom., **17** (1982), 357-453.

[FS1] R. Fintushel and R.J. Stern, *Immersed spheres in 4-manifolds and the immersed Thom conjecture*, Turkish J. Math., **19** (1995), 145-157.

[FS2] R. Fintushel and R.J. Stern, *Knots, links and 4-manifolds*, Inventiones, to appear.

[Fu] M. Furuta, *Monopole equations and the 11/8 conjecture*, preprint.

[Go1] R. Gompf, —it A new construction of symplectic manifolds, Ann. Math., **142** (1995), 527-595.

[Go2] R. Gompf, *Killing the Akbulut-Kirby 4-sphere, with relevance to the Andrews-Curtis and Schoenflies problems*, Topology, **30** (1991), 97-115.

[Gr] M. Gromov, *Pseudo-holomorphic curves in symplectic manifolds*, Invent. Math., **82** (1985), 307-347.

[HL] M. Hutchings and Y-J. Lee, *Circle valued Morse theory, R-torsion and Seiberg-Witten invariants of 3-manifolds*, Topology, to appear.

[Ho] E. Horikawa, *Algebraic surfaces of general type with small* c_1^2, I, Ann. Math., **104** (1976), 357-387.

[IP1] E.-N. Ionel and T.H. Parker, *The Gromov invariant of Ruan-Tian and Taubes*, preprint.

[IP2] E.-N. Ionel and T.H. Parker, *Gromov invariant and symplectic maps*, preprint.

[KKM] D. Kotschick, P.B. Kronheimer and T.S. Mrowka, in preparation.

[KMT] D. Kotschik, J.W. Morgan and C.H. Taubes, *Four-manifolds without symplectic structures but with non-trivial Seiberg-Witten invariants*, Math. Res. Letters, **2** (1995), 119-124.

[K] P.B. Kronheimer, *Minimal genus in* $S^1 \times M^3$, Inventiones, to appear.

[KM] P.B. Kronheimer and T.S. Mrowka, *The genus of embedded surfaces in the projective plane*, Math. Res. Letters, **1** (1994), 797-808.

[Ko1] D. Kotschick *The Seiberg-Witten invariants of symplectic 4-manifolds (after C.H. Taubes)*, Seminaire Bourbaki, 48eme annee (1995-96), no. 812.

[Ko2] D. Kotschick, *On irreducible 4-manifolds*, preprint.

[LM] F. Lalonde and D. McDuff, *The classification of ruled, symplectic 4-manifolds*, preprint.

[Liu] A. Liu, *Some new applications of the wall crossing formula and Gompf's conjecture and its applications*, Math. Res. Letters, **3** (1996), 569-586.

[LL] T.J. Li and A. Liu, *On the equivalence between* SW *and* GT *in the case* $b^+ = 1$, preprint, 1997.

[LL2] T.J. Li and A. Liu, *General wall crossing formula*, Math. Res. Letters, **2** (1995), 797-810.

[LL3] T.J. Li and A. Liu, *Symplectic structure on ruled surfaces and generalized adjunction formula*, Math. Res. Letters, **2** (1995), 453-471.

[Mc1] D. McDuff, *Lectures on Gromov invariants for symplectic 4-manifolds*, to appear in the Proceedings of the NATO Summer School, Montreal 1995.

[Mc2] D. McDuff, *The structure of rational and ruled symplectic 4-manifolds*, Journ. Amer. Math. Soc., **3** (1990), 679-712; Erratum: Journ. Amer. Math. Soc., **5** (1992), 987-988.

[Mc3] D. McDuff, *From symplectic deformations to isotopy*, preprint.

[MS] D. McDuff and D. Salamon, *J-Holomorphic Curves and Quantum Cohomology*, American Mathematical Society, Providence, 1996.

[MS2] D. McDuff and D. Salamon, *A survey of symplectic manifolds with* $b^+ = 1$, proceedings of the 1995 Topology Conference at Gökova, Turkey.

[MT] G. Meng and C. H. Taubes, SW= *Milnor torsion*, Math. Res. Letters, **3** (1996), 661-674.

[Mi1] J.W. Milnor, *A duality theorem for Reidemeister torsion*, Ann. Math., **76** (1962), 137-147.

[Mi2] J.W. Milnor, *Lectures on the h-cobordism theorem, notes by L. Seibenman and J. Sondau*, Princeton University Press, 1965.

[Mor] J.W. Morgan, *The Seiberg-Witten Equations and Applications to the Topology of Smooth Four-Manifolds*, Mathematical Notes, **44**, Princeton University Press, 1996.

[MSz] J. Morgan and Z. Szabo, *Embedded genus 2 surfaces in four-manifolds*, Duke Math. Journal, to appear.

[MST] J. Morgan, Z. Szabo and C.H. Taubes, *A product formula for the Seiberg-Witten invariants and the generalized Thom conjecture*, J. Diff. Geom., **44** (1996), 706-788.

[OO] H. Ohta. and K. Ono, *Note on symplectic 4-manifolds with* $b^{2+} = 1$, II, preprint, 1995.

[PW] T.H. Parker and J.G. Wolfson, *Pseudo-holomorphic maps and bubble trees*, J. Geom. Analysis, **3** (1993), 63-98.

[Ru] Y. Ruan, *Symplectic toppology and complex surfaces*, in 'Geometry and Topology on Complex Manifolds', T. Mabuchi, J. Noguchi and T. Ochial, eds., World Scientific Publications, Singapore, 1994.

[Sa] D. Salamon, private communications.

[SW1] N. Seiberg and E. Witten, *Electro-magnetic duality, monopole condensation and confinement in* $N = 2$ *supersymmetric Yang-Mills theory*, Nucl. Phys., **B426** (1994), 19-52.

[SW2] N. Seiberg and E. Witten, *Monopoles, duality and chiral symmetry breaking in* $N = 2$ *supersymmetric Yang-Mills theory*, Nucl. Phys., **B426** (1994), 581-640.

[Sw] S. Wang, *A vanishing theorem for Seiberg-Witten invariants*, Math. Res. Letters, **2** (1995), 305-310.

[Sz] Z. Szabo, *Simply connected, irreducible 4-manifolds with no symplectic structures*, Inventiones, to appear.

[T1] C.H. Taubes, *The Seiberg-Witten and the Gromov invariants*, Math. Res. Letters, **2** (1995), 221-238.

[T2] C.H. Taubes, *Counting pseudo-holomorphic submanifolds in dimension 4*, J. Diff. Geom., **44** (1996), 818-893.

[T3] C.H. Taubes, SW=> Gr: *From the Seiberg-Witten equations to pseudo-holomorphic curves*, J. Amer. Math. Soc., **9** (1996), 845-918.

[T4] C.H. Taubes, Gr=> SW. *From pseudo-holomorphic curves to Seiberg-Witten solutions*, preprint.

[T5] C.H. Taubes, Gr=SW. *Counting curves and connections*, preprint.

[T6] C.H. Taubes, *The Seiberg-Witten invariants and symplectic forms*, Math. Res. Letters, **1** (1994), 809-822.

[T7] C.H. Taubes, *More constraints on symplectic forms from the Seiberg-Witten invariants*, Math. Res. Letters, **2** (1995), 9-13.

[T8] C.H. Taubes, *The Seiberg-Witten invariants and pseudoholomorphic varieties for self-dual, harmonic 2-forms*, preprint, 1998.

[T9] C.H. Taubes, *The structure of pseudoholomorphic subvarieties for a degenerate almost complex structure and symplectic form on* $S^1 \times B^3$, preprint, 1997.

[Tu] V. Turaev, *Reidemeister torsion in knot theory*, Russian Math. Surveys, **41** (1986), 119-182.

[Tu2] V. Turaev, *A combinatorial formula for the Seiberg-Witten invariants of 3-manifolds*, preprint, 1998.

[W1] E. Witten, *Monopoles and 4-manifolds*, Math. Res. Letters, **1** (1994), 769-796.

[W2] E. Witten, *Two dimensional gravity and intersection theory on moduli space*, Surveys in Diff. Geom., **1** (1991), 243-310.

[Ye] R. Ye, *Gromov's compactness theorem for pseudo-holomorphic curves*, Trans. Amer. Math. Soc., **342** (1994), 671-694.

Made in the USA
Monee, IL
03 May 2026

49411524R00195